Intelligent Observer and Control Design for Nonlinear Systems

Springer-Verlag Berlin Heidelberg GmbH

Dierk Schröder (Ed.)

Intelligent Observer and Control Design for Nonlinear Systems

With 178 Figures

Springer

Editor:
Prof. Dr.-Ing. Dr.-Ing. h. c. Dierk Schröder
Technical University of Munich
Institute for Electrical Drive Systems
Arcisstrasse 21
D-80333 München
Germany

Contributors:
Prof. Dr.-Ing. Dr.-Ing. h. c. Dierk Schröder
Dr.-Ing. Ulrich Lenz
Dipl.-Ing. Michael Beuschel
Dipl.-Ing. Franz D. Hangl
Dr.-Ing. Thomas Frenz
Dr.-Ing. Dieter Strobl
Dr.-Ing. Stephan Straub
Dr.-Ing. Kurt Fischle
Dipl.-Ing. Martin Rau
Dipl.-Ing. Anne Angermann

Library of Congress Cataloging-in-Publication Data applied for
Die Deutsche Bibliothek - Cip-Einheitsaufnahme
Intelligent observer and control design for nonlinear systems / Dierk Schröder (ed.). -
Berlin ; Heidelberg ; New York ; Barcelona ; Hong Kong ; London ; Milan ; Paris ; Singapore ;
Tokyo : Springer, 2000

ISBN 978-3-642-08346-4 ISBN 978-3-662-04117-8 (eBook)
DOI 10.1007/978-3-662-04117-8

Originally published by Springer-Verlag Berlin Heidelberg New York in 2000.
Softcover reprint of the hardcover 1st edition 2000

Production: ProduServ GmbH Verlagsservice, Berlin
Cover design: MEDIO GmbH, Berlin
Typesetting: Camera-ready by editor
Printed on acid -free paper SPIN:10653279 62/3020PT - 5 4 3 2 1 0

Preface

Research is a continuous effort. Engineers and research groups are creating new strategies and solutions by using results of other scientists who have been working on the same topic for a long time, thus aquiring a deeper understanding. Indeed we are dependent upon one another and should support each other.

This book is a contribution in this continuous line of scientific efforts and is a result of the research by PhD–students at my institute. We would like to present our ideas and results and we hope very much to provide support for other scientists interested in this area.

The starting point of our considerations is: We are engineers, and therefore we have basic knowledge of the system under consideration. But often there is a lack of precise information for a sufficiently accurate model, due to structured or unstructured uncertainties or, more severe, nonlinearities. How can we get this desired information? The idea is to identify unknown parts of the plant by a learning procedure. An idea which was already proposed by others, but we think we are able to contribute some new aspects and extensions to this area.

One aspect of our research is to assume in the first step that we know the linear part of the nonlinear plant, but we do not know the type and the parameters of the nonlinearities. In real life these nonlinearities are not smooth in general, typical nonlinearities in motion control are e.g. friction and backlash. So we concentrated on this topic, the identification of type and parameters of the nonlinearities. This led to dynamic learning structures, providing exact information about existing nonlinearities. With this information we achieved a much more precise model of the nonlinear plant. The next steps are nonlinear observers and the controller design. One of the major guidelines in our work is that the learning process is mathematically proven stable and convergent. Therefore these intelligent strategies could be used off–line and on–line.

A second fascinating idea is to learn the optimal controller even though one has only a very limited amount of knowledge of the nonlinear plant. There have been proposals for such a scenario, but up to now there have been very important restrictions. We reduced these restrictions to some extent, but there is additional research necessary; for example to reduce the learning time or to separate the effects of unknown disturbing inputs to the plant during learning. A combination of the first and second approach leads to a possible design of a nonlinear state space controller, where existing nonlinearities are taken into account already during the procedure of controller design.

We noticed that these methods are applicable in different areas of motion control, e.g. electrical drives, machine tools, processing machines with continuous moving webs (rolling mills, printing machines), or even identification and control in combustion engines. Therefore we decided to gather our results up to now in this book and thus provide an easy access for other researchers. We also hope to get information from their experiences and new results and to start a fruitful discussion. Thank you in advance.

München, October 1999

Dierk Schröder

Contents

1 Introduction — Control Aspects **1**

Dierk Schröder

1.1 Linear Plants 4

1.2 Linear Plants with Uncertain Parameters 5

1.3 Linear Plants and Nonlinear Controllers 8

1.4 Nonlinear Plants 9

1.5 Our Conceptions of Nonlinear Control Strategies and Observation 13

1.6 References 15

2 Motion Control **19**

Dierk Schröder

2.1 Control of Electromechanical Systems 19

2.1.1 Introduction 19

2.1.2 Cascaded Control 20

2.1.3 State–Space Control 27

2.1.3.1 Proportional State–Space Controller 27

2.1.3.2 State–Space Control with Integrating Contribution 34

2.1.4 Generalized Considerations for Electromechanical Systems . . . 40

2.2 Actuator, Mechanical System and Process 46

2.3 Objectives of this book (Example Motion Control) 59

2.4 Conclusions 63

2.5 References 64

3 Learning in Control Engineering **67**
Ulrich Lenz

3.1 Intelligent Control as Artificial Intelligence 67
3.2 Artificial Intelligence Realized by a Non–Biologic Structure . . . 68
3.3 Basic Structures for Control 69
3.3.1 Open–Loop Control 69
3.3.2 Closed–Loop Control 70
3.3.3 "Conditional Feedback" Control Structure 71
3.4 Scopes for Intelligent Control 72
3.4.1 Methods of Intelligent Control 73
3.4.2 Application of Learning in Control Engineering 73
3.4.2.1 Example: Direct and Indirect Approach 74
3.5 Requirements for Adaptive Methods 75
3.5.1 Stability 76
3.5.2 Improving the Controller's Performance 78
3.5.3 Expandable Knowledge 78
3.6 Classification due to System's Structure or Restrictions 78
3.7 References 80

4 Nonlinear Function Approximators **83**
Michael Beuschel

4.1 Nonlinear Function Approximation 83
4.1.1 Concepts of Function Approximation 84
4.1.2 Basis Functions for Function Approximation 85
4.1.3 Universal and Convergent Function Approximation 86
4.2 Neural Networks as Function Approximators 88
4.2.1 Radial Basis Function (RBF) Network 88
4.2.2 General Regression Neural Network (GRNN) 89
4.2.2.1 GRNN at Multidimensional Input Space 91
4.2.2.2 Polynomial Activation (DANN) 92
4.2.2.3 Restricted Update Area 92
4.2.3 Other Neural Network Approaches 93
4.3 Neuro–Fuzzy Systems as Function Approximators 94

4.3.1 Principles of Neuro–Fuzzy Systems 94
4.3.2 Neuro–Fuzzy Example: Tuning of Output Fuzzy Sets 96
4.4 Example 98
4.5 Conclusion 101
4.6 References 102

5 Systematic Intelligent Observer Design 105
Ulrich Lenz

5.1 Definitions: Dynamic Systems Containing an Isolated Nonlinearity 107
5.1.1 Dynamic System with an Isolated Nonlinearity 108
5.1.2 Approximation of a Static Nonlinearity 110
5.2 Hybrid Notation of Signals in the Time and Frequency Domain 111
5.3 Systematic Observer Design 111
5.3.1 Conditions 111
5.3.2 Observer Design for Identification 113
5.3.2.1 Observer Approach 113
5.3.2.2 Dimensioning the Observer Feedback Matrix L 113
5.3.2.3 Specification of the Error Transfer Function $H(s)$ 114
5.3.2.4 Deriving a Stable Adaptation Law Using Known Error Models . 115
5.3.2.5 Reflections on Parameter Convergence 121
5.3.2.6 Simplifying the Observer Design 124
5.4 Intelligent Observer Design Following the Luenberger Approach 124
5.4.1 Prerequisites 124
5.4.2 Systematic Observer Design 126
5.4.2.1 Deriving the Error Transfer Function 126
5.4.2.2 Adaptation Law 127
5.5 Summary 131
5.6 References 132

6 Identification of Separable Nonlinearities 135
Franz Hangl

6.1 Plants with Separable Nonlinearities 135
6.2 Nonlinear Observer Approach 136

6.2.1 Identification with Accessible States 136
6.2.1.1 Adaptive Observer According to the Luenberger Observer . . . 136
6.2.2 Identification of Nonlinearities in Plants with Unknown Internal States 139
6.2.2.1 Neural Observer Approach 139
6.2.3 The Error Decoupling Filter 141
6.2.3.1 The Adaptive Law 143
6.3 Implementation of A–Priori Knowledge 144
6.3.1 Additive A–Priori Knowledge 145
6.3.2 Multiplicative A–Priori Knowledge 146
6.4 References 148

7 Identification and Compensation of Friction 149

Thomas Frenz

7.1 Introduction 149
7.2 Design of Hardware 155
7.3 Implementation: Learning of Friction Characteristic 157
7.4 Application: Compensation of Friction Influence 160
7.5 Conclusion 163
7.6 References 165

8 Detection and Identification of Backlash 167

Dieter Strobl

8.1 Introduction 167
8.2 Example System for Backlash Identification 168
8.2.1 Model of an Elastic Two–Mass System 168
8.2.2 Identifiability of the Backlash Characteristic 169
8.2.3 State Space Description of the Nonlinear System 170
8.3 Identification of Backlash 172
8.3.1 Representation of Backlash for the Identification with a Neural Network 172
8.3.2 Load–Side Backlash Observer (LBO) 174
8.3.2.1 State Space Representation 174

8.3.2.2 Observer Design and Error Model 174
8.3.3 Motor–Side Backlash Observer (MBO) 178
8.3.3.1 State Space Representation 178
8.3.3.2 Observer Design and Error Model 179
8.4 Simulation Examples 180
8.5 Experimental Validation 183
8.5.1 Experimental Set–Up and Parameters 183
8.5.2 Results of Online Backlash Identification 184
8.6 Conclusion 187
8.7 References 188

9 Identification of Isolated Nonlinearities in Rolling Mills 189
Stephan Straub

9.1 Introduction 189
9.2 Neural Networks in Rolling Mills 189
9.2.1 Plant Description 189
9.2.2 Compensation of Winder Eccentricities 191
9.2.3 Identification of the Roll Bite 198
9.3 Experimental Results 204
9.3.1 Plant Description 204
9.3.2 Identification Results 207
9.3.3 Compensation Results 209
9.4 Conclusion 214
9.5 References 215

10 Input–Output Linearization: an Introduction 217
Kurt Fischle

10.1 A Useful Canonical Form for Nonlinear Systems 217
10.2 Basic Concept of Input–Output Linearization 223
10.3 Simplified Ideal Control Law 228
10.4 Short Summary 232
10.5 References 233

11 Stable Model Reference Neurocontrol **235**

Kurt Fischle

11.1 Introduction 235

11.2 Description of the Concept 237

11.3 Application Example: Nonlinear Second–Order System 240

11.3.1 Simulation Results 240

11.3.2 Experimental Results 244

11.4 Modifications 247

11.4.1 Modification for Plants with Control Saturation 249

11.4.2 Modification for Plants with $L_g L_f^{n-1} h(\underline{x}) \neq$ const. 249

11.4.3 Method with Differentiation of y 250

11.4.4 Modifications for Reduction of the Learning Times 250

11.5 Short Summary 251

11.6 References 253

12 Dynamic Neural Network Compositions **255**

Stephan Straub

12.1 Introduction 255

12.2 Classification of Identification Methods 256

12.2.1 Motivation 256

12.2.2 Different Net Structures 257

12.2.3 Nonlinear Observer Structures and Dynamic Identificators . . . 260

12.3 Identification of Systems with Unknown Structure Using a Dynamic Identificator 265

12.3.1 Motivation and Theoretical Approach 266

12.3.2 Design of a Dynamic Identificator 269

12.3.3 Possible Control Concepts 272

12.3.4 Simulation 1: Example 274

12.3.5 Simulation 2: Two–Mass System 275

12.3.6 Simulation 3: Inverse Control 278

12.4 Conclusion 280

12.5 References 281

13 Further Strategies for Nonlinear Control with Neural Networks 283

Martin Rau, Anne Angermann

13.1 Introduction . . . 283
13.2 Compensation and State–Space Control Strategies for a Class of Nonlinear Systems . . . 284
13.2.1 Systems with Isolated Nonlinearities and Nonlinear Observer . . 285
13.2.2 Exact Compensation of Isolated Nonlinearities . . . 285
13.2.2.1 Transfer Function Description of the Nonlinear System . . . 286
13.2.2.2 Compensation Algorithm . . . 287
13.2.2.3 Realization of the Compensation Filter $K(s)$ . . . 288
13.2.3 State–Space Control of the Compensated System . . . 291
13.2.3.1 Simulation Example . . . 293
13.2.4 Conclusions . . . 297
13.2.5 Alternate Compensation and Control Design . . . 297
13.3 Nonlinear Control with a Controllable Canonical Form . . . 302
13.4 Nonlinear Control Design with Neural Networks and Numerical Optimization . . . 304
13.4.1 Model Reference Neuro Control for Systems with Isolated Nonlinearity . . . 305
13.4.2 Considerations on Numerical Optimization in Nonlinear Control 307
13.5 Time–Optimal Tension Control of Continuous Moving Webs Systems . . . 308
13.5.1 Introduction . . . 308
13.5.2 Controller Design . . . 313
13.5.3 Experimental Validation . . . 321
13.5.4 Conclusion . . . 323
13.6 References . . . 326

List of Figures 336

Index 337

1 Introduction — Control Aspects

Dierk Schröder

When considering nonlinear control, the variety of control aspects is extremely wide. Due to this situation it is necessary to provide an overview of this book's objectives. However, this introduction cannot cover every detail of all existing linear and nonlinear control strategies. In figure 1.1 and figure 1.2 basic structures of control methods are shown and these structures will enable us to identify the scope of this book.

Figure 1.1 covers control approaches for linear plants, figure 1.2 for nonlinear plants. The shaded boxes in figure 1.2 show the aspects which are discussed in this book.

In figure 1.1 we distinguish between plants which are linear and precisely known in structure and parameters (left side), and plants which are linear and imprecisely known (right side).

The mature linear control theories like cascaded control or state space control can be used for the plants which can be seen on the left hand side [8, 9, 12, 13, 26, 27, 39, 41].

But normally the parameters of the plant are not precisely known or are time–variant due to a large variety of reasons; for example aging or temperature dependent behaviour of the plant. They can also be unknown depending on the process being used or if there are structural uncertainties. In this situation adaptive or robust control is advantageous [1, 2, 14, 15, 19, 24, 32, 44].

Additionally we can use nonlinear controllers like two–step or hysteresis–band control for linear plants, other solutions are controllers for time–optimal behaviour [37].

All these control strategies are not the scope of this book. This also holds for the control of nonlinear plants, when there are no parametric and structural uncertainties and the nonlinearities are smooth, the states are measurable and the output signal can be differentiated as often as necessary [12, 22, 28, 43, 44]. Additionally we will not discuss the strategy of "local linear models" (e.g. LOLIMOT [33], NARX or NOE [25], [16, 30]). The approach for this interesting strategy is the assumption, that we can approximate the real behaviour of the plant by

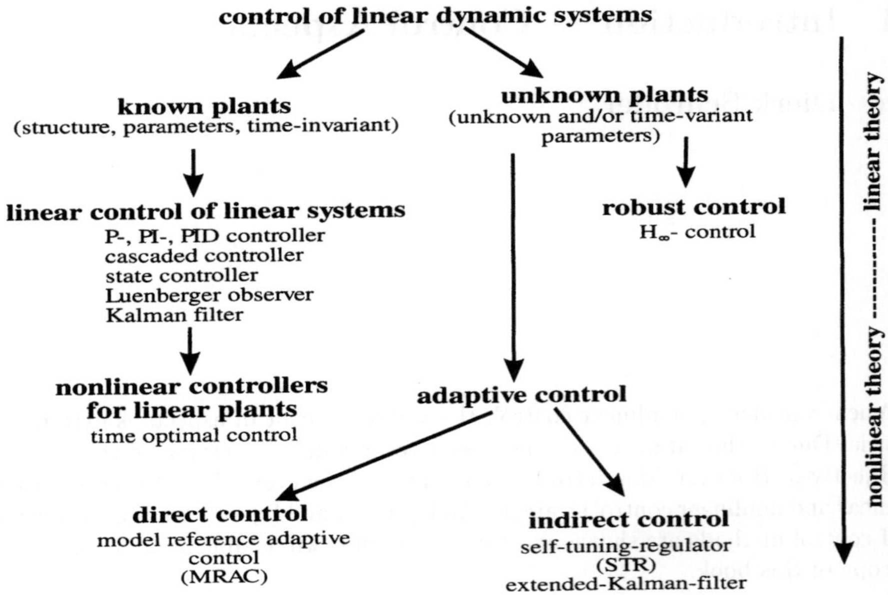

Figure 1.1: Fundamental structures of control methods: linear plants

different linear models in different operating sections. That means, at least the parameters of the plant are different in these sections. Therefore, the objective of this strategy is the identification of the parameters of the plant in different operating sections; this problem can be solved with standard methods [17, 18]. The step ahead of this new approach "local linear models" is the opportunity to use additional and independent signals for the definition of the sections to achieve a more complete system representation. These independent signals are not control variables or states of the plant. The main benefit of "local linear models" is their ability to approximate multidimensional nonlinear dynamic systems. This black box approach is useful especially for plants without knowledge about their structure, but often leads to results that are hardly interpretable physically.
Here, we will consider plants of known structure with a linear part and one or more nonlinearities. Our approach enables us to to identify these nonlinearities, and thus create a more detailled model of the system under consideration and take advantage of the improved model for control purposes. In mechanical systems or motion control, the system's structure is usually known, but the nonlinearities friction, backlash or a real hysteresis loop are unknown. This is the typical field of application of our approach.

We begin our considerations in this book, with a nonlinear, imprecisely known plant; by "imprecise" we mean, the linear part is exactly known, but the nonlin-

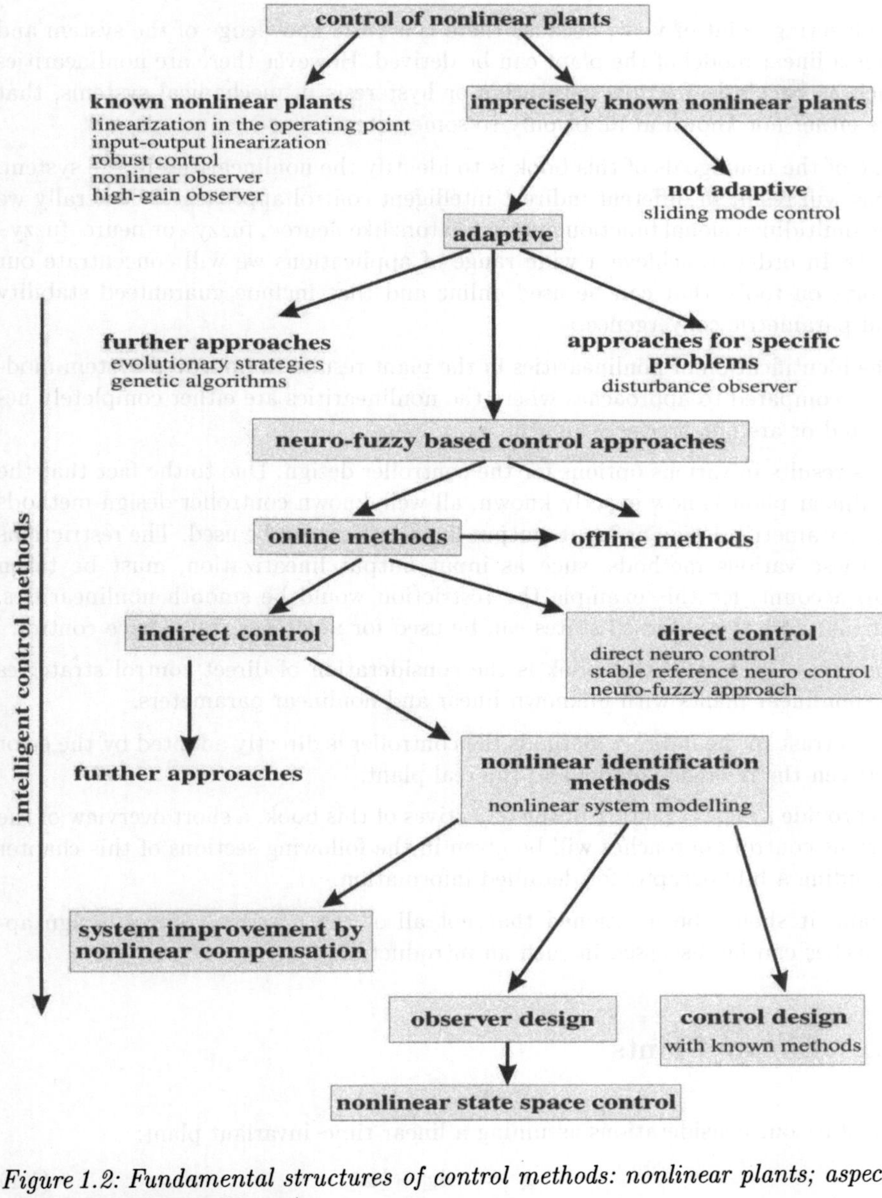

Figure 1.2: Fundamental structures of control methods: nonlinear plants; aspects of this book (shaded boxes)

earities are not known. We define such nonlinear systems as nonlinear, separable plants. This assumption is very often valid in real control applications from the engineering point of view, because there is a basic knowledge of the system and thus a linear model of the plant can be derived. However there are nonlinearities such as backlash, friction, saturation or hysteresis in mechanical systems, that are either not known at all or only to some extent.

One of the main goals of this book is to identify the nonlinearities in the system. This will result in different indirect intelligent control approaches. Generally we use multidimensional function approximators like neuro–, fuzzy– or neuro–fuzzy–tools. In order to achieve a wide range of applications we will concentrate our efforts on tools, that can be used online and that include guaranteed stability and parametric convergence.

The identification of nonlinearities in the plant results in improved system modeling compared to approaches where the nonlinearities are either completely neglected or are not precisely known.

This results in various options for the controller design. Due to the fact that the nonlinear plant is now exactly known, all well–known controller design methods like parametric design or input–output linearization can be used. The restrictions of these various methods, such as input–output linearization, must be taken into account; for this example the restriction would be smooth nonlinearities. Furthermore the observed states can be used for nonlinear state space control.

Another objective of this book is the consideration of direct control strategies for nonlinear plants with unknown linear and nonlinear parameters.

In contrast to the indirect methods the controller is directly adapted by the error between the reference model and the real plant.

To provide a clearer picture of the objectives of this book, a short overview of the various control approaches will be given in the following sections of this chapter including a bibliography for detailled information.

Again, it should be mentioned that not all of the possible control design approaches can be discussed in such an introduction.

1.1 Linear Plants

We start our considerations assuming a linear time–invariant plant:

$$\dot{\underline{x}} = \underline{A}\underline{x} + \underline{b}u \tag{1.1}$$

$$y = \underline{c}^T\underline{x}\,, \tag{1.2}$$

where $\underline{x}$ are the states of the system and the system matrices $\underline{A}$, $\underline{b}$ and $\underline{c}$ are known and time–invariant; y is the measurable output signal of the system. Under these assumptions linear control strategies can be used to control the plant. The linear control theory is well–known, has a variety of powerful methods and a long history of successful applications.

Normally these assumptions are valid neither during the whole period of operation nor in the full range of operation, because the system has time–variant parameters due to, for example, aging or a nonlinear static characteristic like saturation. If we can neglect these effects, we can use cascaded control, conditional feedback structures for the controller or state space control. Each of these techniques has specific advantages (figure 1.1) [8, 9, 12, 13, 26, 27, 39, 41].

As examples:

- Cascaded control is used very often, because the plant is divided from inner to outer control loops which can be put into operation one after the other. The influence of nonlinearities in the inner sections of the plant like pulse width modulation (PWM) for actuators will be attenuated. Disturbance rejection is favorable because the disturbances are detected at the next feedback signal in the cascaded system. However, the dynamic behaviour will decrease with an increasing number of cascaded control loops, where the innermost loop will have the best dynamic response.
- The response to set point variations and the disturbance rejection behaviour can be set independently by conditional feedback.
- State space control is very advantageous, because it will provide very good dynamic behaviour, as will be shown in the next chapter.

1.2 Linear Plants with Uncertain Parameters

Adaptive or robust control starts, when the parameters of the linear plant are either not exactly known or time–variant. The basic idea in adaptive control is first the identification of the uncertain or variant parameters; second the decision of how to change the parameters of the controller; and third the modification of these parameters.

There are two main approaches for adaptive controllers [32, 44]: first the "self–tuning controller" (STC) shown in figure 1.3 and second the "model reference adaptive control" (MRAC) shown in figure 1.4. Using the self–tuning method the procedure mentioned above is used, since the parameters a_i and b_i as elements of $\underline{A}$ and $\underline{b}$ are estimated. The control law is known; therefore the decision process is fixed, the controller parameters can be calculated based on the identified knowledge and the modification of the controller parameters follows as the last step.

This self–tuning method is an indirect control method, because identification (estimation) and controller adaptation are separated; moreover, this method provides information of the plant parameters.

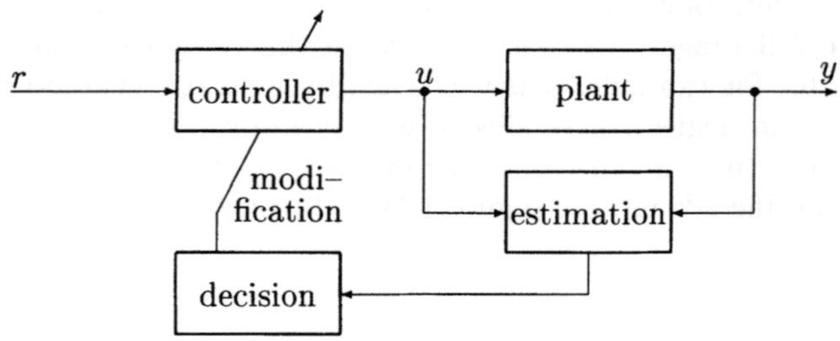

Figure 1.3: Self–Tuning Controller (STC, indirect control method)

For linear plants there are many techniques to estimate the unknown parameters like the least–squares–method and its extensions. For the controller there are many design techniques such as PID, pole placement, linear quadratic control, minimum variance control or H_∞–control. A wide variety of self-tuning systems can be obtained by combining different estimation and control techniques.

A model reference adaptive control system (figure 1.4) is composed of four parts: the plant with parameters which are time–variant or not exactly known, the reference model, which represents the desired input–output behaviour of the controlled system, the controller with adjustable parameters and an adaptation mechanism.

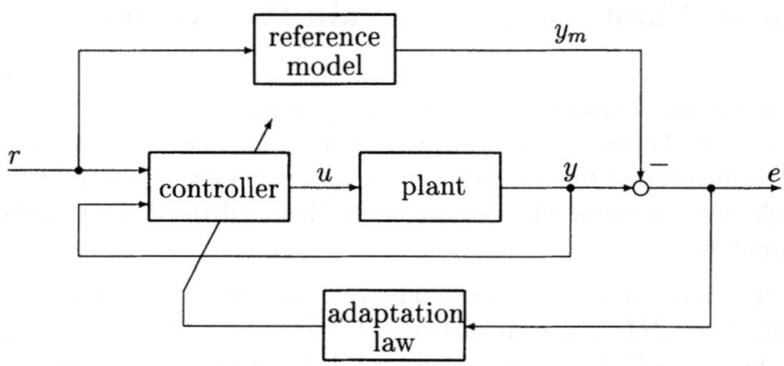

Figure 1.4: Model Reference Adaptive Control (MRAC, direct control method)

In this control method the adaptation process is stimulated by the error e. It should be considered that the constraints of the real plant must be taken into account in the design of the reference model.

MRAC is a direct control method, because the controller parameters are directly adapted from the error signal e; in this selection no data of the varying plant parameters are available.

As described above, STC and MRAC control start from different perspectives. In STC the parameters of the plant itself are estimated, in MRAC the error between the reference model output and the real plant is minimized. Very important for both strategies are the stability and convergence of the adaptation process. Generally STC design is more flexible, but the proof of stability and convergence can be difficult. Furthermore the input signals of the plant must be "rich" during the adaptation process. "Rich" means the input signals must perturb all states of the plant in such a way that the adaptation process is possible. Under certain circumstances MRAC systems can adapt without "rich" signals.

It must be noted here that it is necessary to use nonlinear criteria like the Lyapunov theory to guarantee stability during the adaptation process. Both control structures can be extended to nonlinear control design, if either neural nets or fuzzy logic are integrated [11, 31].

When the structure of the plant is known and STC provides information of the parameters of the plant, then the states of the system can be observed.

In real applications the results of the observers are influenced by noisy measurement signals and additional time–variant disturbances. In these cases the Kalman–Bucy–Filter is used for linear systems to get the optimal observer result (observed states) according to a defined criterion function (mean square estimation error). This function is based on the knowledge of the relevant time–invariant or time–variant system matrices. The filter model must reflect the real system exactly. An extension of the proposed method is the Extended–Kalman–Filter which can be used for state observation in nonlinear systems on the one hand and for linear systems with unknown parameters on the other hand. The second approach is based on an augmented state space where the unknown parameters represent additional states. The adaptation is done via the error between the measured and observed outputs. This method can be extended to the identification of neural net parameters [38, 35] if nonlinear functions have to be identified. The identification result can be used for control purposes (online or control design). One disadvantage is that the parameter convergence cannot be guaranteed if the Extended–Kalman–Filter is used for the identification of parameters in linear and nonlinear systems [29, 37].

In some cases robust approaches are sufficient to reach the desired control purposes, especially when possible adaptive methods cannot be applied for different reasons [1, 14, 15]. Robust control methods are used if systems, linear or nonlinear, are not known exactly. One approach to get the desired control behaviour in every operating point is the H_∞–control [26]. The controller is designed in such a way, that additive and multiplicative deviations in the reference and disturbance transfer functions are taken into account. The control parameters are determined such that the sensitivity of the whole transfer function is low against parameter deviations (frequency response method). In contrast to this, the parametric space method [1] is based on the mapping of the desired poles into the space of the uncertain parameters with the limits of the assumed variations. Further explanations can be found in literature.

Many books and publications concentrate on these control problems. The following books that are related to the subject should be mentioned [2, 10, 19, 21, 23, 24, 32, 44].

1.3 Linear Plants and Nonlinear Controllers

There is another approach for linear plants, where the structure and the parameters of the plant are known and time–invariant. This approach uses a very specific nonlinear controller, which can be a two–step– or a hysteresis band–controller. Using such controllers results in an actuator signal in the plant, where the actuator normally supplies the maximum positive or the maximum negative output signal (saturation). The output signal of the plant will show a motion inside the hysteresis band or a "chattering" when the response from the output signal of the plant to the output signal of the actuator is delayed and not taken into account during the design.

Such controllers can be used advantageously on the one hand, when the plant has low order, low pass behaviour, and the actuator allows very high switching frequencies, and, on the other hand, in an extension of this strategy, when the parameters of the plant are not exactly known (parametric uncertainties). Due to the saturated actuator signal the output signal of the plant will respond with the fastest transient slope and the imprecise knowledge of the parameters of the plant will vary this slope only to some extent. Furthermore the high switching frequency of the actuator limits the overshooting of the output signal. A reduction of the switching frequency of the actuator or comparable to this of the bandwidth of the control loop can be achieved when, instead of the bang–bang–controller, a hysteresis–band–controller is used. The wider the hysteresis band and the greater the time constants of the plant, the lower the desired switching frequency of the actuator will be.

If there is imprecise knowledge of the plant's order, then this will be acceptable to some extent as long as these unstructured uncertainties are at much higher frequency bands than the control loop.

Another type of nonlinear "controller" are control loops where the actuator uses pulse width modulation (PWM). PWM means that the ouput signal of the actuator is switched for a period of time to the positive maximum and another period of time to the negative maximum of the output signal. (A two–step actuator; there are also three– and multi–step actuators in power electronics.) By varying the two periods of time and keeping the sum of both periods of time constant, a mean value of the output signal can be achieved. During transients the maximum positive or the maximum negative output signal can be switched by the actuator for a longer period of time, thus again the fastest possible transient slope results. Due to the two different periods of time a chattering of the output signal cannot be avoided; this chattering will be attenuated the higher the order of the plant and the shorter the sum of both periods will be. More complex PWM–strategies are known for three–phase systems [42].

The basic idea of using a nonlinear controller with an actuator driven to positive or negative saturation for a linear, time–invariant plant can be extended — for example — to a third control strategy: the time optimal control strategy. where the controller forces the actuator for specific periods of time to positive or negative maximum output signal, to achieve the fastest possible transient response. The duration of these specific periods are determined by the order and the parameters of the plant [37].

A fourth type of switching controller strategy is the sliding mode control [12, 36, 44, 46, 47]. The fundamental idea for this control strategy is to design in

- a first step a sliding manifold — or in the $\underline{x}$, $\dot{\underline{x}}$–phase plane of a second order system a sliding surface;
- second to find a control signal, which forces the trajectories of the plant that the sliding manifold is attractive and stability is achieved near the sliding manifold, and
- third remaining near the sliding manifold, when it is reached [46, 47].

Again "chattering" will occur, when the sliding manifold is reached due to imperfections in the control loop and/or unstructured uncertainties. There are various approaches to reduce or avoid chattering.

1.4 Nonlinear Plants

As mentioned in the figures 1.1 and 1.2 and explained in the introduction the objectives of this book are online identification of nonlinearities, nonlinear observation of the states and related controller design strategies. We will assume an engineering point of view during the first steps of considerations. Engineering point of view means that we have a basic knowledge of the structure and the parameters of the linear part of the plant. This combination of requirements will be reduced later step by step.

Thus we have a list of combined assumptions:

1. a plant which has a known structure, known parameters of the linear part of the plant and a static nonlinearity, that is precisely known;

2. a plant which has a known structure, known parameters of the linear part of the plant and a static nonlinearity, that is not known;

3. the same situation as in the first example with the "property" that the error–transfer–function is not "strictly positive real" (SPR); this term will be explained later on;

4. as the first and second example but there are several nonlinearities, which are not known;

5. there is a restriction, that the number of measurable states is smaller than the number of nonlinearities;

6. a plant which has a known structure, the parameters of the linear part and the parameters of the nonlinearities are not exactly known; and

7. the structure, the parameters and the nonlinearities are not exactly known, but the order of the linear system is known and — a very important requirement — the nonlinearity should be smooth; this includes all plants, where the whole system is described by the equation $\dot{\underline{x}} = f(\underline{x},\ \underline{u})$.

For the first type of plant, there are many possible solutions to design the controller. The most simple case would be a nonlinear plant, where the static nonlinearity is at the input of the plant (Hammerstein–type model). In this very particular situation the known static nonlinearity can be compensated by an inverse nonlinearity — if it exists — at the entrance of the plant. Due to this compensation of the nonlinearity, the resulting system is linear and the mature linear design strategies for the controller can be used.

The controller design is more difficult, if the known nonlinearity is not at the entrance of the plant. Again, there are many different solutions for the controller design. Linearization at specific operating points to achieve a linear input–output behaviour of the plant, using these linear input–output representations of the nonlinear plant and designing the controller for only these specific operating points is an approach. During this strategy we considered only the operating points and the first derivatives at the operating points (method of the first approximation) [20], [34].

Another family of linearization methods are extended– or the gain–scheduling–linearization [4, 5, 10, 32, 48]. The main idea of this approach is the concept of a linearization at relevant equilibrium points of the nonlinear plant (extended linearization). Then the controller is designed for the linearized equilibrium points to achieve in the equilibrium points the same pole–zero–combination for the closed loop, thus the characteristic equation is invariant to the chosen operating point. Therefore in a narrow area around these equilibrium points the same transient behaviour is obtained.

It must be noted, that the mentioned linearization procedures will provide a chance for the application of linear control strategies for the specific operating points. But during transient operations the plant will leave the operating points and thus the system can operate between these operating points. Due to this divergence from the assumptions during the controller design, no absolute statements for the transient behaviour are possible in this situation. Therefore it seems necessary to use only such states of the plant, which have a low dynamic behaviour and also give a representation of the nonlinearity of the plant. The same constraints must be accepted when considering the proof of stability.

Up to now linearization at the operating points and a controller design with linear design strategies have been considered. This method tries to use linear methods for a nonlinear plant and requires smooth nonlinearities.

A different approach to control a nonlinear, precisely known plant with smooth nonlinearities is the feedback linearization (input–output linearization, global linearization, input state linearization) [7, 11, 21, 32, 40, 44, 45, 48]. This approach starts with the idea to differentiate the output signal y of the nonlinear input–output equation of the plant as often as necessary in order to achieve an equation where the input signal u of the plant appears for the first time — this equation represents an explicit relationship $y^{(n)} = f(\underline{x}, \underline{u})$ between the input u and the output signal y. That means in this context that the nonlinear plant can be represented by a chain of integrators and a static nonlinear function $f(x, u)$ at the input. When this nonlinear static function $f(x, u)$ is compensated by the inverse function, the linear integrators remain. The controller design for such a linear chain of integrators is possible with linear or nonlinear (for example time–optimal [37]) control strategies. Another possible approach would be a linear transfer function of a desired system behaviour instead of the integrator chain (former desired control behaviour after linearization). This will result in a different explicit relationship between u and y.

The input–output linearization needs several important requirements: first — as mentioned above — the structure and the parameters of the linear and nonlinear components must be precisely known, second the nonlinearity must be smooth and third all states of the plant must be available, because they are necessary for the design of the modified input signal v. There is another difficulty, which must be taken into account. When we consider a linear system and its transfer function, we have to study whether the transfer function of the plant has zeros of the numerator of the plant in the left (stable, minimum phase system) or in the right (unstable, non–minimum phase system) half plane of the s–plane. It is well–known, that a right–half–plane–zero can result in an unstable pole of the linear system. For nonlinear systems the situation is basically the same; instead of the zeros of the transfer function, the so–called "zero dynamics" has to be considered here. The concept of zero dynamics is discussed in chapter 10.

The mathematical tools for the input–output linearization are shown in a short summary in chapter 10 of this book; we refer to numerous publications.

The concept of input–output linearization for SISO–systems can be extended to MIMO–systems. But it must be mentioned, that any imprecise knowledge of the structure and moreover the parameters in the MIMO–plant will result in a failure of decoupling of the different subsystems.

Let us summarize the most important limitations of the very valuable input–output linearization:

- it cannot be used for all nonlinear systems (smooth nonlinearities or local smooth nonlinearites are required),
- the full state has to be measured,
- no robustness is guaranteed in the presence of imprecise parameters or unmodelled dynamics.

Up to now we have considered a nonlinear plant with smooth nonlinearities and well–known structure and parameters. The input–output linearization is not applicable when the parameters are imprecisely known or the nonlinearities are not smooth. There are many other proposed strategies which cannot be mentioned here.

Some of the mentioned nonlinear controller design methods need the states of the plant, but these states are not always fully measurable. These non–measurable states must be reconstructed by an observer. It must be stated that the design of these observers for nonlinear plants is as complex as the controller design. As it has been discussed above there is no generally applicable controller design for nonlinear plants; a comparable situation exists for the observer design.

To provide a short introduction to this subject the following statements can be found in literature [10, 29]:

- Extended–Kalman–Filters need the complete plant data for design, they minimize the mean square estimation error, and there is no a–priori performance or stability guarantee.
- A constant gain Extended–Kalman–Filter has an intrinsic robustness against modeling errors, but the margins can be quite restrictive, e.g. when one deals with hard nonlinearities.
- A global linearization observer is derived from the global linearization method, the same restrictions are valid.

- A pseudo–linearization method transforms the original plant into an observer canonical form; the design is easy when the canonical form is derived, but only local properties can be guaranteed and all data of the plant is again required.

- Extended linearization observer design starts from gain–scheduling techniques, the same advantages and disadvantages as mentioned for the controller design must be accepted.

- Thau's observer design method provides sufficient conditions for estimation convergence for some classes of nonlinear systems. The fundametal assumption — as in some of our approaches — is to assume separable nonlinearities, which must be known (which is not our presumption), use the Lipschitz Criterion to achieve an assymptotically convergence of the dynamic error function; but the drawback is that it might not be able to handle modelling errors.

- "Local linear models" can be used for the identification of nonlinear plants [33, 16, 30].

- There are other approaches like a design starting from the "set theoretic approach" or "sliding mode observer", these methods should not be discussed here.

This list of different control and observer design strategies shows very interesting and valuable solutions for nonlinear plants. It can be seen, that there are some restrictions and due to this many research groups are actively working on reducing or avoiding these restrictions in order to improve the strategies for nonlinear systems. A small list of different books and publications is given at the end of this introduction to give an improved view of these activities.

We would like to present some of our ideas and results.

1.5 Our Conceptions of Nonlinear Control Strategies and Observation

As mentioned before, we will use an engineering point of view. This often results in a linear part of the plant, which is known more or less precisely and additional nonlinearities, which are not known at all or known only to some extent. That means the system in general is a nonlinear one, not precisely known in parameters and perhaps in terms of the structure.

The list of different cases of nonlinear system configurations, that were discussed above, is the starting point in the following chapters.

An obvious idea in such a situation is to improve the model of the nonlinear plant by extraction of additional knowledge. To achieve this improvement, we decided to use intelligent strategies like fuzzy–logic, neural or neural–fuzzy–learning approaches.

When we have obtained an improved nonlinear model of the plant and thus have gained a deeper system understanding then we can decide on various extensions like nonlinear observer design, compensation of the nonlinear effects, or nonlinear state space approach as examples. For more details refer to figure 1.2.

To avoid a too complex learning process we decided to use as much a–priori knowledge of the system as possible. This coincides with the engineering point of view and results in the use of every item concerning the structure or parameters of the plant. This presupposition influences the learning strategy and structure of the identification unit. It should be noted here, that using as much a–priori knowledge as possible supports the learning process. Considering this, we additionally place great value in intelligent strategies with guaranteed stable and convergent learning, which is in our opinion an inalterable requirement in nonlinear control engineering. This enables us furthermore to use these methods for online applications.

We must state that this activity "nonlinear plant identification, observation and nonlinear control" is a very wide area and will be an extremely interesting subject for future research, which cannot be fully covered by a small group of researchers. Therefore in this book only the state of the art at the Institute is presented.

1.6 References

[1] Ackermann, J.:
Robuste Regelung.
Springer–Verlag, Berlin, Heidelberg, 1993.

[2] Aström, K.J.:
Adaptive Control.
Addisson Wesley, 1989.

[3] Aström, K. J.:
Adaptive Control around 1960.
IEEE Transactions on Control Systems, Vol. 111, 1989, pp. 44–49.

[4] Baumann, W.T. und Rugh, W.J.:
Feedback Control of Nonlinear Systems by Extended Linearization.
IEEE Transactions on Automatic Control, 1986, Vol. 31, pp. 40–46.

[5] Baumann, W.T.:
Feedback Control of Multi–input Nonlinear Systems by Extended Linearization.
IEEE Transactions on Automatic Control, 1988, Vol. 33, pp. 193–197.

[6] Bushnell, L.G.:
On the History of Control.
IEEE Transactions on Control Systems, June, 1996, pp. 14–17.

[7] Byrnes, Isidori, Willems:
Passivity, Feedback Equivalence, and the Global Stabilization of Minimum Phase Nonlinear Systems.
IEEE Transactions on Automatic Control, Vol. 36, No. 11, Nov. 1979, pp. 1228–1240.

[8] D'Azzo, J.J.:
Linear Control System Analysis and Design.
McGraw–Hill, New York, 1995.

[9] Driels, M.:
Linear Control Systems Engineering.
McGraw–Hill, New York, 1996.

[10] Engell, S.:
Entwurf nichtlinearer Regelungen.
R. Oldenbourg Verlag, München, 1995.

[11] Fischle, K.:
Ein Beitrag zur stabilen adaptiven Regelung nichtlinearer Systeme.
Dissertation, TU München, 1998.

[12] Föllinger, O.:
Regelungstechnik.
Hüthig Buch Verlag, 6. Auflage, 1990.

[13] Föllinger, O.:
Nichtlineare Regelungen I, II.
R. Oldenbourg Verlag, München, 1993, 7th Edition.

[14] Francis, D.A.:
Robust Control Theory.
Springer–Verlag, Berlin, Heidelberg, 1995.

[15] Green, M., Limebeer, D.J.N.:
Linear Robust Control.
Prentice Hall New York, 1995.

[16] Hunt, K., Irwin, G. R., Warwick, K.:
Neural Network Engineering in Dynamic Control Systems.
Springer–Verlag, Berlin, Heidelberg, New York, 1995, pp. 61–82.

[17] Isermann R.:
Identifikation dynamischer Systeme Band 1.
Springer–Verlag, Berlin, Heidelberg, 1987.

[18] Isermann R.:
Identifikation dynamischer Systeme Band 2.
Springer–Verlag, Berlin, Heidelberg, 1988.

[19] Isermann R., Lachmann, K.H., Matko, D.:
Adaptive Control System.
Prentice Hall New York, 1992.

[20] Isidori, A.:
Nonlinear Control Systems.
Springer–Verlag, Berlin, Heidelberg, 1989.

[21] Kaufmann, H., bar–Kana, I.:
Direct Adaptive Control Algorithms.
Springer–Verlag, Berlin, Heidelberg, 1994.

[22] Khalil, H.K.:
Nonlinear Systems.
Prentice Hall, USA, 1996, 2nd Edition.

[23] Kokotovic, P.:
Foundations of Adaptive Control.
Springer–Verlag, Berlin, Heidelberg, 1991.

[24] Landau, J.D.:
Adaptive Control.
Marcel Dekker, Inc., New York, USA, 1979.

[25] Ljung, L.:
System Identification—Theory for the User.
Prentice–Hall, Englewood Cliffs, N.J., 1987.

[26] Ludyk, G.:
Theoretische Regelungstechnik 1.
Springer–Verlag, Berlin, Heidelberg, 1995.

[27] Ludyk, G.:
Theoretische Regelungstechnik 2.
Springer–Verlag, Berlin, Heidelberg, 1995.

[28] Marino, R., Tomei, P.:
Nonlinear Control Design.
Prentice Hall, London, 1995.

[29] Misawa, E. A., Hedrick, J. K.:
Nonlinear Observers – A State–of–the–Art Survey.
Journal of Dynamic Systems, Measurement and Control, Vol. 111, 1989, pp. 344–352.

[30] Murray–Smith, R.:
A Local Model Network Approach to Nonlinear Modelling.
Ph.D. Thesis, University of Strathclyde, UK, 1994.

[31] Miller, W.T., Sutton, R.S., Werbos, P.J.:
Neural Networks for Control.
Bradford Book, The MIT Press, Cambridge, Massachusetts, London, England, 1990.

[32] Narendra, K.S., Annaswamy, A.M.:
Stable Adaptive Systems.
Prentice–Hall International Edition, Englewood Cliffs, New Jersey, 1989.

[33] Nelles, O.:
LOLIMOT—Lokale, lineare Modelle zur Identifikation nichtlinearer, dynamischer Systeme.
Automatisierungstechnik 45, Vol. 4, Oldenbourg–Verlag, 1997.

[34] Nijmeijer, H., von der Schaft, A.J.:
Nonlinear Dynamic Control.
Springer–Verlag, Berlin, Heidelberg, 1990.

[35] Obradovic, D.:
On–line Training of Recurrent Neural Networks with Continuous Topology Adaptation.
Technical Report Siemens AG, Corporate Research and Development, ZFE ST SN 41, Munich, Germany, 1994.

[36] Palm, R.:
Model Based Fuzzy Control.
Springer–Verlag, Berlin, Heidelberg, 1997.

[37] Papageorgiou, M.:
Optimierung — Statische, dynamische, stochastische Verfahren für die Anwendung.
R. Oldenbourg Verlag, München, 1991.

[38] Puskorius, G.V., Feldkamp, L.A.:
Neurocontrol of Nonlinear Dynamical Systems with Kalman Filter Trained Recurrent Networks.
IEEE Transactions on Neural Networks, Vol. 5, No. 2, 1994, pp. 279–297.

[39] Rohrs, C.E., Melsa, J.L., Schultz, D.G.:
Linear Control Systems.
McGraw–Hill, New York, 1993.

[40] Schäffner, C.:
Analyse und Synthese neuronaler Regelungsverfahren.
Herbert Utz Verlag Wissenschaft, 1996.

[41] Schmidt, G.:
Grundlagen der Regelungstechnik.
Springer–Verlag, Berlin, Heidelberg, 1994.

[42] Schröder, D.:
Elektrische Antriebe 2.
Springer–Verlag, Berlin, Heidelberg, 1995.

[43] Schwarz, H.:
Nichtlineare Regelungssysteme.
R. Oldenbourg Verlag, München, 1991.

[44] Slotine, J.-J.E., Li, W.:
Applied Nonlinear Control.
Prentice–Hall International Editions, New Jersey, 1991.

[45] Tzirkel–Hancock, E., Fallside, F.:
A Direct Control Method For a Class of Nonlinear Systems Using Neural Networks.
Report CUED/F-INFENG/TR.65, Cambridge University Engineering Department, Trumpington Street, Cambridge CB2 1PZ, March 1991.

[46] Utkin, V.I.:
Variable Structure System with Sliding Mode in Discontinuous Systems I, II.
Automation and Remote Control, 1972.

[47] Utkin, V.I.:
Sliding Modes in Control and Optimization.
Springer–Verlag, Berlin, Heidelberg, 1992.

[48] Wang, J., Rugh, W.J.:
Feedback Linearization Families for Nonlinear Systems.
IEEE Transactions on Automatic Control, 1987, Vol. 32, pp. 935–940.

[49] Zeitz, M., Schaffner, J.:
Dezentraler Entwurf nichtlinearer Beobachter.
Workshop GMA–Ausschuss 1.4. Theoretische Verfahren der Regelungstechnik, Interlaken, Switzerland, 29.9. - 2.10. 1996.

2 Motion Control

Dierk Schröder

2.1 Control of Electromechanical Systems

2.1.1 Introduction

Up to now, research and development in motion control have focussed on the electrical drive only, without taking into account the potential of optimization of the total system consisting of the actuator and the mechanical and electrical components. However, due to these research activities it was possible to replace DC–machines by robust AC–machines which are controllable in torque and speed.

The first step in the achievement of this change was the replacement of the naturally–commutated inverters by self–commutated inverters consisting of switchable power semiconductors (e.g. GTO). Thus, a controllable voltage source for AC–machines became available. The second difficulty that had to be overcome was the control of the AC–drives. The two existing theoretical approaches, decoupling and field orientation, were well–known. Their practical application was possible by implementing them in digital microcontrollers.

Meanwhile, AC–drives with voltage source inverters and digital field oriented control are state of the art. These AC–drives achieve at least the same or better static and dynamic characteristics than DC–drives. Furthermore new considerations are emerging to achieve higher switching frequencies for the voltage source inverters. Therefore, much better characteristics can be achieved in the near future [1]. Is it sensible to put so much effort in the optimization of one component, without looking at the total system?

If we consider many practical applications of electrical drives, the cascaded control loop configuration is the solution up to now. In these cascaded control loop configurations a current control loop is normally the most inner loop, superimposed by a speed control loop which is again superimposed by a position control loop. The parameters of these control loops are usually optimized to avoid excitation of resonant effects in the mechanical system. Therefore, the cross–

over–frequencies ω_D of the open loops have to be kept low and the dynamic possibilities of the drive can often not be fully exploited.

Of course, cascaded control could be replaced by state–space control. This holds especially when the mechanical system is a linear two–mass system with an elastic shaft. State–space control requires the exact knowledge of the system's structure and parameters. However, the damping factor d of the elastic shaft is usually not known exactly and the poles of the current control loop cannot be influenced by pole placement as a result of the already existing optimization of this control loop. Therefore, some parameters are difficult to determine. Moreover, to get the speed of the second mass, an additional sensor or an observer is necessary. This holds for the difference angle of the shaft as well.

The linear state–space approach fails, if nonlinearities like friction and backlash occur in the mechanical system. Additionally, in most practical applications we have to deal with multi–mass systems. Thus the complexity of the mechanical system increases considerably and consequently the design of the controller also becomes more complex.

Furthermore, the actuator (electrical drive) and the mechanical system are usually part of a production plant and process, whose additional complexity must be taken into account. In the following sections we will have a closer look at these problems with the help of two examples.

2.1.2 Cascaded Control

In real applications the actuator and the mechanical system are often delivered from different partners. These partners agree to define specific characteristics of the components to guarantee that the requirements of the client are met. If these components are mismatched, their strong interdependence can generate undesired effects which could impose servere restrictions. The following example shall be a first step into the problems of motion control.

Figure 2.1 shows the signal flow graph of a two–mass system, where not only the behaviour of the elastic shaft is taken into account but also the usually occurring nonlinearities: backlash and friction. In most applications such a model of the plant is sufficient for the controller design.

For the following considerations, the dynamic transfer function $G_M(s)$ of the actuator (elctrical drive) is approximated by a first order delay

$$G_M(s) = \frac{M_M(s)}{M_M^*(s)} = \frac{1}{1 + sT_\sigma} \tag{2.1}$$

T_σ: time constant of torque control loop

which is not shown in figure 2.1. The mechanical part of the actuator is represented by the moment of inertia Θ_1.

Now we want to design a cascaded speed control for the considered system; therefore, we have to neglect both nonlinearities shown in the signal flow graph. In many practical applications this cannot be done and we need special nonlinear observers and controllers. Approaches like these will be presented in chapters 7, 8, 9.

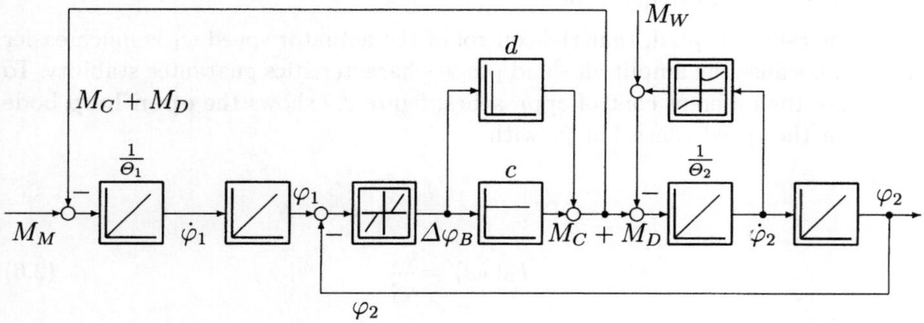

Figure 2.1: Signal flow graph of a two–mass system with friction and backlash effects

It has been a well–known fact for many years and it is a standard example in advanced lecture books, that due to this resonant mechanical system very severe problems can appear.

Let us consider both transfer functions of the linear mechanical system

$$G_{S1}(s) = \frac{\dot{\varphi}_1(s)}{M_M(s)} = \frac{1}{s(\Theta_1 + \Theta_2)} \frac{1 + s\frac{d}{c} + s^2\frac{\Theta_2}{c}}{1 + s\frac{d}{c} + s^2\frac{\Theta_1\Theta_2}{(\Theta_1 + \Theta_2)c}} \tag{2.2}$$

$$G_{S2}(s) = \frac{\dot{\varphi}_2(s)}{M_M(s)} = \frac{1}{s(\Theta_1 + \Theta_2)} \frac{1 + s\frac{d}{c}}{1 + s\frac{d}{c} + s^2\frac{\Theta_1\Theta_2}{(\Theta_1 + \Theta_2)c}} \tag{2.3}$$

with the input value actuator torque M_M and the output values actuator speed $\dot{\varphi}_1$ and load speed $\dot{\varphi}_2$. For the resonance frequency Ω_{0N} of the mechanical system we obtain

$$\Omega_{0N} = \sqrt{\frac{(\Theta_1 + \Theta_2)c}{\Theta_1\Theta_2}} \tag{2.4}$$

which does not depend on the damping factor d. In both transfer functions the first part

$$G_{\text{ideal}} = \frac{1}{s(\Theta_1 + \Theta_2)} \tag{2.5}$$

describes the transfer function of an ideally coupled two–mass system. The second contributions in G_{S1} and G_{S2} show the influence of the nonideal shaft. Depending on the output variables $\dot{\varphi}_1$ and $\dot{\varphi}_2$ the bode plots differ considerably. In G_{S1} the conjugate complex pair of zeros is located at a lower frequency than the complex pair of poles ($\omega_{0Z} < \Omega_{0N}$). Therefore, the phase angle φ of the plant increases, starting from $-90°$, first which reduces the total phase shift of the system. In G_{S2}, there is only one zero and one conjugate complex pair of poles. Therefore, the phase angle φ will decrease starting from $-90°$ to $-270°$ and returning to $-180°$ at very high frequencies.

It is generally accepted, that the control of the actuator speed $\dot{\varphi}_1$ is much easier indeed, because the amplitude– and phase–characteristics guarantee stability. To illustrate the different control approaches, figure 2.2 shows the **open loop** bode plots for the speed control of $\dot{\varphi}_1$ with

$$F_0(j\omega) = \frac{\dot{\varphi}_1}{\dot{\varphi}_1^*} \tag{2.6}$$

for different resonance frequencies Ω_{0N} of the mechanical system. Figure 2.3 shows the **open loop** bode plots for the control of $\dot{\varphi}_2$.

In both control loops a cascaded speed–torque–control–configuration and a PI–speed–controller are assumed. The parameters of that speed–controller $G_{\dot{\varphi}}(s)$ can be optimized by the symmetrical optimization criterion (SO). Hence, we obtain for the transfer function of the controller

$$G_{\dot{\varphi}}(s) = K_R \frac{1 + sT_R}{s} \tag{2.7}$$

with the parameters

$$T_R = 4T_\sigma \tag{2.8}$$

$$K_R = \frac{T_{\Theta N}}{8T_\sigma^2} \tag{2.9}$$

$$T_{\Theta N} = \frac{(\Theta_1 + \Theta_2)\Omega_{0N}}{M_{iN}} \tag{2.10}$$

Ω_{0N} angular nominal motor speed

M_{iN} inner nominal torque

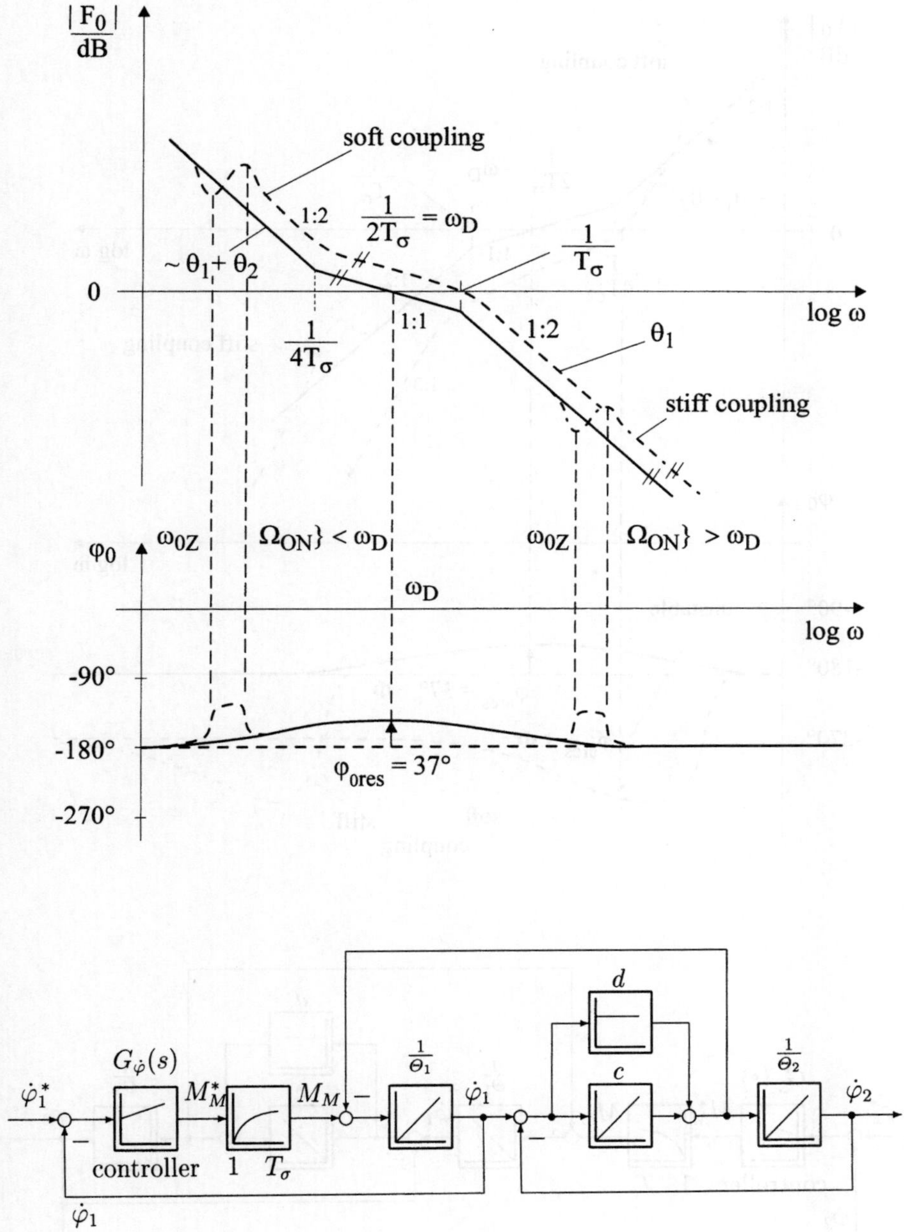

Figure 2.2: Bode plot of the open $\dot{\varphi}_1$ control loop (solid line: one–mass system, dashed line: two–mass system) and corresponding signal flow graph

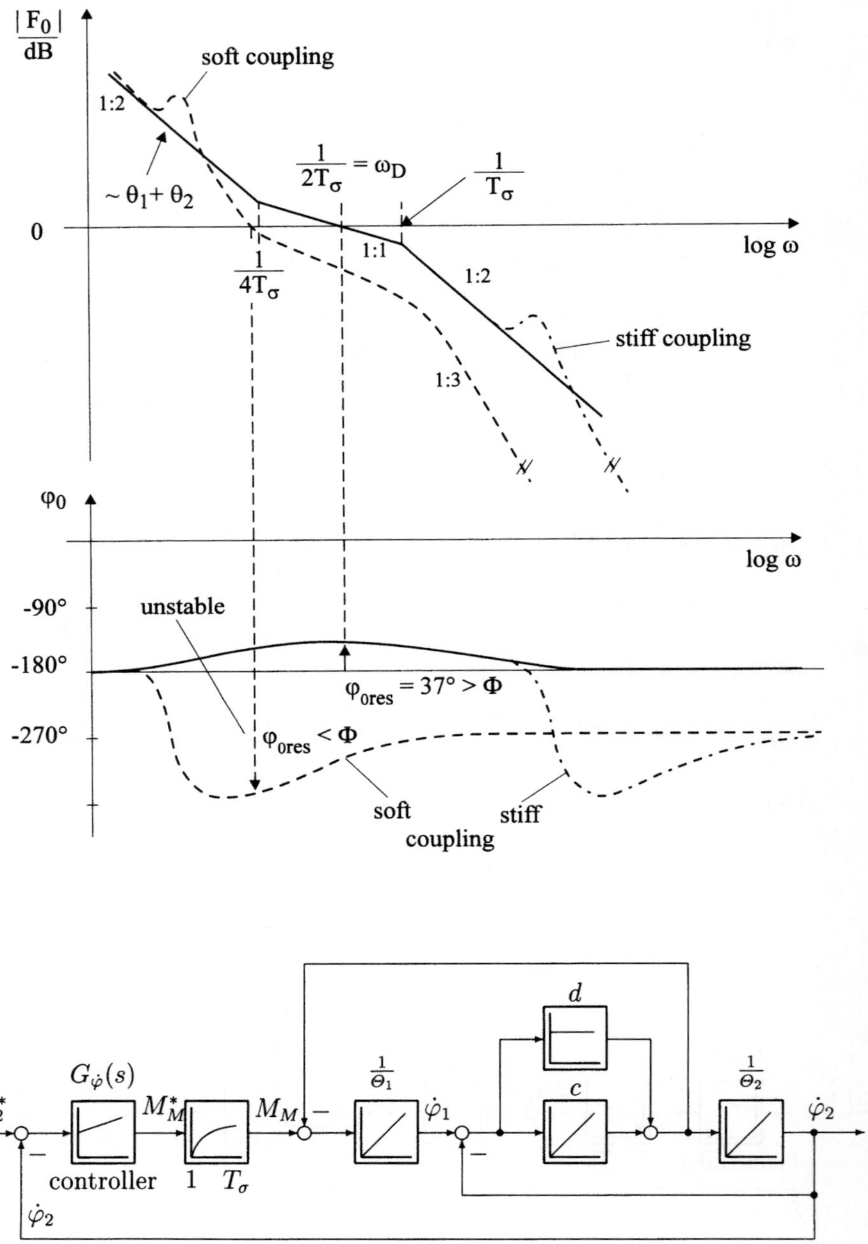

Figure 2.3: Bode plot of the open $\dot{\varphi}_2$ control loop (solid line: one–mass system, dashed line: two–mass system) and corresponding signal flow graph

If we discuss figure 2.2 first and presume a very high mechanical resonance frequency Ω_{0N} in relation to ω_D (stiff coupling), it can be stated that the second part of the transfer function G_{S1} will not influence the optimization process of the speed controller at all.

If we decrease the mechanical resonance frequency Ω_{0N} more and more — this means an increasing influence of the elasticity of the shaft — the optimization is also possible, because the increase of the phase angle (zeros) always occurs at lower frequencies than the decrease (poles). Thus, the optimization of the speed controller and the control of the electromechnical system seems to be without any negative effects.

If we consider figure 2.3 now, we have to discuss the effects of the resonant mechanical system in a similar approach. At very high resonance frequencies Ω_{0N} (stiff coupling) the speed controller for $\dot{\varphi}_2$ can be optimized, because the gain of the **open loop** system is very low within the region, where the phase angle φ_0 passes the critical $-180°$ phase line. That means

$$|F_0(j\omega)| \quad = \quad \left|\frac{\dot{\varphi}_2}{\dot{\varphi}_2^*}\right| \ll 0 \quad \text{at } \varphi_0 = -180°$$

But the optimization of the controller is impossible, if the mechanical resonance frequency Ω_{0N} is lower or in the order of ω_D (soft coupling); this can easily be derived from figure 2.3. Hence, we are able to control the system in $\dot{\varphi}_2$ and at the same time fully exploit the dynamic possibilities (T_σ) of the actuator, only in case the system is very stiff. Otherwise we have to reduce the system's dynamics by the choice of a very low cross–over frequency $\omega_D \ll 1/2T_\sigma$.

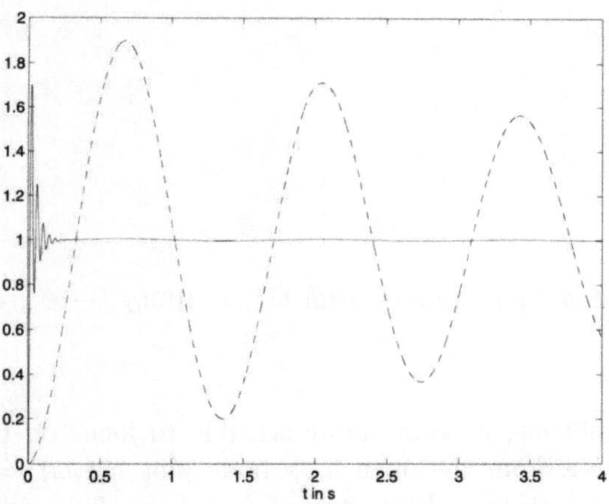

Figure 2.4: Step response with $\Omega_{0N} = 0,1\,\omega_D$ (— $\dot{\varphi}_1$, - - - $\dot{\varphi}_2$)

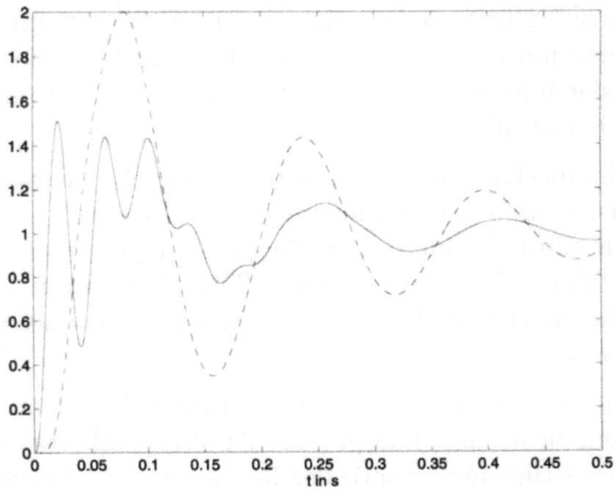

Figure 2.5: Step response with $\Omega_{0N} = \omega_D$ (— $\dot{\varphi}_1$, - - - $\dot{\varphi}_2$)

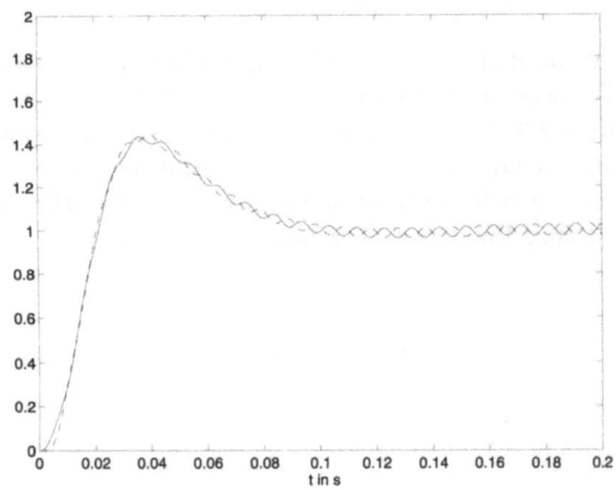

Figure 2.6: Step response with $\Omega_{0N} = 10\,\omega_D$ (— $\dot{\varphi}_1$, - - - $\dot{\varphi}_2$)

Due to these problems, it seems more sensible to focus on the control of the actuator speed $\dot{\varphi}_1$. From the open loop bode plot $F_0(j\omega) = \dot{\varphi}_1/\dot{\varphi}_1^*$ shown in figure 2.2 we have seen that the control of $\dot{\varphi}_1$ is basically possible. However, the value of real interest is the load speed $\dot{\varphi}_2$, and therefore has to be included in our considerations. In the case of very soft coupling the actuator speed can be controlled nearly ideally, whereas the load speed will oscillate with roughly the resonance frequency Ω_{0N} (see figure 2.4). Even in the case of a moderate elastic shaft ($\Omega_{0N} \approx \omega_D$), the situation does not improve. As shown in figure 2.5, we get dissatisfying behaviour of both the actuator and the load speed. Only if there is a stiff coupling (see in figure 2.6) the control of $\dot{\varphi}_1$ and $\dot{\varphi}_2$ shows acceptable results.

At this point a very important effect should be mentioned. Considering equation (2.2), the nominator and the denominator will be approximately equal, if $\Theta_1 \gg \Theta_2$. In other words, we can control $\dot{\varphi}_1$ in case that we either have a stiff mechanical shaft or a relation $\Theta_1 \gg \Theta_2$. This relation between the moments of inertia can e.g. be achieved by a gear box between Θ_1 and Θ_2 with a gear box ratio i. Then the two–mass system is approximately reduced to the ideal transfer function $G_{\text{ideal}} = 1/\left[(\Theta_1 + \Theta_2)\, s\right]$. The same considerations are, of course, valid also for $G_{S2}(s)$.

2.1.3 State–Space Control

The cascaded control of linear electromechanical systems, discussed in section 2.1.2, showed severe restrictions in case of missing coordination between the design of the electrical and mechanical components of the considered system.

2.1.3.1 Proportional State–Space Controller

The cascaded control loop configuration can be replaced by state–space control which is shown in figure 2.7 in normalized representation. In figure 2.7 a proportional state–space controller is used. Due to this controller the normalized original equation of the plant

$$\begin{pmatrix} \dot{n}_1 \\ \dot{\alpha}_{12} \\ \dot{n}_2 \end{pmatrix} = \begin{pmatrix} -\frac{d_{12}}{T_{\Theta 1}} & -\frac{c_{12}}{T_{\Theta 1}} & \frac{d_{12}}{T_{\Theta 1}} \\ \frac{1}{T_N} & 0 & -\frac{1}{T_N} \\ \frac{d_{12}}{T_{\Theta 2}} & \frac{c_{12}}{T_{\Theta 2}} & -\frac{d_{12}}{T_{\Theta 2}} \end{pmatrix} \cdot \begin{pmatrix} n_1 \\ \alpha_{12} \\ n_2 \end{pmatrix} + \begin{pmatrix} \frac{1}{T_{\Theta 1}} \\ 0 \\ 0 \end{pmatrix} \cdot m_1 \tag{2.11}$$

$$\underline{\dot{x}} \quad = \quad A \quad \cdot \quad \underline{x} \quad + \quad \underline{b} \quad \cdot m_1$$

is converted into

$$\underline{\dot{x}} = A \cdot \underline{x} + \underline{b} \cdot \underbrace{\left(n_2^* \cdot K_V - \begin{pmatrix} r_1 & r_2 & r_3 \end{pmatrix} \cdot \begin{pmatrix} n_1 \\ \alpha_{12} \\ n_2 \end{pmatrix} \right)}_{\text{new input variable}} \tag{2.12}$$

$$\underline{\dot{x}} = A \cdot \underline{x} + \underline{b} \cdot K_V \cdot n_2^* - \underline{b} \cdot \underline{r}^T \cdot \underline{x} \tag{2.13}$$

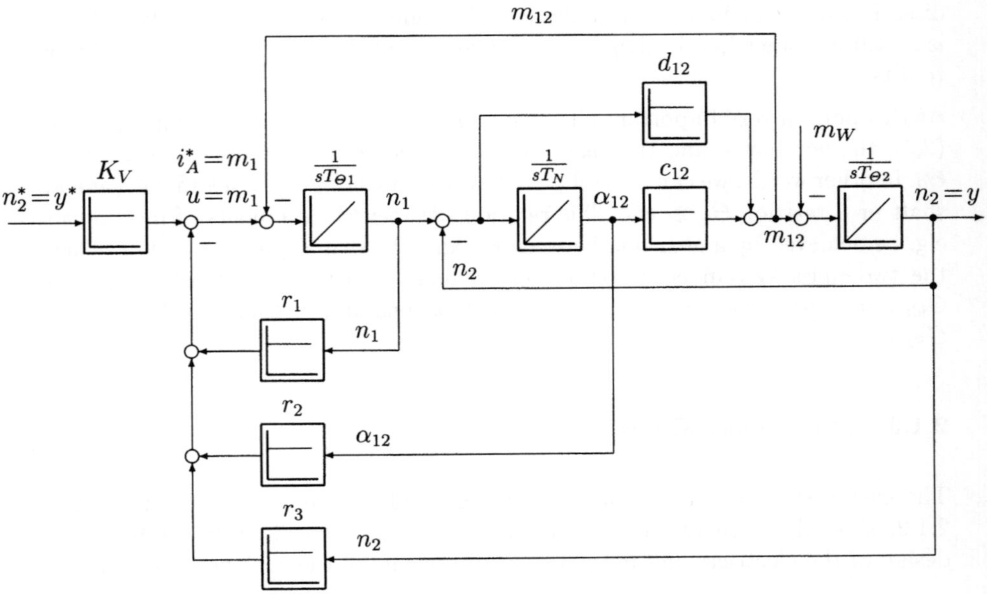

Figure 2.7: State–space control of $\dot{\varphi}_2$ without integrating contribution ($T_N = 1sec$)

$$\underline{\dot{x}} = (A - \underline{b} \cdot \underline{r}^T) \cdot \underline{x} + \underline{b} \cdot K_V \cdot \underbrace{n_2^*}_{\substack{\text{new} \\ \text{reference signal}}} \tag{2.14}$$

Consequently the original plant matrix A is transformed to A_{ZR}

$$A_{ZR} = A - \underline{b} \cdot \underline{r}^T = \tag{2.15}$$

$$= \begin{pmatrix} -\frac{d_{12}}{T_{\Theta 1}} - \frac{r_1}{T_{\Theta 1}} & -\frac{c_{12}}{T_{\Theta 1}} - \frac{r_2}{T_{\Theta 1}} & \frac{d_{12}}{T_{\Theta 1}} - \frac{r_3}{T_{\Theta 1}} \\ \frac{1}{T_N} & 0 & -\frac{1}{T_N} \\ \frac{d_{12}}{T_{\Theta 1}} & \frac{c_{12}}{T_{\Theta 1}} & -\frac{d_{12}}{T_{\Theta 1}} \end{pmatrix} \tag{2.16}$$

One should pay close attention to a very important simplification in figure 2.7: The reference values of the current, i_A^*, and the torque, m_1^*, are identical and the actual values i_A and m_1 are also identical to the reference values i_A^* and m_1^*. This simplification results from one of the following reasons:

- if the control loop for the actuator current (torque, respectively) is very fast, its behaviour can be assumed as ideal.
- in case this assumption cannot be made, we have to take into account the new state value i_A (m_1, respectively) by an additional feedback channel with a proportional gain (e.g. r_0). However, if the current control loop is optimized, its poles are fixed and can no longer be shifted into the desired position; that means the dynamic charcteristic of the current control loop must be taken for granted during the design of the state–space controller gains r_1 to r_3.

If this simplification is made we get a third order transfer function of the closed speed control loop, in contrast to a fourth or fifth order transfer function if the dynamic response of the current (torque) loop is taken into account.
The poles of the denominator equation of this transfer function can be determined by solving

$$N_{ZR}(s) = \det(sI - A_{ZR}) = 0 \tag{2.17}$$

resulting in

$$\begin{aligned} N_{ZR}(s) &= s^3 + \left[d_{12}\left(\frac{1}{T_{\Theta 1}} + \frac{1}{T_{\Theta 2}}\right) + \frac{r_1}{T_{\Theta 1}}\right] \cdot s^2 + \\ &+ \left[\frac{c_{12}}{T_N}\left(\frac{1}{T_{\Theta 1}} + \frac{1}{T_{\Theta 2}}\right) + \frac{d_{12}(r_1 + r_3)}{T_{\Theta 1}T_{\Theta 2}} + \frac{r_2}{T_{\Theta 1}T_N}\right] \cdot s^1 + \\ &+ \frac{c_{12}(r_1 + r_3)}{T_{\Theta 1}T_{\Theta 2}T_N} \end{aligned} \tag{2.18}$$

or

$$P^*(s) = s^3 + p_2 s^2 + p_1 s^1 + p_0 \tag{2.19}$$

The controller parameters can be calculated by pole placing or more directly by using the "damping optimization criterion" [9, 10]. This results in the following controller parameters

$$r_1 = \left[p_2 - d_{12}\left(\frac{1}{T_{\Theta 1}} + \frac{1}{T_{\Theta 2}}\right)\right] \cdot T_{\Theta 1} \tag{2.20}$$

$$r_2 = \left[p_1 - \frac{c_{12}}{T_N}\left(\frac{1}{T_{\Theta 1}} + \frac{1}{T_{\Theta 2}}\right) - \frac{d_{12}}{c_{12}} \cdot \frac{p_0}{T_N}\right] \cdot T_{\Theta 1}T_N \tag{2.21}$$

$$r_3 = p_0 \cdot \frac{T_{\Theta 1}T_{\Theta 2}T_N}{c_{12}} - r_1 \tag{2.22}$$

with

$$p_0 = \frac{1}{D_3 \cdot D_2^2 \cdot T_{ersn}^3} \quad (2.23)$$

$$p_1 = \frac{1}{D_3 \cdot D_2^2 \cdot T_{ersn}^2} \quad (2.24)$$

$$p_2 = \frac{1}{D_3 \cdot D_2 \cdot T_{ersn}} \quad (2.25)$$

$$D_2 = D_3 = 0,5 \qquad \text{damping optimization criterion} \quad (2.26)$$

If, for example, unity gain is assumed for the closed speed control loop in the stationary state we obtain the filter factor K_V

$$K_V = r_1 + r_3 \quad (2.27)$$

The parameter T_{ersn} of the closed speed control loop cannot be chosen completely freely, as we must consider

- the neglected dynamic behaviour of the closed current control loop,
- the stresses, which the electrical and mechanical system components are able to endure and
- the neglected nonlinearities.

Under these restrictions we can achieve the following results for the three different mechanical systems listed in the following table and depicted in the figures 2.8–2.10.

				command variable control			disturbance rejection		see
case	fig.	Ω_{0N}	T_{ersn}	m_{12max}	α_{12max}	i_{1max}	$\alpha_{12\infty}$	$n_{2\infty}$	fig.
		$1/s$	ms	1	°	1	°	1	
1	2.8	628,3	4	0,572	0,1	1,55	0,0175	$8,28 \cdot 10^{-4}$	2.6
2	2.9	62,83	40	0,0568	1,0	0,15	1,75	$8,22 \cdot 10^{-3}$	2.5
3	2.10	6,283	400	0,00571	10,0	0,014	175,1	$8,33 \cdot 10^{-2}$	2.4

The figures 2.8–2.10 show an acceptable dynamic response for a step change of the reference value of the speed (compared to figures 2.4–2.6). Yet the disturbance rejection is not perfect due to the stationary error resulting from the influence of the load torque that cannot be compensated by the proportional state–space controller.

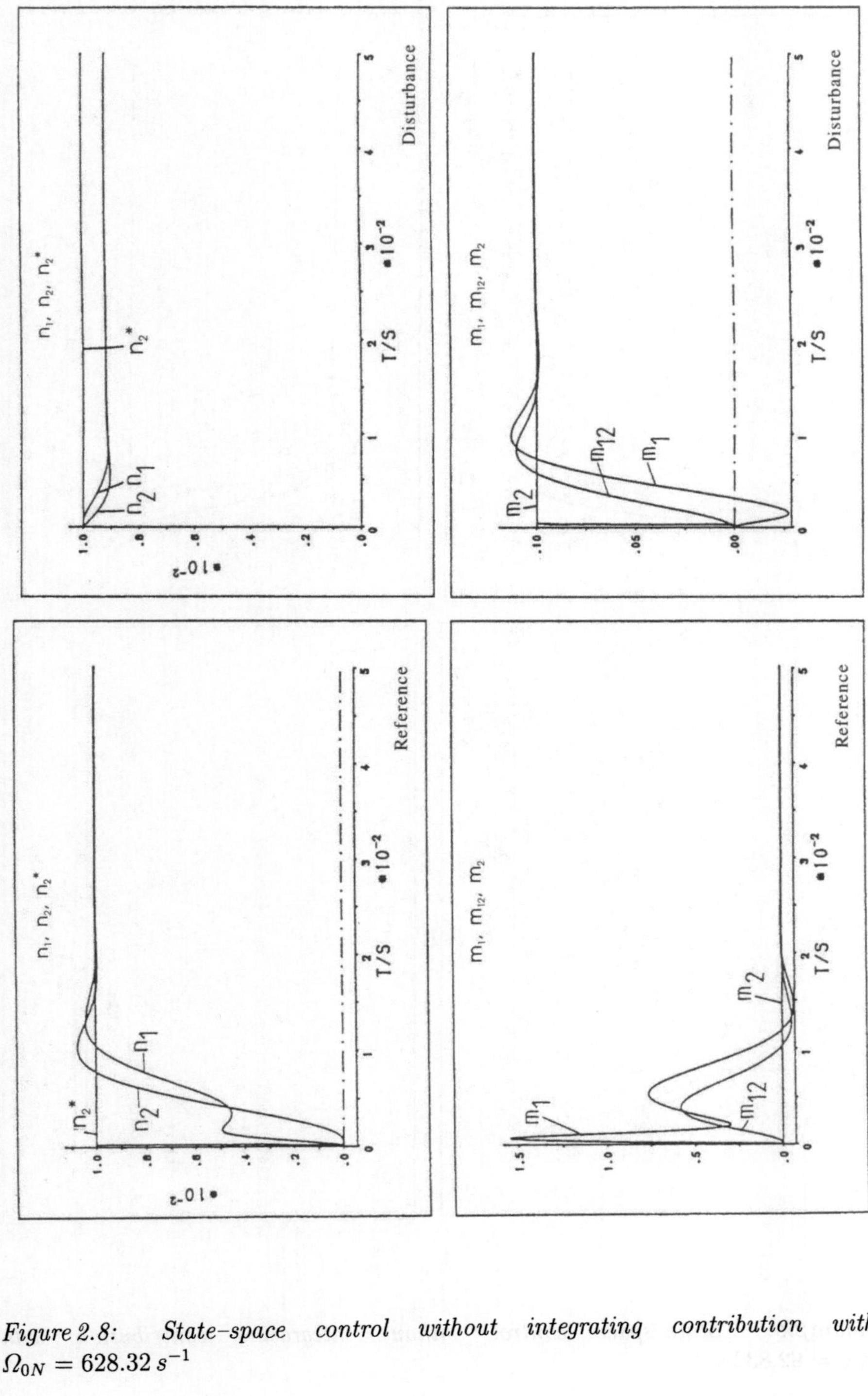

Figure 2.8: *State–space control without integrating contribution with* $\Omega_{0N} = 628.32\, s^{-1}$

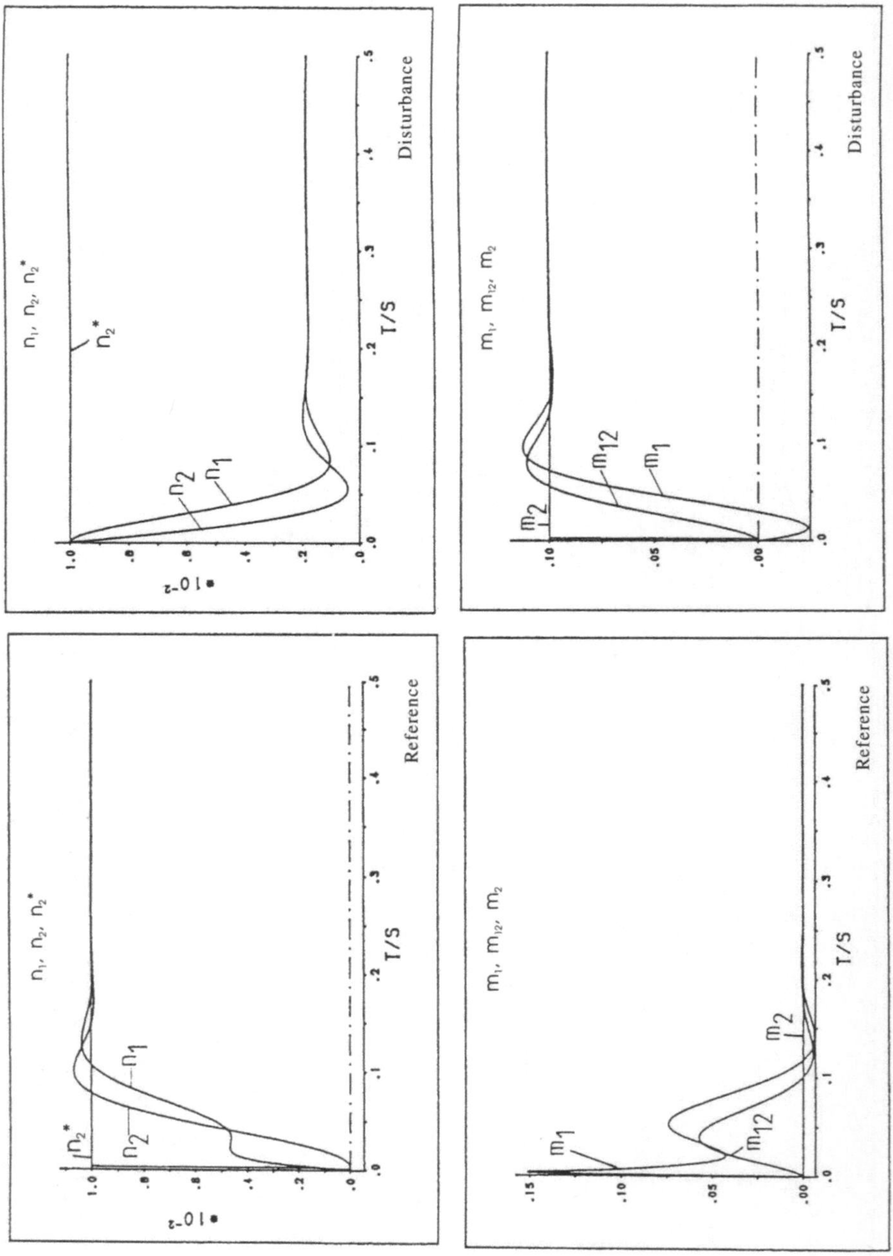

Figure 2.9: *State–space control without integrating contribution with* $\Omega_{0N} = 62.832\, s^{-1}$

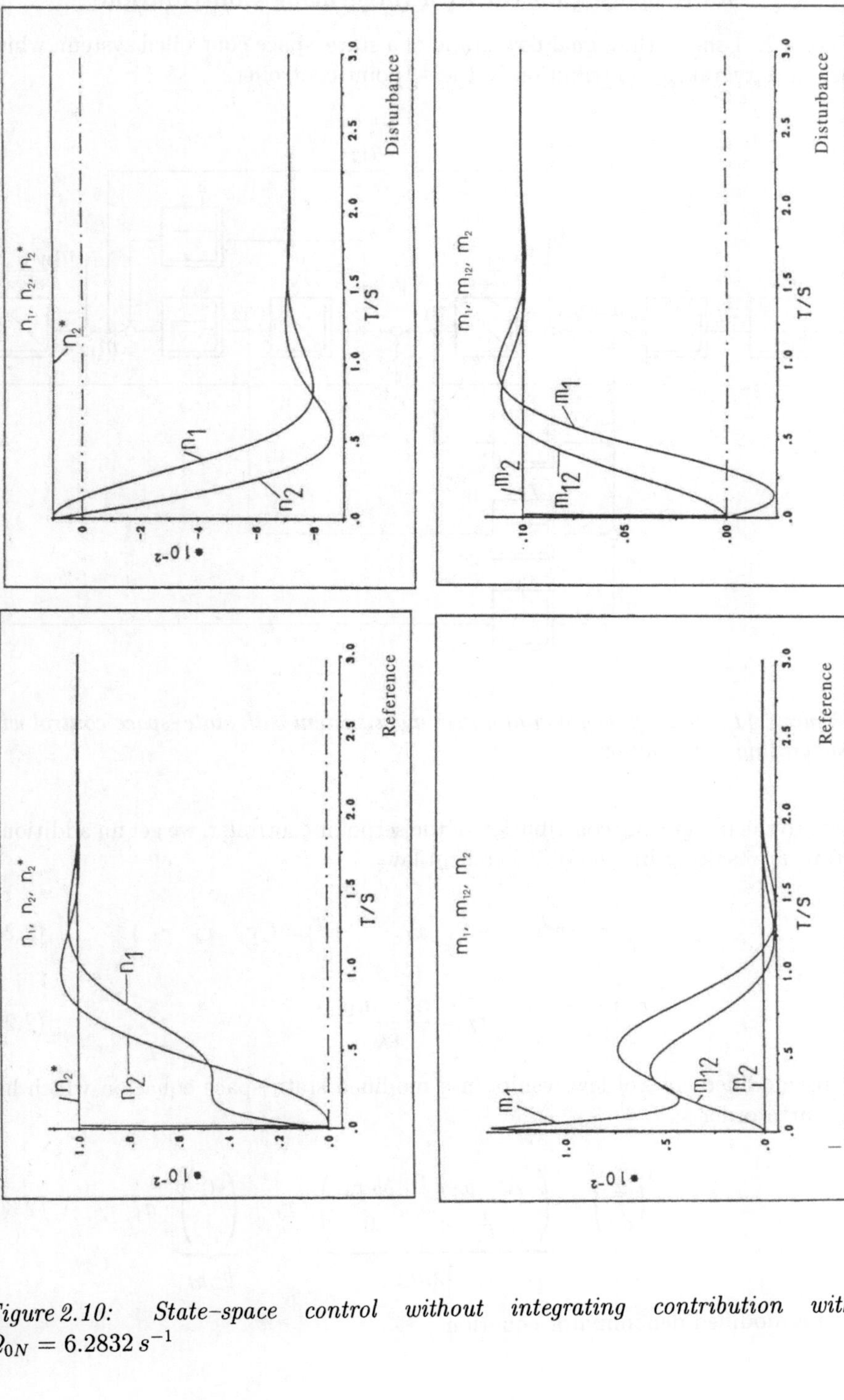

Figure 2.10: *State–space control without integrating contribution with* $\Omega_{0N} = 6.2832\,s^{-1}$

2.1.3.2 State–Space Control with Integrating Contribution

Figure 2.11 shows the signal flow graph of a state–space controlled system, which has an integrating contribution in the setpoint controller.

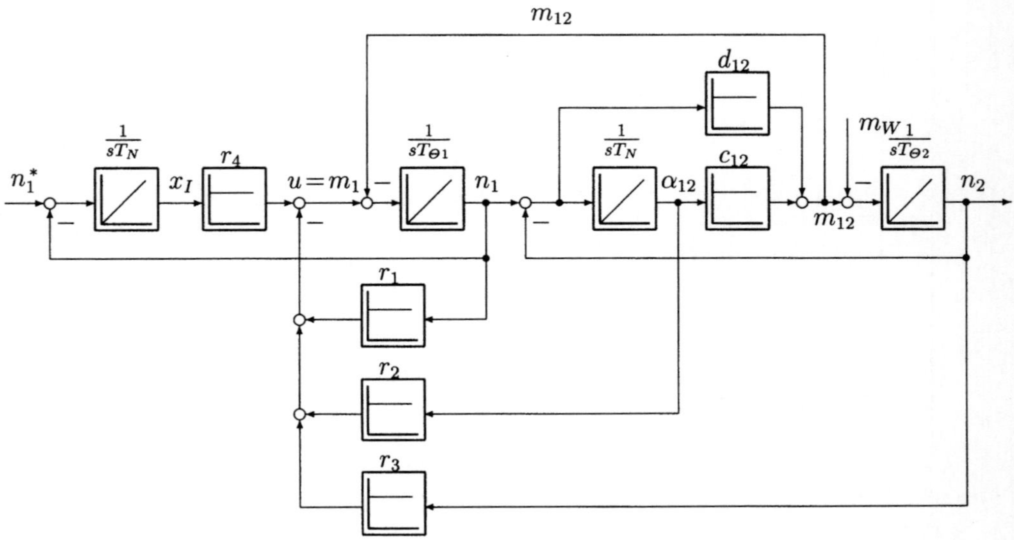

Figure 2.11: Signal flow graph of a two–mass system with state–space control with integrating contribution

Due to the integrating contribution of the setpoint controller, we get an additional state x_I resulting in a modified control law

$$u = m_1 = -\underline{r}^T \cdot \underline{x} + r_4 \cdot x_I; \qquad \underline{r}^T = \begin{pmatrix} r_1 & r_2 & r_3 \end{pmatrix} \tag{2.28}$$

$$\dot{x}_I = \frac{n_1^* - n_1}{T_N} \tag{2.29}$$

This modified control law results in a modified state–space equation which has a fourth order

$$\dot{\underline{x}}_I = \begin{pmatrix} \dot{\underline{x}} \\ \dot{x}_I \end{pmatrix} = \underbrace{\begin{pmatrix} A - \underline{b} \cdot \underline{r}^T & \underline{b} \cdot r_4 \\ -\frac{1}{T_N} \cdot \underline{c}^T & 0 \end{pmatrix}}_{A_{ZRI}} \cdot \underline{x}_I + \underbrace{\begin{pmatrix} 0 \\ \frac{1}{T_N} \end{pmatrix}}_{\underline{b}_{ZRI}} \cdot n_1^* \tag{2.30}$$

and a modified denominator equation

$$N_{ZRI}(s) = \det(sI - A_{ZRI}) = N_{Norm} \tag{2.31}$$

The feedback gains $r_1 - r_4$ of the controller can be calculated by

$$r_1 = T_{\Theta 1} \cdot p_3 \tag{2.32}$$

$$r_2 = -\frac{T_{\Theta 1} T_{\Theta 2} T_N^2}{c_{12}} \cdot p_0 + T_{\Theta 1} T_N \cdot p_2 - c_{12}\left(1 + \frac{T_{\Theta 1}}{T_{\Theta 2}}\right) \tag{2.33}$$

$$r_3 = \frac{T_{\Theta 1} T_{\Theta 2} T_N^2}{c_{12}} \cdot p_1 - r_1 \tag{2.34}$$

$$r_4 = \frac{T_{\Theta 1} T_{\Theta 2} T_N^2}{c_{12}} \cdot p_0 \tag{2.35}$$

and using the "damping optimization criterion" we get

$$p_0 = \frac{1}{D_4 D_3^2 D_2^3 T_{ersn}^4} = \frac{64}{T_{ersn}^4} \tag{2.36}$$

$$p_1 = \frac{1}{D_4 D_3^2 D_2^3 T_{ersn}^3} = \frac{64}{T_{ersn}^3} \tag{2.37}$$

$$p_2 = \frac{1}{D_4 D_3^2 D_2^2 T_{ersn}^2} = \frac{32}{T_{ersn}^2} \tag{2.38}$$

$$p_3 = \frac{1}{D_4 D_3 D_2 T_{ersn}} = \frac{8}{T_{ersn}} \tag{2.39}$$

The considerations of the section above hold again, and taking these considerations into account we get the results shown in the following table and in the figures 2.12–2.14 (compared to figures 2.4–2.6).

case	fig.	Ω_{0N}	T_{ersn}	command variable control m_{12max}	α_{12max}	i_{1max}	disturbance rejection $\alpha_{12\infty}$	$n_{2\infty}$	see fig.
		$1/s$	ms	1	°	1	°	1	
1	2.12	628,3	4,5	0,592	0,1	1,18	0,0175	0	2.6
2	2.13	62,83	45,0	0,059	1,0	0,092	1,75	0	2.5
3	2.14	6,283	450,0	0,0059	10,0	0,009	175,1	0	2.4

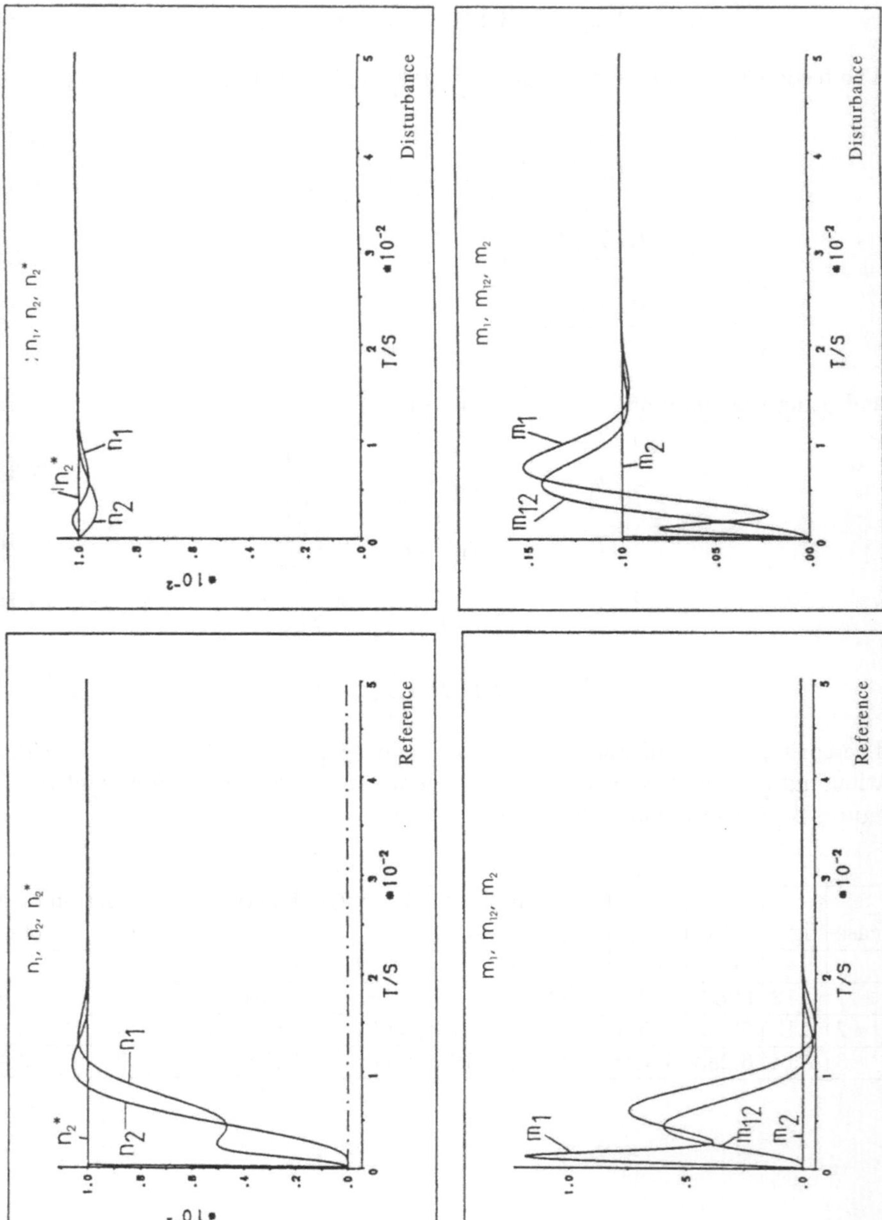

Figure 2.12: State–space control with integrating contribution with $\Omega_{0N} = 628.32\,s^{-1}$

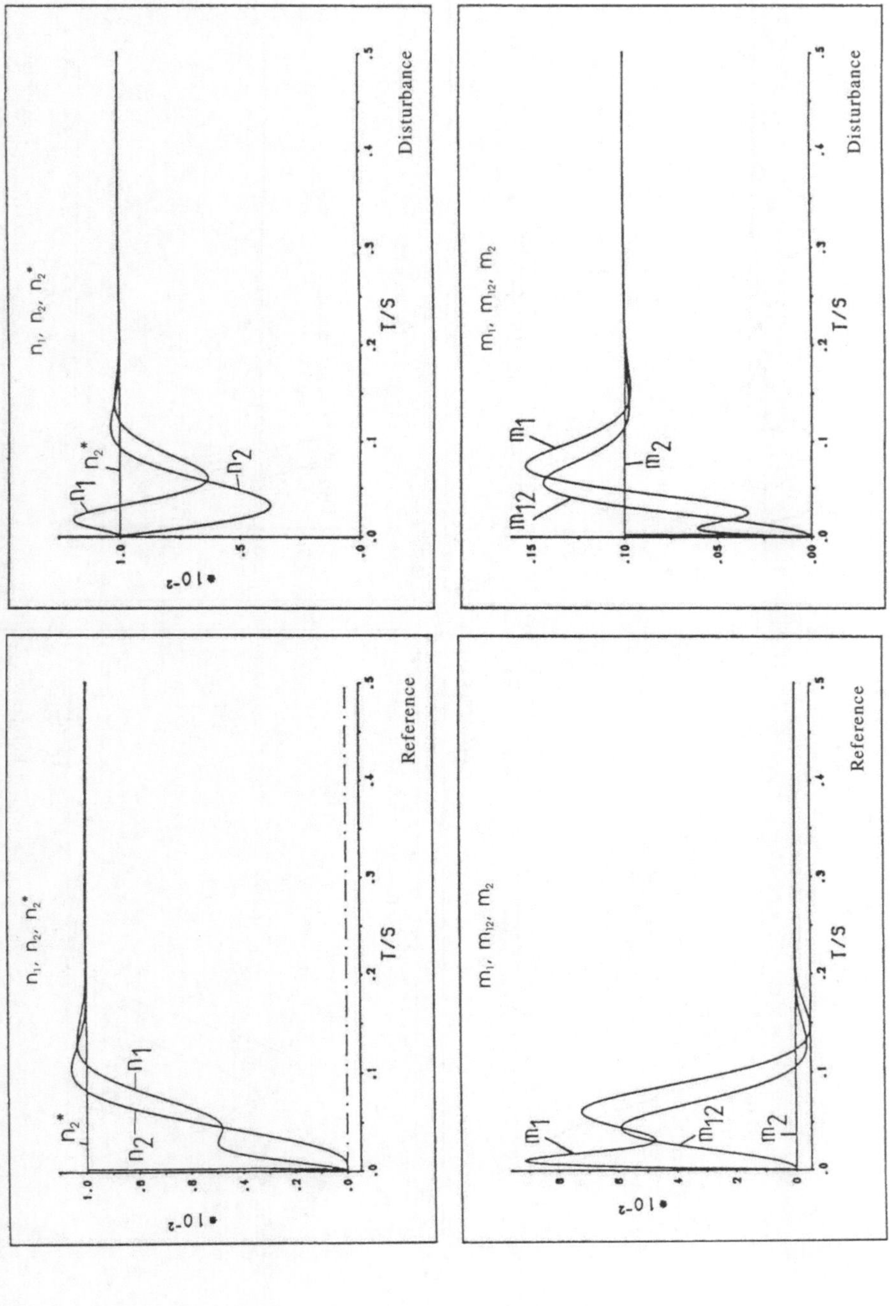

Figure 2.13: State–space control with integrating contribution with $\Omega_{0N} = 62.832\, s^{-1}$

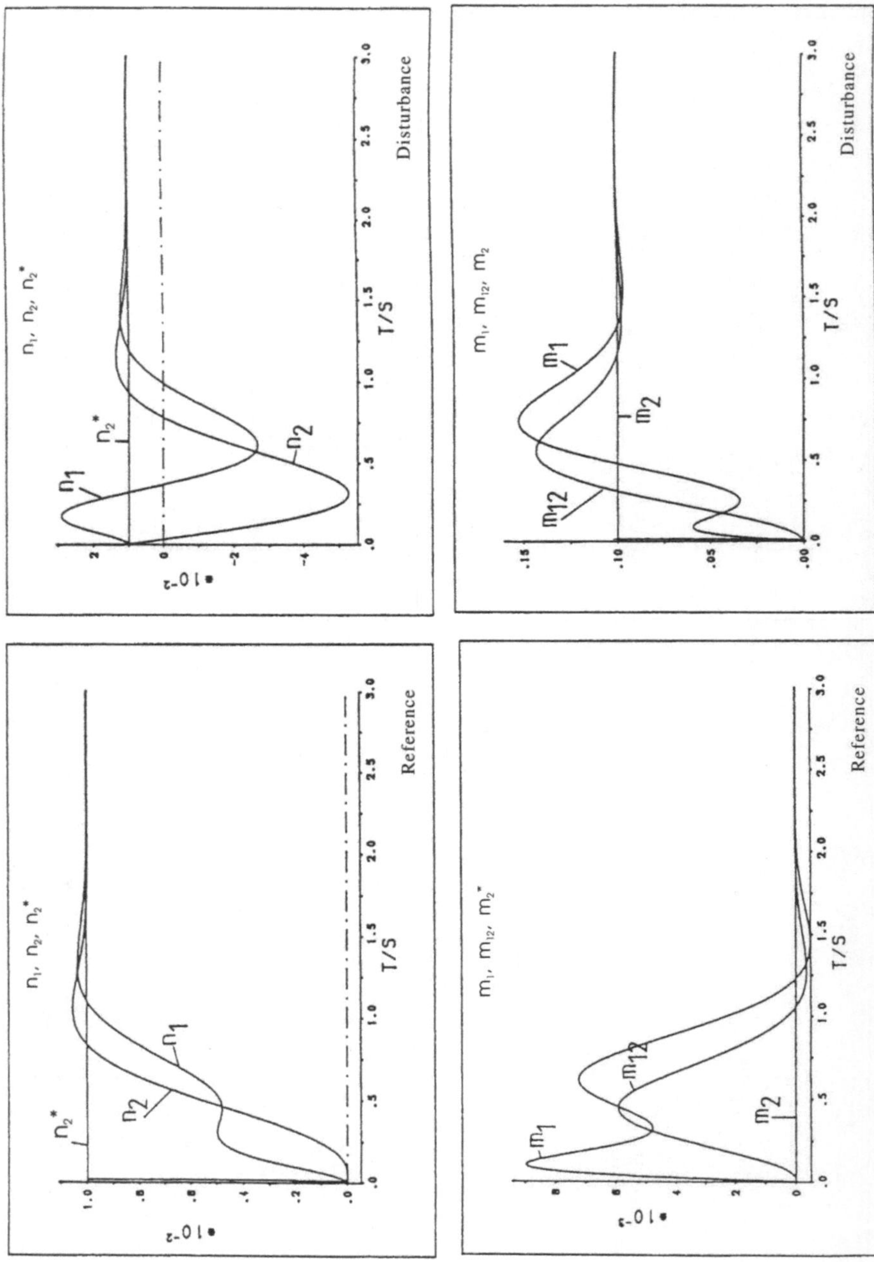

Figure 2.14: *State–space control with integrating contribution with* $\Omega_{0N} = 6.2832\,s^{-1}$

These results should be discussed now:

- all results are very acceptable,
- the mechanical frequency Ω_{0N} determines the dynamic response,
- if the mechanical system is stiff, the current (torque) has a high peak value which generally cannot be accepted (cost),
- if the mechanical system is soft, then the difference angle α_{12} (m_{12}) of the shaft is the limit for the design,
- state–space control with integrating contribution results in a slower dynamic response compared to proportional state–space control.

These results are promising; yet some difficulties remain. We assumed a model of the mechanical system which can be represented by two moments of inertia, elasticity and damping. But

- the behaviour of the current (torque) control loop was neglected,
- nonlinearities were neglected,
- all parameters were supposed to be known and time–invariant,
- we assumed the existance of sensors for the measurement of all states without delay, noise and error.

Usually not all states are measureable which requires state–space observers; for further information the control books dealing with state–space control can be used.

There is a remaining difficulty while bringing the system into operation. As we have seen, all controller parameters must be calculated in advance. The control structure and the parameters have to be implemented and then the entire system must be put into operation completely. There is no opportunity to put the system into operation gradually as it is usually done in cascaded control loops.

Up to now, the nonlinearities in the mechanical system have been neglected. But friction and backlash are common in all mechanical systems. In the last few years the system in figure 2.1 was under investigation to reduce or avoid the negative effects of such nonlinearities. Figure 2.15 shows dynamic results of a position state–space controlled system with the nonlinearity friction alone.
Due to this nonlinearity, stick–slip limit cycles can occur.

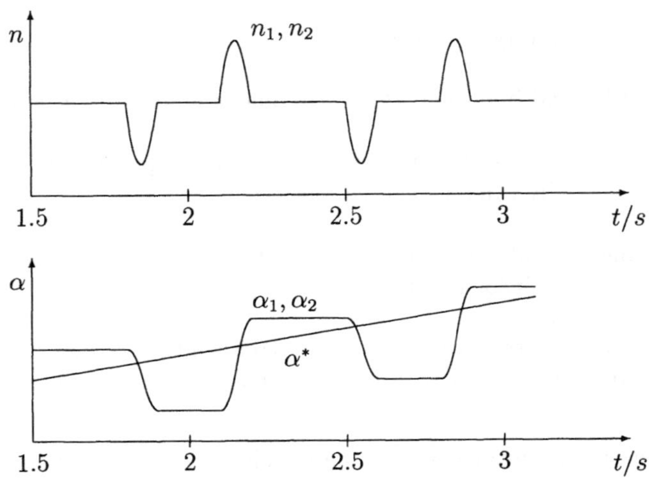

Figure 2.15: Position state–space controlled two mass system with friction

2.1.4 Generalized Considerations for Electromechanical Systems

Up to now, we have concentrated on a linear two–mass system where the elasticity and the damping of the shaft are taken into account. Generally though, such a model is not an adequate representation of a real system. Therefore, in many practical applications it is necessary to develop more detailled models for the controller design. An approach that is often used in mechanical engineering is the finite element analysis. For linear systems, such an analysis gives a perfect representation of the plant, yet there are some important disadvantages of this method:

- The representing model is of relatively high order, requiring much computational power during the analysis.
- Due to its high order, the model can no longer be used for the controller design.
- Therefore, a reduction of the order of the linear system is necessary, for example by the method of Litz [8]. Nevertheless, severe problems will remain.
- The interpretation of the reduced system in relation to the real physical system is not possible. Therefore, the design of the control strategies is

difficult when unacceptable stresses of the mechanical components and instability should be avoided in the real system.

- Furthermore, investigations have shown that the simulation results of the reduced system can no longer be compared with the behaviour of the real system. This is also an effect of one: the non-interpretability of the reduced system, and two: the nonlinearities which have been neglected up to now.

Therefore, a Lagrange approach seems more suitable from the engineering and control point of view, because there is both basic knowledge about the principal structure of the system and the fundamental parameters of the electrical and mechanical components in general. By this approach, the order of the linear model of the system can be kept as low as possible. Figure 2.16 shows the linear signal flow graph of a three–mass system, where the elasticity and damping of the shaft are taken into account. This configuration can be treated in the same way as the two–mass system discussed above, after the signal flow graph of the shaft and the third moment of inertia have been added. We get a fifth order model then, with one integrating block and two complex pole pairs.

It can easily be derived from the figures 2.2 and 2.16, that the addition of one shaft and one moment of inertia to the former system results in an additional section in the signal flow graph and has to be repeated as often as there is an increase of these types of mechanical components. This fundamental procedure forms the starting point for the object–oriented modelling approach and languages [12] – [17] and is the mathematical basis of the descriptor representation. These aspects will be discussed further in section 2.2 and 2.3.

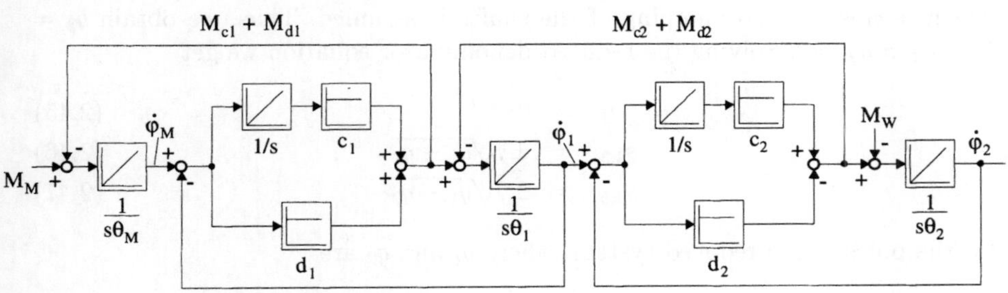

Figure 2.16: Three–mass system

The model shown in figure 2.16 can be described by the following equations

$$\ddot{\varphi}_2 = \frac{1}{\Theta_2}\left(c_2(\varphi_1-\varphi_2)+d_2(\dot{\varphi}_1-\dot{\varphi}_2)-M_L\right) \quad (2.40)$$

$$\ddot{\varphi}_1 = \frac{1}{\Theta_1}\left(c_1(\varphi_M-\varphi_1)+d_1(\dot{\varphi}_M-\dot{\varphi}_1)-c_2(\varphi_1-\varphi_2)-d_2(\dot{\varphi}_1-\dot{\varphi}_2)\right) \quad (2.41)$$

$$\ddot{\varphi}_M = \frac{1}{\Theta_M}\left(-c_1(\varphi_M - \varphi_1) - d_1(\dot{\varphi}_M - \dot{\varphi}_1) + M_M\right) \tag{2.42}$$

For the transfer function we obtain

$$G_{S3}(s) = \frac{\dot{\varphi}_2}{M_M} = \frac{b_2 s^2 + b_1 s + b_0}{a_5 s^5 + a_4 s^4 + a_3 s^3 + a_2 s^2 + a_1 s} \tag{2.43}$$

with the parameters

$$\begin{aligned}
b_2 &= d_1 d_2 \\
b_1 &= c_1 d_2 + c_2 d_1 \\
b_0 &= c_1 c_2 \\
a_5 &= \Theta_M \Theta_1 \Theta_2 \\
a_4 &= d_1 \Theta_2 \left(\Theta_M + \Theta_1\right) + d_2 \Theta_M \left(\Theta_1 + \Theta_2\right) \\
a_3 &= c_1 \Theta_2 \left(\Theta_M + \Theta_1\right) + c_2 \Theta_M \left(\Theta_1 + \Theta_2\right) + d_1 d_2 \left(\Theta_M + \Theta_1 + \Theta_2\right) \\
a_2 &= \left(d_1 c_2 + d_2 c_1\right)\left(\Theta_M + \Theta_1 + \Theta_2\right) \\
a_1 &= c_1 c_2 \left(\Theta_M + \Theta_1 + \Theta_2\right)
\end{aligned}$$

After the integrating part

$$\frac{1}{(\Theta_1 + \Theta_2 + \Theta_3)s} \tag{2.44}$$

which represents an ideally coupled three–mass system, has been separated from the denominator equation, a fourth order equation remains which can be solved analytically. A more simple approach results from the assumption of $d_1 = d_2 = 0$, which means that no damping of the shafts is assumed. Thus, we obtain $b_2 = b_1 = a_4 = a_2 = 0$. Solving the reduced denominator equation we get

$$s_1 = 0 \tag{2.45}$$

$$s_{2,3} = \pm j\sqrt{q_1 + q_2} \tag{2.46}$$

$$s_{4,5} = \pm j\sqrt{q_1 - q_2} \tag{2.47}$$

for the poles of the reduced system, where q_1 and q_2 are

$$q_1 = -\frac{c_1\Theta_1(\Theta_M + \Theta_2) + c_2\Theta_M(\Theta_1 + \Theta_2)}{2\Theta_M\Theta_1\Theta_2} \tag{2.48}$$

$$q_2 = \frac{\sqrt{[c_1\Theta_1(\Theta_M + \Theta_2) + c_2\Theta_M(\Theta_1 + \Theta_2)]^2 - 4c_1c_2\Theta_M\Theta_1\Theta_2(\Theta_M + \Theta_1 + \Theta_2)}}{2\Theta_M\Theta_1\Theta_2} \tag{2.49}$$

This result seems to be quite complex, but we will see, that we can nevertheless derive simple rules to understand such systems.

Let us briefly return to the two–mass system and have a look at equation 2.3 and its corresponding bode plot in figure 2.17. As already discussed, such a system has only one mechanical resonance frequency which causes a resonant rise in amplitude–frequency response.

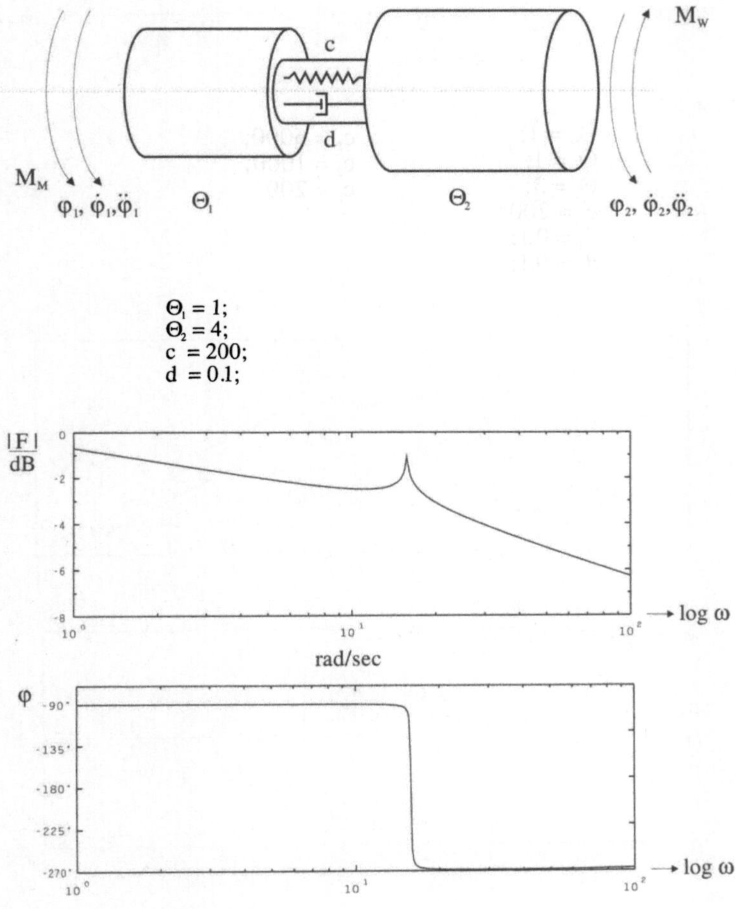

Figure 2.17: Bode plot of a two–mass of inertia system

Now, a three–mass system is assumed in figure 2.18 with $(\Theta_1 + \Theta_2) = \Theta_L$ and $\Theta_1 < \Theta_2$. It is interesting to notice, that the variation of the higher mechanical frequency does not influence the lower mechanical frequency very much. This result holds especially, if c_2 is relatively high compared to c_1. Only if c_2 is also in same range of c_1, then there is a small reduction of the lower mechanical

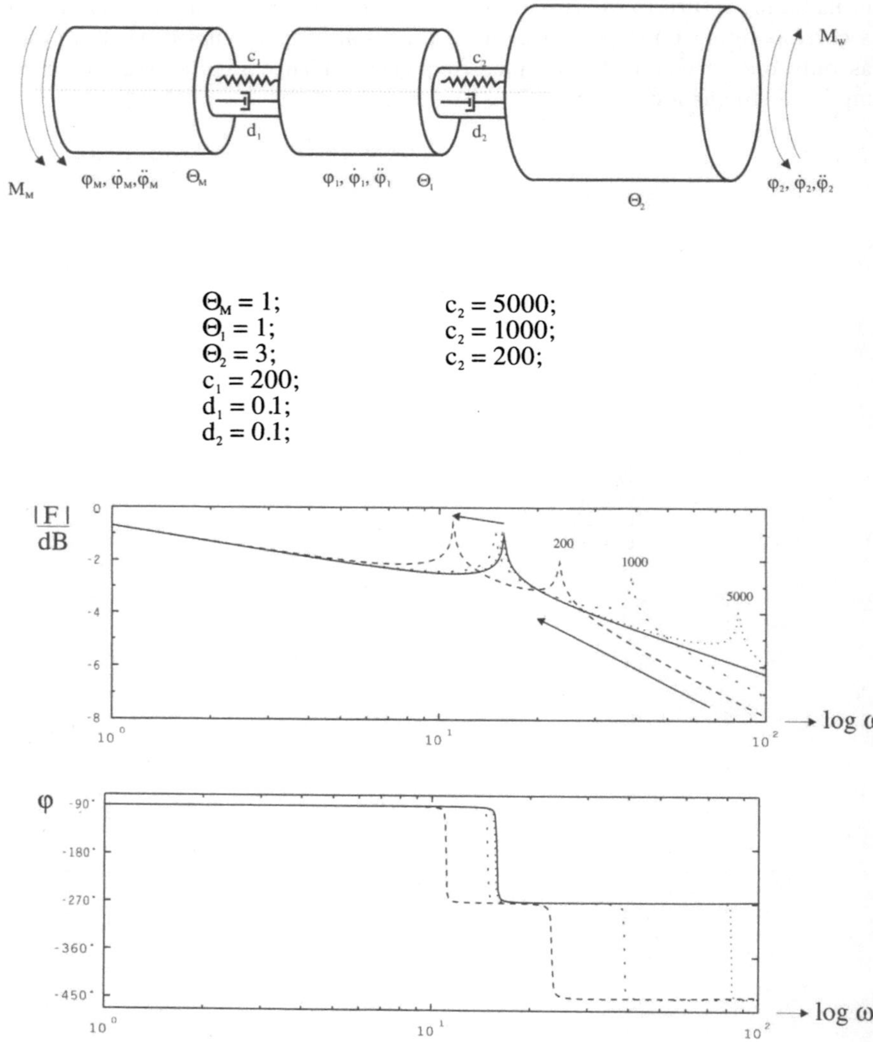

Figure 2.18: Bode plot of a three–mass system with $(\Theta_1 + \Theta_2) = \Theta_L$ *and* $\Theta_1 < \Theta_2$

frequency. For comparison, the bode plot also shows the frequency response of the two–mass system (solid line).

In figure 2.19 the same system as in figure 2.18 is shown, only that now $\Theta_1 > \Theta_2$ and $(\Theta_1 + \Theta_2) = \Theta_L$. Again we achieve results very similar to figure 2.18.

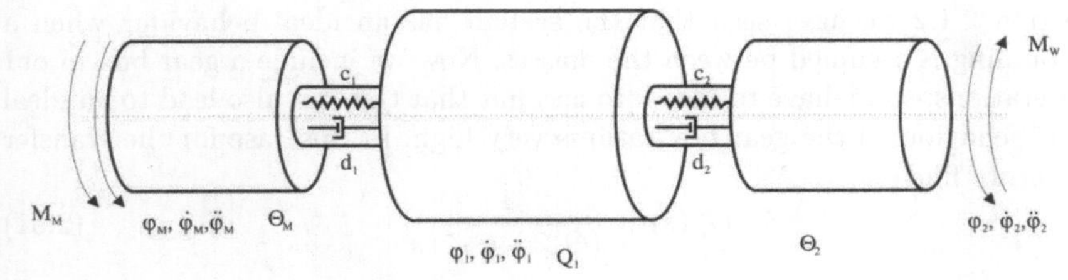

$\Theta_M = 1;$
$\Theta_1 = 3;$
$\Theta_2 = 1;$
$c_1 = 200;$
$d_1 = 0.1;$
$d_2 = 0.1;$

$c_2 = 5000;$
$c_2 = 1000;$
$c_2 = 200;$

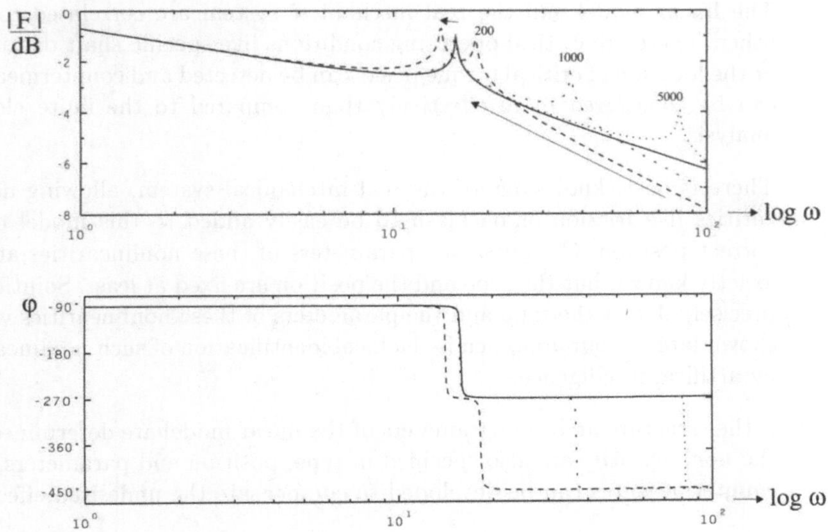

Figure 2.19: Bode plot of a three–mass system with $(\Theta_1 + \Theta_2) = \Theta_L$ *and* $\Theta_1 > \Theta_2$

Due to these results, it seems acceptable to study nonlinear effects in resonant two–mass systems.

In case there is a gear box between actuator and load, for the relevant load inertia we obtain

$$\Theta_L{}^* = \frac{\Theta_L}{i^2} \quad (i = \text{gear box ratio}) \tag{2.50}$$

In section 2.1.2 we have seen that the system has an ideal behaviour when a stiff coupling is assumed between the masses. Now we include a gear box in our considerations, so we have to take into account that this can also lead to an ideal system behaviour if the gear box ratio is very high. In this case for the transfer function we find

$$G_S(s) = \frac{1}{(\Theta_M + \Theta_L{}^*)\, s} \tag{2.51}$$

This also holds true for mechanical systems of higher order.

The Lagrange approach of modeling the linear part of mechanical systems has the following advantages:

1. Due to the basic engineering knowledge of the real mechanical system a low order can be chosen for the model and most of the linear parameters can easily be determined.

2. The linear model and the real mechanical system are correlated to each other. Therefore critical operating conditions like specific shaft distortions or the location of critical torque peaks can be detected and countermeasures can be considered more effectively than compared to the finite element analysis.

3. There is basic knowledge of the real mechanical system, allowing nonlinearities like friction or backlash to be easily added to this model at the correct position. Of course the parameters of these nonlinearities are not exactly known, but the type and the position are fixed at least. Solutions to precisely detect the type and the parameters of these nonlinearities will be shown later — our approach is the local identification of such nonlinearities by artifical intelligence.

4. If the structure and the parameters of the linear model are determined and the nonlinearities are also specified in type, position and parameters, then countermeasures can be developed to compensate the undesired effects.

Up to now, we only discussed the electrical (i.e. actuator) and the mechanical section of the systems. In the next section we will therefore include into our considerations the technological process, where these systems are integrated. As an example we will use production plants with continuous moving webs.

2.2 Actuator, Mechanical System and Process

In figure 2.20 the signal flow graph of a plant with continuous processing of material is shown. For means of easier understanding let us assume that the following components in figure 2.20 are ideal.

- the electrical drives, i.e. the actuators: They are modelled as first order delays with the parameters k_m and T_e.
- the mechanical system: The elasticity of the shaft and the moments of inertia of the rollers at the nip sections are neglected; therefore the mechanical system of one section reduces to a single integrating block ($T_{\Theta N}$).
- the process: Different idealizations have been made here; the most important is the linearization at the operating point. Furthermore, a linear stress–strain behaviour (i.e. Hooke's law is valid) and a constant thickness and width of the processed material is assumed. Additionally, any disturbances from the nip sections are neglected. The identifier ε_N in figure 2.20 denotes the nominal strain.

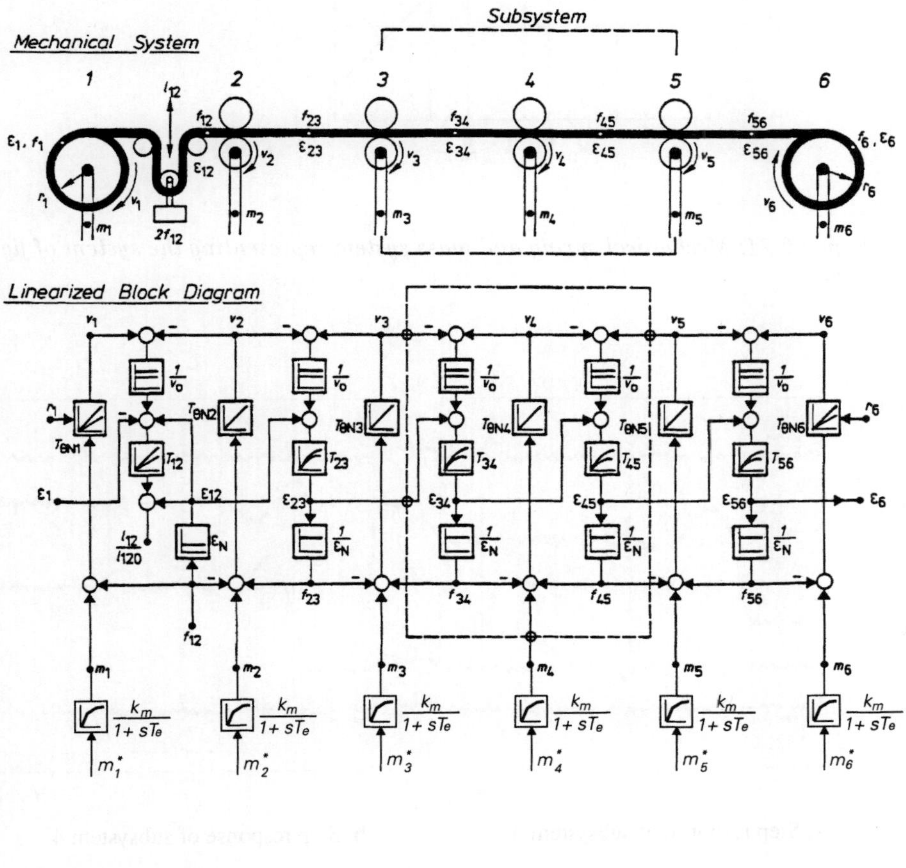

Figure 2.20: Multi variable process system with continuous processing of material

From figure 2.20 we can see that the plant is composed of identical sections connected in series, leading to a modularly structured signal flow graph. Each section is coupled by the web with both the preceding and the following sections. Therefore, we must distinguish between a coupling in upstream and downstream direction of the machine. The coupling of the sections is also reflected in the transfer functions.

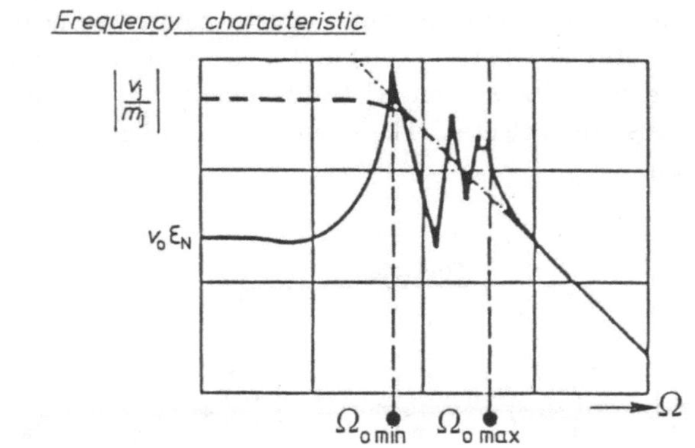

Figure 2.21: Mechanical spring and mass system representing the system of figure 2.20

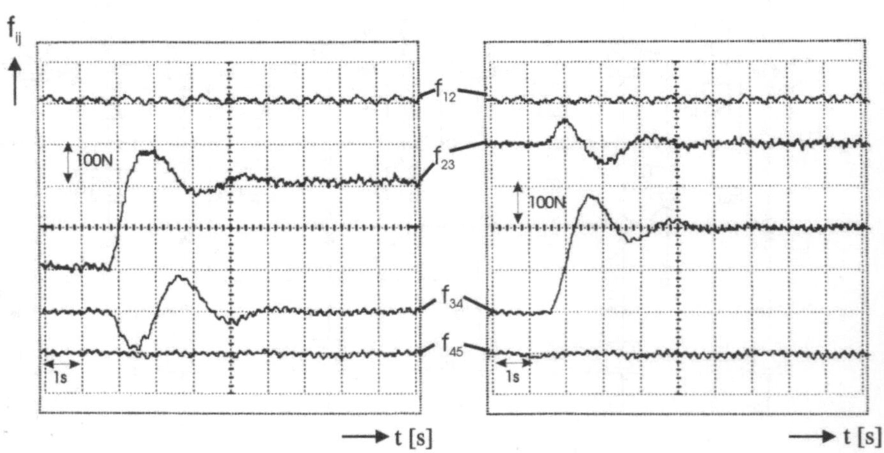

a, Step response of subsystem 3 b, Step response of subsystem 4

Figure 2.22: Step response of f_{23} and the coupling to f_{34} and vice versa

As mentioned in section 1.1.4 we can state, that an increase of the number of the same type of sections (or processing stations) will result in an increase of the same number of signal flow graphs of the total system. Therefore we have the same option like before, namely to use object oriented modelling approaches.

Figure 2.21 and figure 2.22 show the frequency characteristic and the coupling effects of such a multiresonant system. Due to the coupling of the speeds and the web–forces, a desired variation of the production speed or the web–force in one production section will also produce undesired speed and web–force variations in the other production sections.

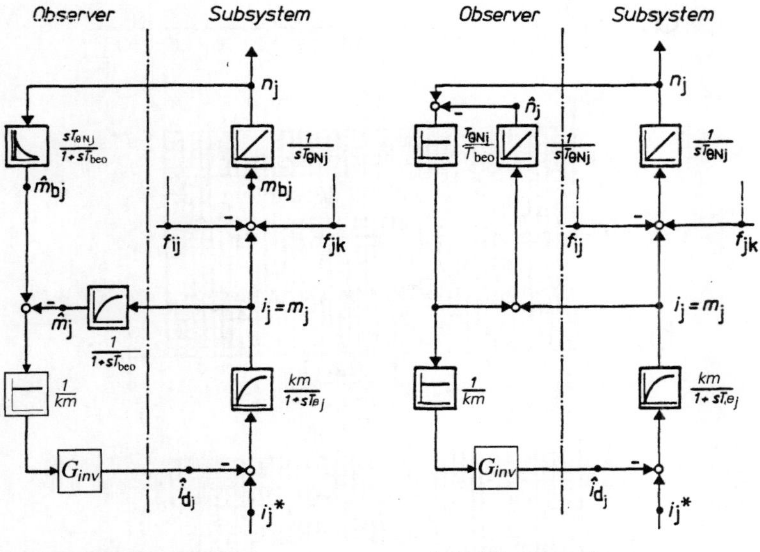

a, Realized observer *b, Luenberger observer*

Figure 2.23: Reduced decoupling observer

A general idea is to decouple the processing stations. Decoupling can be started by using an observer for the web–forces as shown in figure 2.23. The realized observer in 2.23a can be transformed into a Luenberger observer, shown in figure 2.23b. This observer is able to estimate the web–forces in the sections before $(j-1)$ and after $(j+1)$ the processing station (j) itself and use the estimated web–force signal as an additional reference value for the current (torque) control loop of the processing station j. By applying such a strategy, the undesired coupling is compensated and each subsystem can be approximated much better as an integrator representing the moment of inertia of the processing station.

Figure 2.24 shows the improvements, which can be achieved by implementing a linear web-force observer.

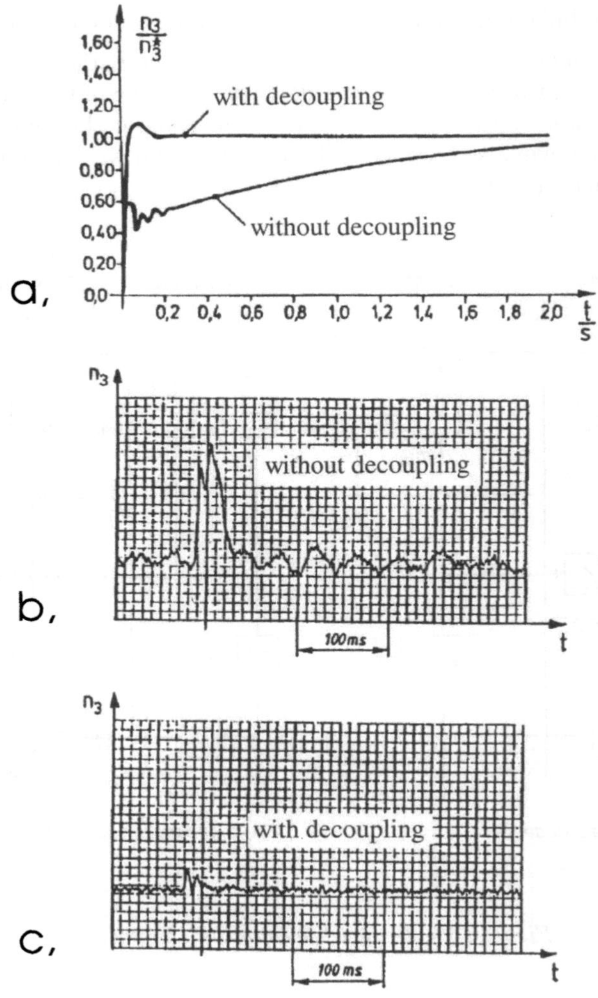

Figure 2.24: Step response of n_3 *due to* n_3^* *(a) and response of* n_3 *due to a* f_{34}*-variation (b and c)*

Today the state of the art control strategy for the winders is the so-called feed-forward control (FFC). With the knowledge of the flux in the machine this can be realized by the control of the armature current (figure 2.25). This conventional control concept calculates the reference values for the motor torque m_M^* and the flux ψ^*, where the motor torque is a function of the desired reference value of the web–force, f^*, the acceleration dv^*/dt, the friction torque m_R and the actual radius r of the winder (WR). The radius is determined by the ratio of the pulse frequency f_{pk} of the speed of the cylinder k next to the winder and the pulse frequency f_{pw} of the winder itself (RR).

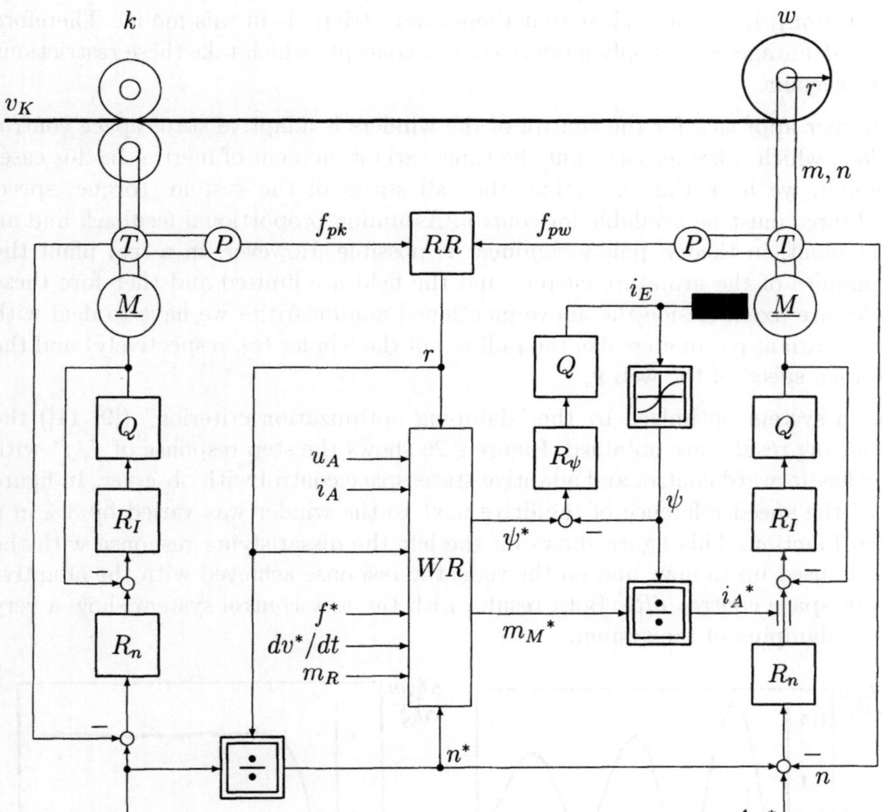

Figure 2.25: Schematic feedforward control of axial winders

The control of the armature current is realized by a limitation of the output signal of the speed controller to the variable saturation limit $i_A^* = m_M^*/\psi$. The reference value of the speed control loop is increased over the reference value of the average speed, n^*, by an additional Δn^*. This leads to the effect that in steady state the speed controller output reaches the saturation limit and the speed control loop is opened. Hence, the speed controller is active only during the start of the processing or if the material tears. During normal operation the speed controller has no influence on the stationary and dynamic state of operation.

The realization of the control with the two signal processing blocks WR and RR is complicated and not very precise if analogue signal processing is used. It should be kept in mind that the equations are valid only, if small disturbances occur and the system is in a state near stationary operation. Considering disturbances occurring in the system (e.g. friction, backlash, the nonlinearity of the excitation), it becomes clear that there are restrictions in this model. Therefore it is advantageous to apply modern control concepts which take these restrictions into account.

Another approach for the control of the winders is adaptive state–space control ($ZR1$) which takes into account the time–variant moment of inertia. In this case, though, we have the restriction, that all states of the system, torque, speed and force must be available for control. Assuming proportional feedback and an LTI plant, in theory, pole assignment is possible. However, in a real plant the dynamics of the armature current and the field are limited and therefore these poles are fixed. Beside the above mentioned nonlinearities we have to deal with time–variant parameters like the radius r of the winder (Θ, respectively) and the average speed of the web v_o.

For a system optimized by the "damping optimization criterion" ([9, 11]) the following results are obtained. Figure 2.26 shows the step response of f/f^* with the feedforward control and adaptive state–space control with observer. In figure 2.27 the speed reference of the drive next to the winder was varied by 3% in a step function. This figure shows on the left the dissatisfying response with the FFC used up to now, and on the right the response achieved with the adaptive state–space control $ZR1$. Both results with the new control system show a very good damping of the system.

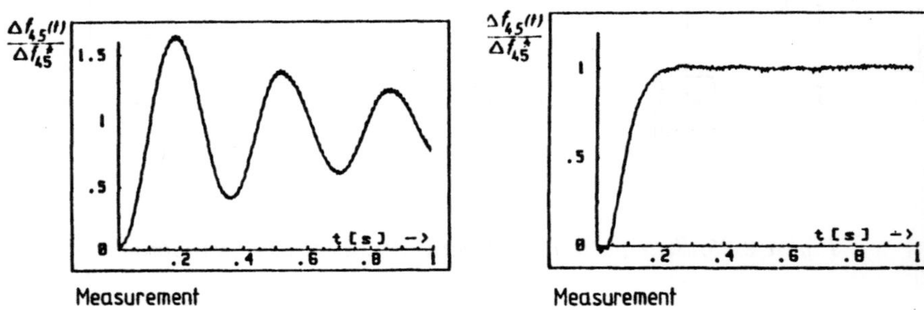

feedforward control *adaptive state–space control*

Figure 2.26: Step responses due to reference input

A very undesired resonant effect occurs, if the roll of the winder is not circular. The resonant speed of the winder is

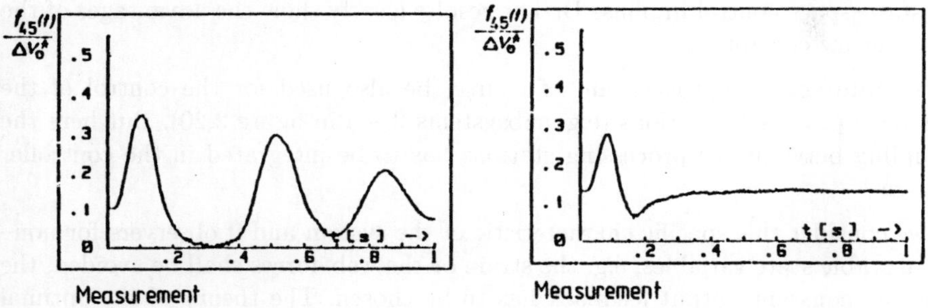

feedforward control *adaptive state–space control*

Figure 2.27: Step responses due to disturbance

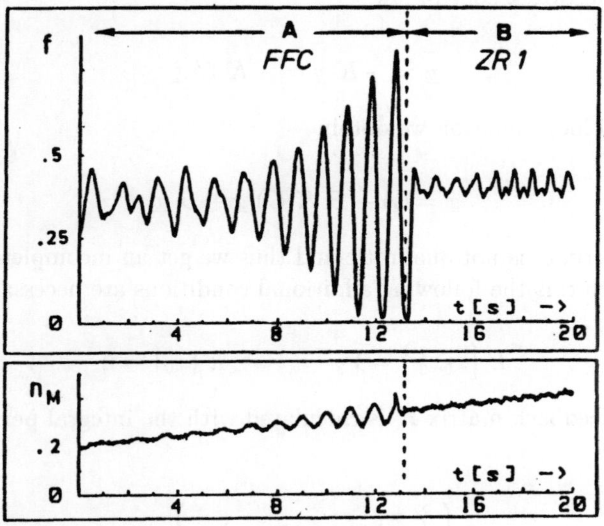

Figure 2.28: Web force and winder speed for a noncircular winder

$$N_W = \frac{\Omega_{0W}}{2\pi} \tag{2.52}$$

Figure 2.28 shows the results with FFC and with an adaptive state–space controller $ZR1$.

In the experiment the winder had a very small deviation from circularity. The speed n_M of the winder was increased slowly until it nearly reached N_W. In the experimental phase A a conventional feedforward control was used and switched to state–space control in phase B. The results clearly show the advantages of the state–space control.

The state–space approach can, of course, be also used for the control of the system's processing stations (e.g. subsystems 2 – 5 in figure 2.20). But here the coupling between the processing stations has to be integrated in the controller design.

If we consider this specific characteristic of the system and if observers for non–measurable state variables, e.g. the strain or the web–forces shall be avoided, the optimal constant output feedback has to be chosen. The theory of the optimal output feedback was developed in [6] and improved for practical application in [7].

In contrast to complete state–space control, the optimal output feedback uses **only the measurable components of the output vector $\underline{y}$**. Thus we obtain the following feedback control law

$$\underline{u} = -K\ \underline{y} = -K\ C\ \underline{x} \tag{2.53}$$

For the closed loop equation we obtain

$$\dot{\underline{x}} = (A - B\ K\ C)\ \underline{x} + B\ \underline{w} \tag{2.54}$$

where the matrix C is not quadratic and thus we get an incomplete state–space control. Due to this the following additional conditions are necessary

$$E\left\{\underline{x}_0 \underline{x}_0^T\right\} = V; \qquad E\left\{x_0\right\} = 0 \tag{2.55}$$

The optimal feedback matrix K is calculated with the integral performance criterion

$$J = E\left\{\int\limits_0^\infty \left(\underline{x}^T Q \underline{x} + \underline{u}^T R\ \underline{u}\right) dt\right\} \rightarrow Min \tag{2.56}$$

The solution of J depends on the initial value $\underline{x}_0$ because of the incomplete state–space control. Due to this fact the covariance matrix V is necessary. Thus, the average value of the integral performance criterion J is calculated with all possible initial values $\underline{x}_0$.

This method has been applied to the system shown in figure 2.20. In this application the optimized current control loop was represented by a simple time-delay element with a gain V_{ij} and a time constant T_{ij}. The design of the speed control was achieved with regard to the multi variable system.

The control structure is shown in figure 2.29. Each speed controller uses the actual value of the speed of the preceding processing station which leads to a decoupling effect. To avoid a steady–state error, a feedback with an integrator is added. The gains K_{lm} and the integral time constants T_j of the speed feedback control loop depend on the solution of the integral performance criterion J and a special Riccati equation.

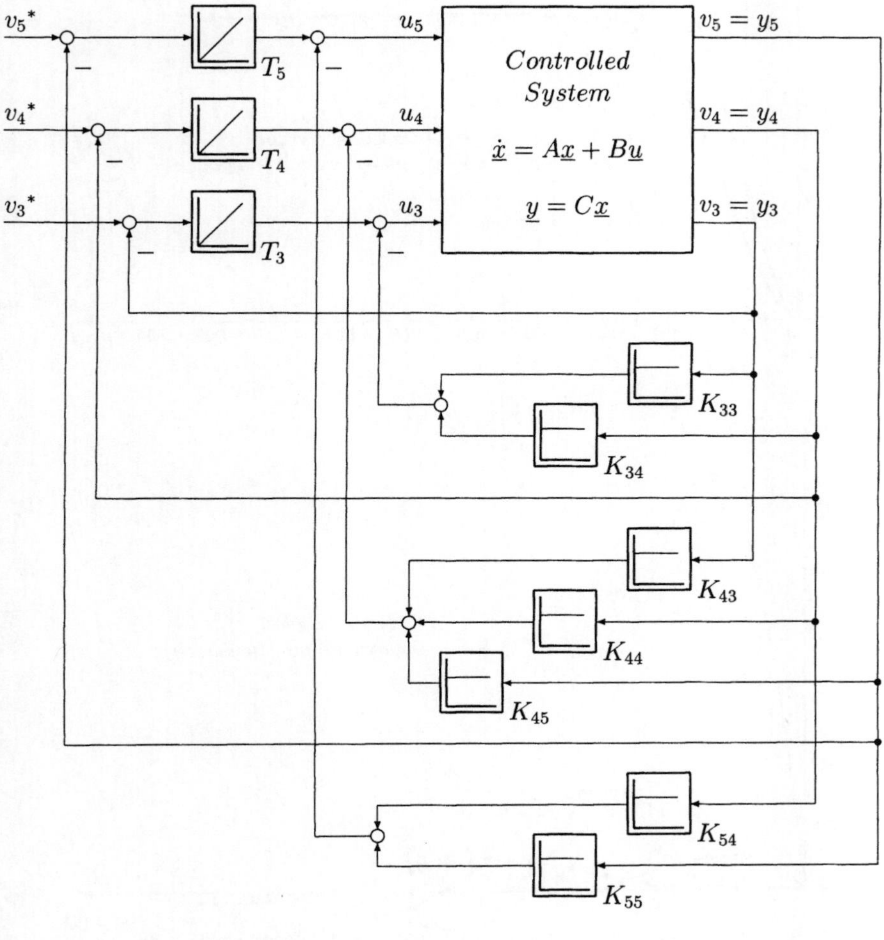

Figure 2.29: Structure of the control with optimal output feedback

A typical control result shows figure 2.30. Compared to a cascade control with simple PI–controllers, the decoupling effect can be seen.

As the state–space control of a multi variable system is complex and often unpractical in industrial plants, decentralized control methods should be used, where

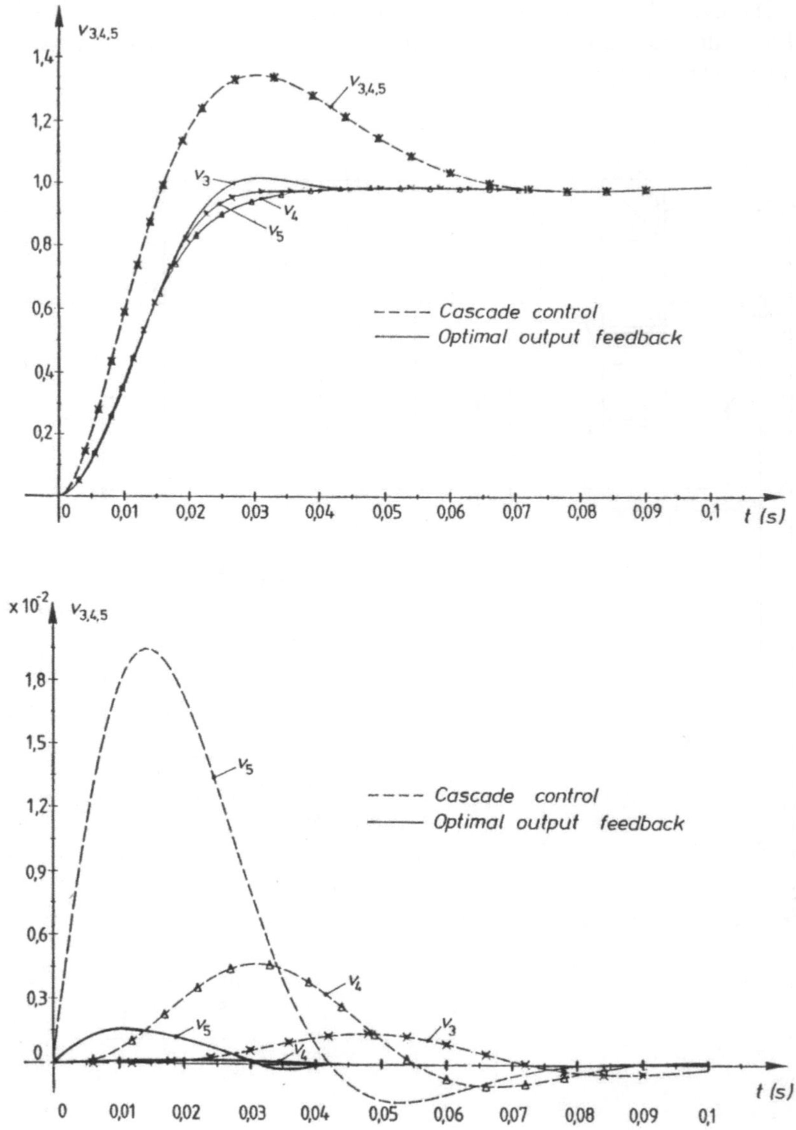

Figure 2.30: Step responses of the speed with optimal output feedback; above: step response to reference value, below: step response to disturbance

the state–space control is designed for subsystems of low order. To design a decentralized control we have to separate the total system into subsystems. As shown in figure 2.20 the subsystem consists of the roller, the electrical drive and the web section on the left side of the roller. Each subsystem can be controlled with a low order state–space control. But if we design the controller for an isolated subsystem, we get a significant deterioration of the dynamic behaviour in the total system as shown in figure 2.22. Due to the influence of the coupling, oscillations occur and the forces of the neighbouring subsystems also have undesired dynamic variations. This is the consequence of the neglect of the coupling during the design of the control.

To get a satisfying dynamic behaviour, the quantities of coupling must be taken into account. To achieve this, three possibilities exist:

- the design of decoupling networks
- the use of so–called equivalent terminating models
- the decentralized decoupling.

The first option requires the design of special decoupling networks and presupposes the measurement of the coupling signals.

The second option requires the design of a low order equivalent model of the controlled neighbouring subsystems. If we have only few subsystems (less than three or four) this method is successful and produces results equivalent to method three. But if the number of subsystems increases this method is not recommended because of many iterations during the design of the control.

The third possibility avoids these disadvantages. The goal of the method is to design a controller which minimizes the influence of the neighbouring systems.

Now the state–space controller has two functions:

- to guarantee the desired dynamic and stability of the total system and
- to minimize the influence of the neighbouring system.

The solution to design such a controller is to consider the sensitivity of the eigenvalues [11].

The advantage of this method is that no measurements of the coupling signals are required. It is only necessary to know where the couplings are active in the subsystem.

Figure 2.31 shows the measured step responses of the web forces for a system with five sections. The decoupling effect can clearly be seen in comparison with figure 2.22.

But there are some questions to all control methods considered above. What will happen, if the parameters inside the system will vary considerably — for example

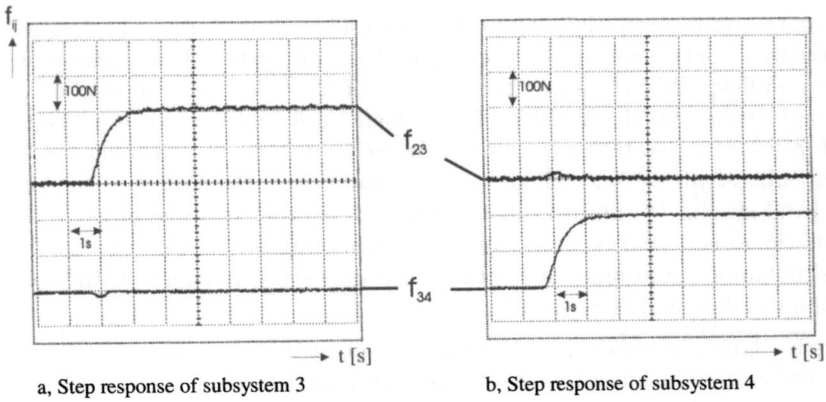

Figure 2.31: Step responses with decentralized control

the ε's — , or if the material changes its characteristics during the processing? Or what will happen during the process of putting the system into operation? Is there any chance to use this approach, if all effects of the mechanical transmission are also added? Furthermore, which dynamic and static effects will occur, if the material in one processing station tears?

The results of this section for motion control are:

- cascaded control loops cannot provide the desired control results;

- state–space controllers can provide much better results than cascaded control loops, but due to time–variant parameters at least adaptive state–space control is necessary;

- even if the adaptive state–space control approach is applied, the controlled system behaviour can show severe limitations due to quickly changing parameters or parameters with high uncertainty;

- it takes much more detailled knowledge of the system to put a state–space control system into operation, than one with cascaded control;

- the coherences in an electromechanical system where the general process is taken into account, usually lead to a signal flow graph of high complexity even if severe simplifications are made.

2.3 Objectives of this book (Example Motion Control)

During the last three decades the control of electrical drives has included the control of torque, speed and position at the shaft of the machine only. The real requirements for motion control are much more complex.

1. In most applications there is no ideal transmission between the load and the actuators, see section 2.1.2 – 2.1.4
2. The plant is not ideal, either. It usually is a complex system often with a nonmodular structure, where a decentralized control design for subsystems (like in the case of continuous processing machines) cannot be done.
3. The interaction between the actuator, the transmission and the load leads consequently to an even more complex system behaviour.

Considering these aspects, it is reasonable to say that new approaches for the understanding of the system, the control, and the optimization of the controller are necessary.

Of course, there are tools for specific problems. For example, if a mechanical system is under investigation, a finite element analysis will result in a set of very high order differential equations. Therefore, the understanding of these equations, the simulation of the system, and the design of the control is very time consuming.

Of course, the order of this set of differential equations can be reduced. But to which order is the reduction necessary? If it is done, is there any chance of understanding the reduced set of equations and comparing these equations and the real physical system?

As discussed in the preceding sections, it is advantageous to use the object oriented approach for the representation of a system. For example, if a multi–mass mechanical plant is under consideration, the Lagrange–approach is suitable and each subsystem consisting of one shaft and one mass can be one object. For the mathematical representation of one object we obtain

$$\dot{\underline{x}} = A\underline{x} + B\underline{u} \tag{2.57}$$

$$\underline{y} = C\underline{x} \tag{2.58}$$

The mathematical representation of a linear multi–mass mechanical plant in descriptor form [18, 19, 20, 21] will be

$$E\dot{\underline{x}} = A\underline{x} + B\underline{u} \tag{2.59}$$

$$\underline{y} = C\underline{x} \tag{2.60}$$

where the first equation is an implicit system of differential equations with a singular matrix E. This first equation in the descriptor form seems very similar to the explicit state–space representation of the object, but in the descriptor form the matrix E is a representation of the connections between the different objects and these are the algebraic equations. Therefore, the first equation of the descriptor form represents the differential equations of the system and the algebraic constraints.

The input–output behaviour of this implicit system of differential equations will be a transfer matrix $T(s)$

$$T(s) := C(sE - A)^{-1}B \tag{2.61}$$

It should be mentioned that not all of the input signals u can be chosen freely, hence the signals $\underline{u}$ and $\underline{x}$ are transformed to $\underline{z}$

$$\underline{z} = \begin{bmatrix} \underline{x} \\ \underline{u} \end{bmatrix} \tag{2.62}$$

and

$$M\dot{\underline{z}} = N\underline{z}; \qquad M = [E, 0]; \qquad N = [A, B] \tag{2.63}$$

The descriptor representation can be extended to systems with imprecise parameters, to parameter identification, observer design, nonlinear systems and structure variable systems by events.

The same considerations are applicable for the technological sections in motion control.

As an important result of these considerations, it seems advantageous to define objects which are subsystems of the plant. Due to this approach, we can define electrical objects, mechanical objects, technological objects, hydraulic objects and information processing objects. These very different objects will be components in a descriptor system which is an implicit system of differential equations and enables a physical understanding of very complex systems. From the engineering point of view this is desired very much, as then critical conditions can be analysed and diminished or even avoided by an adapted information processing object.

Following this object oriented approach for the analysis of complex plants, we have to keep in mind that an object can have a linear dynamic behaviour with known structure and time–invariant parameters, and a static nonlinearity with known characteristics.

However, we discussed, that very often the parameters may be time–variant or known imprecisely and — even more important — the nonlinearity is not known. Therefore, one objective of this book will be the local identification of nonlinearities by intelligent strategies. Local identification means, we will define objects and determine the nonlinearities in this object by intelligent strategies. The result of this procedure will be a nonlinear model of the object.

Due to the learning of the nonlinearity, convergence and stability are important demands. When these demands are met, then on–line learning is achievable and a time–variant model of objects will be the result.

The intelligent identification of nonlinearities will be extended in this book to intelligent observer design (SNO) also.

Additionally, if the order of the plant is not too high in SISO–systems, compensation of the nonlinear effects are possible, resulting in a linear system and the application of the well–tried linear control strategies are possible. (The discussions of the multi–mass system versus the two–mass system, the unstructured dynamics and the control strategies discussions should be noted.)

We discussed in section 2.2, that in an elastic two–mass system undesired vibrations can occur. By appropriate control concepts these vibrations can be damped effectively; yet the torque dynamics were neglected during the discussions.

Extending these considerations to controlled systems where

- the torque dynamics cannot be neglected,
- we have a digital controller and where the processing and transfer of the information leads to a time–delay in the order of the highest mechanical mode,
- we have a multi–mass mechanical system and the undesired vibrations occur at a mass near the output,

then the damping of the undesired vibrations will be extremely difficult to achieve.

Under these restrictions a diminution of the undesired vibrations can be achieved much better by an active absorber. The theoretical and practical considerations will be discussed in the appendix of this book and are valid for both lateral and rotating mechanical systems.

Up to now, we discussed the modelling and control of nonlinear plants from an engineering point of view. The engineering point of view means we have at least some basic knowledge of the relevant structure and the linear parameters of the plant. In contrast to that point of view the order of the system is generally the only knowledge available to a control design.

In this book we will develop new tools which are adequate to the new challenges. We have to build bridges of better understanding between the technologies, mechanical aspects, and the researcher's interest in drives. Furthermore, we need new tools for the control of such systems. Nonlinear control can be a solution; fuzzy logic and neural systems are discussed.

But these strategies, which till now were not common, show interesting and sometimes promising results. For example, we still do research on the system shown in figure 2.20. We have stated, that cascaded control does not show the desired results (figure 2.22). If we use decoupling or adaptive state–space control, improvements in the static and dynamic behaviour can be achieved (figure 2.31). But a general result of the research is that the better the decoupling, the slower the dynamic response of the system will be. If fuzzy logic control is used for such a system, the same decoupling but a much better dynamic response is obtainable.

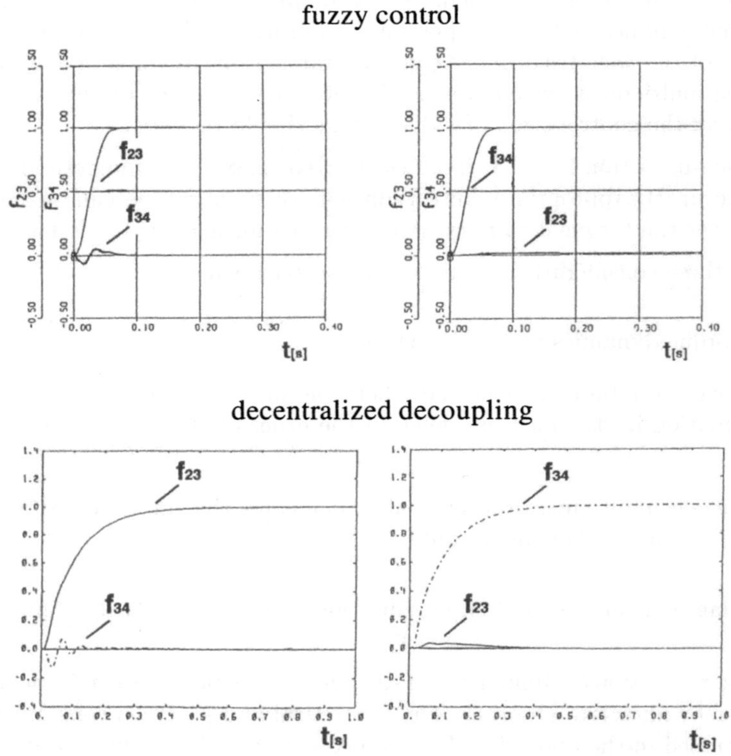

Figure 2.32: Comparison of fuzzy and decentralized decoupling control

Figure 2.32 shows the dynamic behaviour of the web–forces f_{23} and f_{34} when the system is controlled by fuzzy logic (top) and by decentralized decoupling control (bottom). Due to the variation of the reference value of the web–forces, the real value will follow with no overshoot and reach the steady state value in about $0.5s$ with decentralized decoupling control. With fuzzy control we get the steady state value after $0.06s$. In both control methods, the decoupling is equal. So the best result is obtained with fuzzy control.

This example shows some promising features of exotic strategies, which must be investigated further. Again it should be mentioned that motion control is only one of many possible examples to which the strategies discussed in this chapter can be applied to.

2.4 Conclusions

In the first two chapters we discussed the objectives for this book, starting from the theoretical control aspects in the first chapter to an example of applied motion control in this chapter.

When we summarize the results of this chapter first, we have to accept that from both the engineering and the control point of view there generally is some knowledge of the structure of the plant. Therefore, a fundamental signal flow graph of the structure can be designed.

In many cases the plant has a modular structure resulting in the opportunity to define objects which have identical signal flow graphs but different parameters. The objects are coupled by algebraic equations resulting in the descriptor–representation of the plant. Due to this tool of object–oriented modelling very different objects can also be coupled. This is a very important advantage.

Furthermore, there generally is basic knowledge about the most essential parameters of the linear section or objects of the plant. Additionally there can be some knowledge of the positions of nonlinearities in the plant. But the nonlinearities are not generally smooth — especially not in motion control, where backlash and friction are typical hard nonlinearities. However, the parameters of these nonlinearities are difficult to determine exactly. They are often time–variant due to different operating conditions or aging. This also holds for several of the parameters of the linear sections of the plant.

2.5 References

[1] Schröder, D.
Trends in Power Electronics and Drives
Eleventh IFAC World Congress 1990, Tallin/GUS, pp. 215 - 224.

[2] Schäfer, U.
Entwicklung von nichtlinearen Drehzahl– und Lageregelungen zur Kompensation von Coulomb–Reibung und Lose bei einem elektrisch angetriebenen, elastischen Zweimassensystem
Dissertation, TU–München, 1992.

[3] Schröder, D.; Wolfermann, W.
State observer for multi–motor–drives in processing machines with continous moving web
EPE Conference 1985, Brussels, pp. 3.203 - 3.210.

[4] Schröder, D.; Höger, W.
Substantial improvements of the performance of axial winders by adaptive state space control
IFAC–Symposium on "Microcomputer Application in Process Control", Istanbul, 1986, A6.13.

[5] Schröder, D.; Wolfermann, W.
Applications of decoupling and state space control in processing machines with continous moving web
Sixteenth IFAC–Congress 1987, Munich, Vol.3, p. 100-107.

[6] Levin, W.S.; Athans, M.
On the determination of the optimal constant output feedback gain for linear multi variable systems
IEEE–Transactions on Automatic Control AC–15.

[7] Kuhn, U.
Neue Entwurfsverfahren für die Regelung linearer Mehrgrössensysteme durch optimale Ausgangsrückführung
Dissertation, TU–München, 1984.

[8] Litz, L.
Reduktion der Ordnung linearer Zustandsraummodelle mittels modaler Verfahren
Dissertation, Universität Karlsruhe, 1979.

[9] Schröder, D.
Elektrische Antriebe 2: Regelung von Antrieben
Springer–Verlag, Berlin, Heidelberg, 1996.

[10] Naslin, P.:
Essentials of Optimal Control.
Iliffe, 1968.

[11] Wolfermann, W.; Schröder, D.
New Decentralized Control in Processing Machines with Continuous Moving Webs

Proceedings of the Second International Conference on Web Handling IWEB2, Stillwater, Oklahoma, 1993.

[12] Bae, D. S., Hang, I.
A recursive formulation for constrained mechanical system dynamics. Part I: Open loop systems.
Mech. Struc. Mach. Vol. 15, 1987, pp. 359–382.

[13] Anathavaman, M.
Flexible Multibody Dynamics – An Object-Oriented Approach.
Proc. of the Nato ASI on Computer Aided Analysis of Rigid and Flexible Mechanical Systems, Vol. II, 1993, pp. 383–402.

[14] Anderson, M.
Dymola – An Object Oriented Language for Model Representations.
Thesis TFRT–3208, Lund Institute of Technology.

[15] Elmquist, H.
Dymola – User's Manual.
Dynasim AB, Lund, Sweden, 1993.

[16] Kecsheméthy, A.
Objektorientierte Modellierung der Dynamik von Mehrkörpersystemen mit Hilfe von Übertragungselementen.
VDI–Fortschrittsberichte, Reihe 20, Nr. 88, 1993.

[17] Otter, M.
Objektorientierte Modellierung mechatronischer Systeme am Beispiel geregelter Roboter.
VDI–Fortschrittsberichte, Reihe 20, Nr. 147, 1995.

[18] Bryant, R.L., Chern, S.S., Gardner, R.B., Goldsmith, H.L., Briffiths, P.H.:
Exterior Differential Equations.
Springer–Verlag, Berlin, Heidelberg, 1991.

[19] Choquet–Bruhot, Y., Denritt–Morette, C.:
Analysis, Manifolds and Physics.
North Stolland, 1982.

[20] Willems, B.C.:
On Interconnections, Control and Feedback.
IEEE Transactions on Automatic Control, Vol. 42, 1997.

[21] Schupphaus, R.:
Regelungstechnische Analyse und Synthese von Mehrkörpersystemen in Deskriptorform.
VDI Fortschrittsberichte Reihe 8, VDI–Verlag, 1995.

3 Learning in Control Engineering

Ulrich Lenz

Technology in the field of information processing is advancing in large steps, a fast growing area is "Artificial Intelligence".

In recent years intelligent methods have been applied in control engineering. At the same time, complex technical systems consisting of mechanics, electrical drives and control became the subject of an integral view. Combined with the reflection of ideas of computer science and information technology, this has lead to interesting mathematical methods in the field of mechatronics.

In this chapter we do not intend to discuss the state of the art in control engineering with Artificial Intelligence (AI), it is our aim to explain reasonable extensions of control theory based on the methods of AI. AI means learning, and based on the learned knowledge some new approaches of control design become clear.

However, first we have to discuss different interpretations of intelligence.

3.1 Intelligent Control as Artificial Intelligence

Artificial Intelligence — a contradiction in itself? Artificial, non–biologic intelligence? As long as we associate fantasy, creativity, esprit, or senses with the word intelligence, this remains an incompatible contrast. "Intelligence", as it can be artificial, has to be reduced in its meaning to the simple attribute "capability of learning". Yet, for the learning, a complex optimizing functional might be used to learn simple aspects. Learning means also to draw conclusions from the learned knowledge; this leads to intelligence.

Learning needs structures able to generate and store associations. Structures able to learn can be realized by using the state of the art, namely processors coupled with a kind of analogous, digital or hybrid memory. This memory must be able to store associations which represent the learned knowledge. The possibility of a dynamical variation of the learned knowledge by adaptation (learning) with respect to appropriate rules is a substantial characteristic of intelligence.

The technical realization of such intelligent structures is a separate topic in technology and seems to be an infinite field of research.

Here we will discuss in a very short manner three basic approaches for the technical realization of intelligent structures. These attempts are oriented at nature, at the biological representation of learning.

3.2 Artificial Intelligence Realized by a Non–Biologic Structure

Often, the technical research and development are oriented at nature (e.g. bionic). In particular, technology copies three different interpretations of natural/biological intelligence.

Regarding evolution as a learning process of the subject "nature", in which genetic information is learned by selection oriented at the biological success of a species, has lead to several approaches summarized in the topic "genetic algorithms". As in nature, this learning takes a long period of time and is based on trial and error. It is obvious that this success–oriented selection is not very helpful for control engineering since safety aspects always have to be considered.

The result of biologic evolution has lead in its highest stage of development to connected nerve cells to realize the ability of learning reflected by the brain, see figure 3.1.

Figure 3.1: The brain consists of connected neurons. The associations reflected in the connections combine input scheme and action scheme.

The connection of these nerve fibres/cells is done by synapses. During the learning process, the connectivity is adapted to change the stored associations. The processing of a sensoric stimulation is done in parallel by a huge number of neurons at the same time, leading to an extremly fast information processing. The information processing shows the association of a (sensoric) stimulation scheme to a reaction scheme.

All kinds of Artificial Neural Networks (NN) imitate this highly parallel information processing.[1] The learning is achieved by adapting weights, which reflect the biologic synapses. In the following chapter 4 a certain type of NN is introduced with its features. It is a specific RBF–net (radial basis functions) and its selection for control purposes dates back to Schäffner [18].

It seems obvious to copy not only the brain's structure, but also its function as the center of intelligence in life forms for technical approaches. The rule based thinking enables mankind to draw conclusions based on learned rules. The conclusions are drawn considering if–then–decisions, which is imitated by the several (adaptive) fuzzy approaches.[2]

We are interested in Artificial Intelligence, learning in control engineering or "intelligent control". By the increasingly interdisciplinary view in technology, intelligent approaches have found their way into control engineering. **Mechatronics**, the integral view of a system consisting of drive, mechanics and data processing, is the title of this technical evolution. This farseeing thinking has brought forth the idea to apply methods of AI in the field of control engineering.

3.3 Basic Structures for Control

The classical assignment of guidance and control is to affect a system by its input in such a way, that the system's output acts as intended. In principle, we distinguish between open–loop and closed–loop control.

3.3.1 Open–Loop Control

The first approach is the open–loop control, as it is shown in a schematic structure in figure 3.2. Under the stipulation, that the plant and the actuator are known in

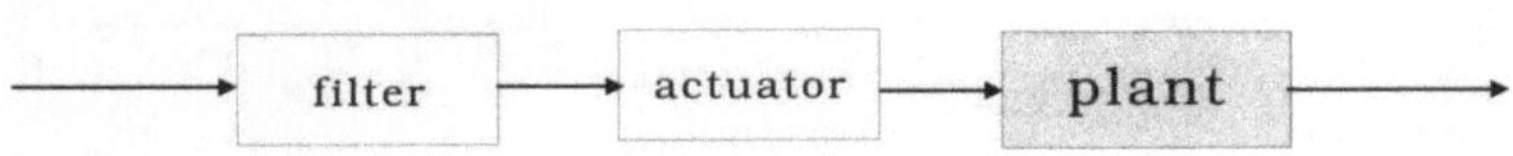

Figure 3.2: Simple structure of open–loop control

their static and dynamic behaviour, and if they are also linear and time–invariant,

[1] Introductions to Neural Networks are e.g. [10, 17, 22]

[2] Basics for fuzzy control see e.g. [22, 3], [2, 14] represent an introduction to neuro–fuzzy approaches.

the open–loop structure is able to synthesize a control law. This affects the system to react as intended within its limitations. This means, the response to setpoint changes can be defined by using the open–loop control.

It is known that if disturbances occur or some parameters of the plant or actuator are time–variant, the open–loop–controlled system's behaviour often becomes unsatisfactuary.

Also the synthesis of a control law seems to be difficult, especially if the system or the actuator is nonlinear and the nonlinearity is not known exactly. Concerning the stability of an open–loop–controlled system is valid, that, if the components of the whole system have a stable transfer behaviour themselves, the whole system is also stable.

3.3.2 Closed–Loop Control

An open–loop structure is not able to correct the system's output deviations from the intended behaviour due to the lack of information.

If the feedback of the system's output y is compared to the reference signal r, the result is a simple feedback or closed–loop control structure. Disturbances affecting the system and leading to a deviation are now recognizable and measures for compensation can be considered by the controller. Such a simple conventional feedback control structure is shown in figure 3.3 in a schematical drawing. Such

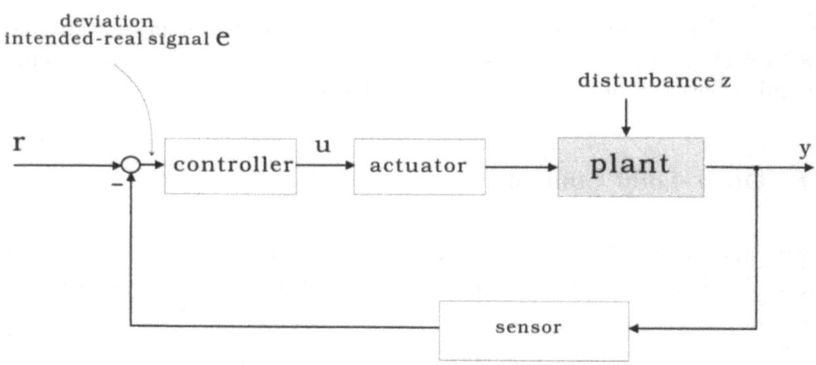

Figure 3.3: Simple control structure. The controller diminishes the deviation of the systems output from the intended behaviour.

closed–loop control structures are characterized by a common feature:
Comparing the system's output y with the reference signal r leads to the control error e; the controller's aim is to diminish this error. Therefore based on the reference signal and the real signal(s) the control law defines a plant's input u which

affects the plant. The control behaviour achieved by this feedback structure is optimizable in view of response to setpoint changes and disturbance suppression.

There exist different closed–loop control approaches. Beside cascaded control, state space control is also well–known. Control theories are developed not only for the linear case but also for some nonlinear applications (chapter 2). As well as for open–loop control, the design of a closed–loop control structure becomes more difficult in the nonlinear or/and time–variant case. A special difficulty arises in the treatment of nonlinear systems with unknown or time–variant nonlinearities.

Due to the feedback or closed–loop structure of such control structures, the stability analysis of the whole structure ranks high. While the stability analysis for linear systems use the well–known theories for linear control, for the treatment of nonlinear systems a generic theory lacks. There are some helpful approaches such as the Popov–criterion, the describing function or the direct method of Ljapunov, however these approaches do not cover the whole spectrum of nonlinear systems.

An appropriate combination of open– and closed–loop control reflects the "conditional–feedback" structure [5, 20].

3.3.3 "Conditional Feedback" Control Structure

The response to variations of the reference signal is realized by using a kind of open–loop control. A closed–loop controller takes action if the system's output differs from the intended behaviour. For the simple linear case such a conditional–feedback structure, also known as "feedforward–feedback" structure, is displayed in figure 3.4. The basic idea in the upper path of this structure is equal to

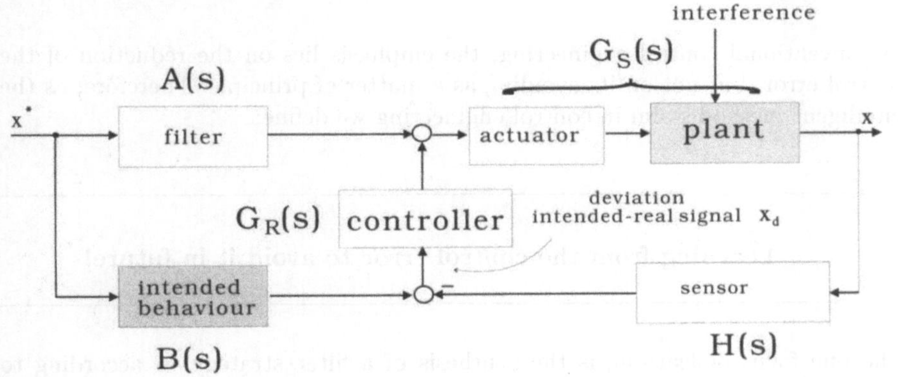

Figure 3.4: Conditional feedback structure: the responses to variations of the reference signal and to disturbances are separately optimizable.

that of an open–loop control; this means, the filter is designed to guarantee the intended response on variations of the reference signal. The design is carried out without reflection of occuring interferences. Disturbances result in a deviation of the system's output from the reference signal, the controller in the feedback path intervenes. If a disturbance $z \neq 0$ occurs, the signal e will be $\neq 0$ and the controller is activated (conditional feedback) to suppress the disturbance.

Based on this proposed structure, the controller response to variations of the reference signal and to disturbances is separately optimizable.

3.4 Scopes for Intelligent Control

From the explanations in sections 3.1 and 3.2 Artificial Intelligence should be explained: it is the control of nonlinear and/or time–variant plants with unknown coefficients.

Conventional control always is characterized by a common fact: the control structure tries to diminish a deviation of the controlled variable if a control error appears! This means, under the same conditions, the control error will always be reproduced. Such errors are caused by

- non–optimal open–loop pre–control,
- disturbances or
- nonlinearities or/and timevariances which have not been considered for the control synthesis.

In conventional control engineering, the emphasis lies on the reduction of the control error, but not on its avoiding as a matter of principle. Therefore, as the intelligent method's aim in control engineering we define:

Learning from the control error to avoid it in future!

The one focus of learning is the synthesis of a filter–strategy — according to open–loop or conditional–feedback control — which leads to the intended static and transient behaviour of the plant, whether it is linear or not.

The other focus is the reduction of the control error due to "deterministic" disturbances. This means, if disturbances occur not only in a stochastic way, if they depend on measurable signals, their dependency can be learned and this knowledge can be used e.g. for compensation purposes.

This intention is a difficult demand, but it is oriented at our philosophical understanding of "learning". We shall characterize a control structure to be intelligent, if by a process of adaptation (learning) a control error principally caused by the structure is minimized within the structure's limitations.

3.4.1 Methods of Intelligent Control

A control structure with adapted control parameters can be considered the first step of intelligent control methods; according to specified algorithms, the controller parameters are adapted to minimize the error.

In literature, the so–called STC–approach (*self tuning controller*, e.g. [12, 21]) is well–known. Another solution is the *model reference adaptive control* (MRAC), which means, that the behaviour of the controlled system is tuned towards that of a reference model adapting an open–loop pre–controller [7]. Such a MRAC–

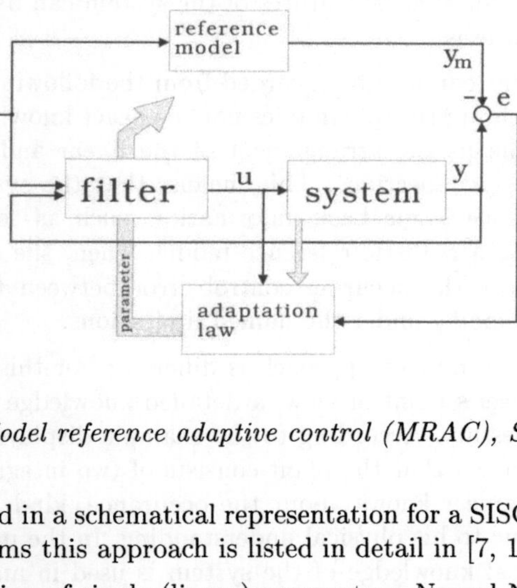

Figure 3.5: Model reference adaptive control (MRAC), SISO–system

structure is displayed in a schematical representation for a SISO–system in figure 3.5. For linear systems this approach is listed in detail in [7, 12, 21].

Besides the adaptation of single (linear) parameters, Neural Networks or Fuzzy Rules are also used as intelligent structures in applications for control engineering [13, 15].

3.4.2 Application of Learning in Control Engineering

In the field of control applications, in principle, learning can be used in two different approaches [21]:

- A control law is adapted to minimize or to avoid the control error. The basis for the learning process is in most cases a defined performance index as a criterion for the adaptation. Intelligent structures are used to synthesize the control law. We call this type of learning in control engineering the "direct approach".

or

- The system is identified by means of these learning structures to provide knowledge as a basis for the controller design. Intelligent structures are used to model the static and dynamic behaviour of the system and to represent the structural or quantitative knowledge in the following step for the controller design. This procedure is called the "indirect approach".

In the direct approach, the intended learning target is a control law which is able to improve the control performance regarding several criteria. An "understanding" of the system in its behaviour is not gained.

Compared to it, the indirect approach produces this understanding. Based on this understanding the specific features of the system can be taken into account for the control synthesis.

These considerations can also be regarded from the following point of view:
The direct approach in principle implies not the exact knowledge of the system's structure, which means the arrangement of the linear and nonlinear dynamic parts and their interconnections. This means, that the system is regarded as a "black box". Hence, some basic information, such as the systems order, is necessary to choose a realistic reference model. Then, the controller has to be adapted to minimize the occuring control error between the system and the intended reference model under the named limitations.

The intention of the indirect approach is different: For this approach, which is closer to the engineer's point of view, a detailed knowledge about the structure of the plant is necessary. According to the example displayed in figure 3.6, this means that it is known that the plant consists of two integrators with a known gain. Also, the engineer knows about the occurance, kind and interconnection of nonlinearities due to his physical understanding. In the indirect approach the physical or technical knowledge of the system is used in an advantageous way, for example to leave the exact shape of a nonlinearity to be identified. The system's modeling with regard to this detailed knowledge leads to the decision, to compensate the influence of such a nonlinearity as a disturbance or to design a control law which considers among the linear parts also the nonlinear ones.

These two different approaches are compared by using a simple example:

3.4.2.1 Example: Direct and Indirect Approach

Given is a rigid drive system with friction according to figure 3.6. This plant has to be controlled in its position; the controller is designed to guarantee an optimal follow–up control within the given limitations.

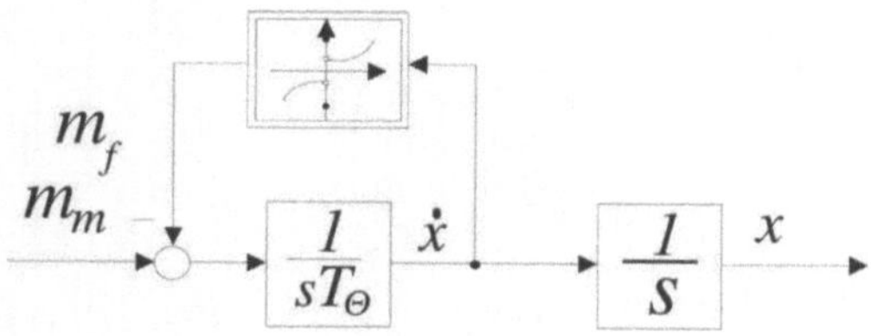

Figure 3.6: Signal flow chart representation of a rigid drive system with speed dependent friction

The direct approach would result in the synthesis of a feedforward control law, to diminish the control error due to set–point changes under certain performance indices. An obvious criterion is the behaviour of such a system without any influence of friction leading to a linear plant for time–optimal control. This could be realized by the operation point depending on the adaptation of an ordinary linear PD–controller whose parameters are stored in maps.

Contrary to this direct approach, the indirect approach gains knowledge about the occuring nonlinearity (friction) to be the basis for compensation purposes to minimize this interfering influence. To do so, we are in need of an intelligent structure which is able to represent the friction in its dependency on rotational speed.

Compensating the friction also leads to a linear I^2–system which can be object to time–optimal control.

Both approaches are compared to each other in figure 3.7. It becomes clear that the indirect approach needs a much more sophisticated understanding of the concerned system: to identify the disturbing influences of friction in that plant, it has to be known exactly in its linear structure.

In this book we present direct as well as indirect approaches. But, first, the requirements for adaptive methods concerning control engineering must be discussed:

3.5 Requirements for Adaptive Methods

All methods to be used in control engineering have to fulfill requirements concerning safety aspects, to have a useful application. Stability is the most important quality of each meaningful control method.

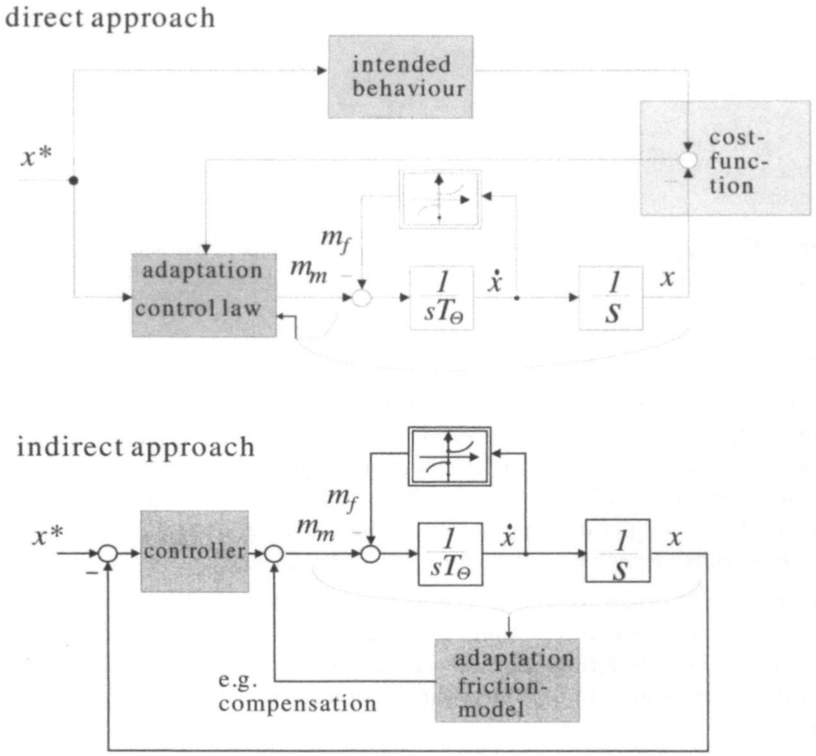

Figure 3.7: Direct and indirect approach for adaptive control of a rigid drive system with speed dependent friction

3.5.1 Stability

If adaptive control methods are used in closed–loop structures, the stability of the system has to be proven mathematically for each operating point. In the field of control engineering there exist different interpretations of stability.

We differentiate between input–output stability, stability in the sense of Ljapunov or asymptotical stability. For autonomous systems we define:

Definition 3.1 Input–Output Stability [19] *A system is called input–output stable, if any limited input signal causes an always limited system's output for any initial state of the system.*

Definition 3.2 Stability in the sense of Ljapunov [8] *The equilibrium point* $\underline{x} = \underline{0}$ *of the system*

$$\dot{\underline{x}} = \underline{X}(\underline{x}), \quad \underline{X}(0) = 0 \tag{3.1}$$

is called stable in the sense of Ljapunov, if for each radius $R > 0$ *a radius* $r > 0$ *exists, so that for* $||\underline{x}(0)|| < r$ *it follows* $||\underline{x}(t)|| < R$ *for all* $t \geq 0$. *If not, the system is unstable.*

Definition 3.3 Asymptotical Stability [21] *The equilibrium point* $\underline{x} = \underline{0}$ *of the system*

$$\dot{\underline{x}} = \underline{X}(\underline{x}), \quad \underline{X}(0) = 0 \tag{3.2}$$

is called asymptotically stable, if it is stable and also a $r > 0$ *exists, that*

$$||\underline{x}(\underline{0}|| < r \rightarrow \lim_{t\to\infty} \underline{x}(t) = \underline{0} \tag{3.3}$$

is valid.

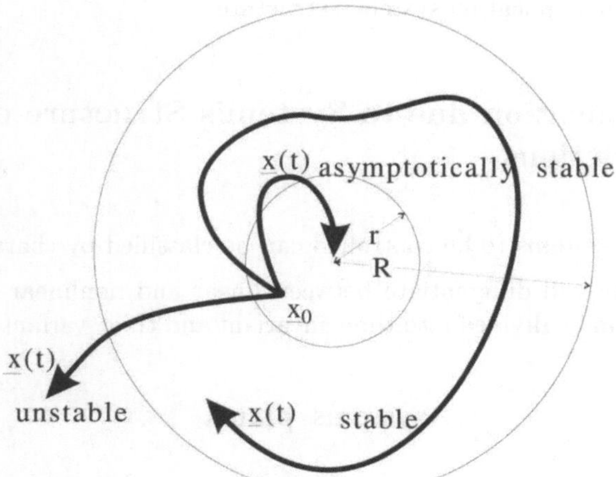

Figure 3.8: Stability of an equilibrium point

These interpretations of stability are illustrated in figure 3.8. Regarding technical systems, stability in the sense of Ljapunov means, that due to a disturbance, the system remains always near its point of operation (trajectory). In difference to that, asymptotical stability leads to the decrease of the disturbance influence.

3.5.2 Improving the Controller's Performance

Due to safety aspects, the application of adaptive systems in control engineering seems to be meaningful, only if the adaptation considerably improves the control performance. Surely, the choice of a reference is difficult. Therefore regarding the system's limitations is always necessary; for example, it is no use to expect an abrupt change of the controlled position of the system displayed in figure 3.6.

3.5.3 Expandable Knowledge

If methods are used for identification, the gained knowledge has to be expoundable, clear and not equivocal. This seems to be a necessary condition for the use of the identified knowledge in an expanded control structure, for example for compensation measures or pre–control.

It is a convenient approach to classify the various methods of intelligent control according to the application system's structure.

3.6 Classification due to System's Structure or Restrictions

Real technical systems to be controlled can be classified by characteristics.

In principle, we will differentiate between linear and nonlinear systems. These main groups can be divided into time–invariant and time–variant systems (figure 3.9).

systems, plants

linear	nonlinear
→ time-invariant	→ time-invariant
→ time-variant	→ time-variant

Figure 3.9: General classification of technical systems

The systems regarded in this book are nonlinear. After a physical modeling, their structure and most of the linear parameters are known more or less precisely. As mentioned before, we distinguish between direct and indirect control approach; the indirect approach has to be based on such a modeling to gain knowledge of the system's structure.

The Engineer's View Against the Black–Box Approach

In cases in which the engineer's examination delivers knowledge, a model of the plant, it is advantageous to use this model as a basis for the controller design. We get such knowledge by a physical modeling, e.g. by discribing a dynamical system using differential equations. Most technical systems are accessible for this modeling.

In comparison to the engineer's view stands the approach used by computer science: A system is not accessible for examination or modeling; regarding the input–output behaviour only leads to an impression of the system's order. This "black–box" approach offers no knowledge e.g. for compensation purposes. Therefore the direct synthesis of a control law according to a defined cost function is in the foreground of the control structure design.

In the following chapter 4 we will introduce a Neural Network as an intelligent structure, which will be used for identification and also for direct synthesis of the control law.

3.7 References

[1] Ackermann, J.:
Robuste Regelung.
Springer–Verlag, Berlin, Heidelberg, 1993.

[2] Brown, M., Harris, C.:
Neurofuzzy Adaptive Modelling and Control.
Prentice Hall, 1994.

[3] Driankov, D., Hellendoorn, H., Reinfrank, M.:
An Introduction to Fuzzy Control.
Springer–Verlag, Berlin, Heidelberg, 2. Auflage, 1996.

[4] Fischle, K.:
Ein Beitrag zur stabilen adaptiven Regelung nichtlinearer Systeme.
Dissertation, TU München, 1998.

[5] Geering, H.P.:
Regelungstechnik.
Springer–Verlag, Berlin, Heidelberg, 4. Auflage, 1996.

[6] Kuschewski, J.G., Hui, S., Zak, S.H.:
Application of Feedforward Neural Networks to Dynamical System Identification and Control.
IEEE Transactions on Control Systems Technology, Vol. 1, No.1, March 1993.

[7] Landau, Y.:
Adaptive Control: The Model Reference Approach.
Marcel Dekker Inc., New York Basel, 1979.

[8] La Salle, Lefschetz:
Stabilitätstheorie von Ljapunov.
B. I. Hochschultaschenbücher, 1967.

[9] Mamdani, D.:
Application of Fuzzy Algorithm for Control of Simple Dynamic Plant.
Proc. of the IEEE, 1974, pp. 1585–1588.

[10] Miller, W., Sutton, R.S., Werbos, P.J.:
Neural Networks for Control.
The MIT Press, 1992.

[11] Narendra, K.S.:
Neural Networks for Control: Theory and Practice.
Proceedings of the IEEE, Vol. 84, No. 10, October 1996, pp. 1385–1406.

[12] Narendra, K.S., Annaswamy, A.M.:
Stable Adaptive Systems.
Prentice–Hall, Inc., Englewood Cliffs, NJ, USA, 1989.

[13] Narendra, K.S., Parthasaranthy, K.:
Identification and Control of Dynamical Systems Using Neural Networks.
IEEE Transactions on Neural Networks, pp. 4–27, Vol. 1, No. 1, March 1990.

[14] Nauck, D., Klawonn, F., Kruse, R.:
Neuronale Netze und Fuzzy–Systeme.
Vieweg Verlag, Braunschweig, 2. Auflage, 1996.

[15] Parasani, T., Zoppoli, R.:
Neural Networks for Feedback Feedforward Nonlinear Control Systems.
IEEE Transactions on Neural Networks, pp. 436–449, Vol. 5, No. 3, May 1994.

[16] Park, Y., Choi, M., Lee, K.:
An optimal tracking Neuro Controller for Nonlinear Dynamic Systems.
IEEE Transactions on Neural Networks, Vol. 7, No. 5, September 1996.

[17] Ritter, H., Martinetz, Th., Schulten, K.:
Neuronale Netze. Eine Einführung in die Neuroinformatik selbstorganisierender Netzwerke.
Addison–Wesley, 1991.

[18] Schäffner, C.:
Analyse und Synthese neuronaler Regelungsverfahren.
Dissertation, TU München 1996.

[19] Schmidt, G.:
Grundlagen der Regelungstechnik.
Springer–Verlag, Berlin, Heidelberg, 1989.

[20] Schröder, D.:
Elektrische Antriebstechnik 2, Regelung von Antrieben.
Springer–Verlag, Berlin, Heidelberg, 1995.

[21] Slotine, K., Li, W.:
Applied Nonlinear Control.
Prentice–Hall, Inc., Englewood Cliffs, NJ, USA, 1991.

[22] White, D.A., Sofge, S.A.:
Handbook of Intelligent Control, Neural, Fuzzy and Adaptive Approaches.
Van Nostrand Reinhold, 1992.

[23] Widrow, B.:
The Original Adaptive Neural Net Broom–Balancer.
Proceedings IEEE International Symposium on Circuits and Systems, 1987, pp. 351–357.

[24] Zadeh, L.A.:
Fuzzy Sets.
Information and Control, Vol. 8, 1965.

4 Neural Networks and Fuzzy Controllers as Nonlinear Function Approximators

Michael Beuschel

Considering nonlinear control, identification of nonlinearities has become essential. Thus, a universal function approximator has to be found which is able to approximate an unknown static[1] nonlinear function. For this objective, various neural network and also fuzzy approaches are offered.

To use one of these methods for adaptive control, two requirements have to be met: First, a stable self–tuning algorithm has to be established (which will be the topic of the next chapter). And second, this method has to perform universal function approximation with required accuracy and convergence, which will be discussed in this chapter. Typical neural networks and a common fuzzy approach will be analyzed and compared to each other to find an optimal choice for this type of control application. Closing, the above approaches will be applied to an example.

4.1 Nonlinear Function Approximation

Nonlinear function approximation is a common approach in identifying nonlinear behaviour that is not accessible analytically but only empirically. In general, this involves a limited number of parameters which can be tuned to fit a given function optimally within a compact set of its input space.

First, we will define a class of nonlinear functions, to which all subsequent identification concepts can be applied.

Definition 4.1 *Any continuous, bounded and time–invariant scalar function* $\mathcal{NL} : \mathbb{R}^P \rightarrow \mathbb{R}$ *of the P–dimensional input vector* $\underline{x}$ *with the scalar output* y *is called a* **nonlinearity** $\mathcal{NL}$

[1] In this chapter we will restrict ourselves to the case where the control law can be separated into *static* nonlinear functions and linear dynamic components

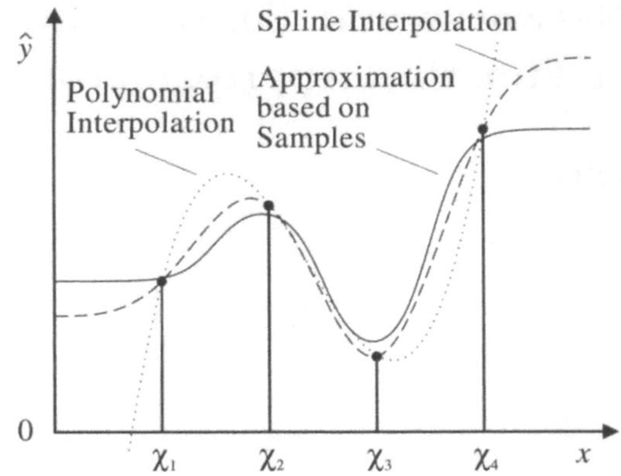

Figure 4.1: Approximation and interpolation using given sample points

$$y = \mathcal{NL}(\underline{x})$$

with $\underline{x} = [x_1, x_2, \ldots x_P]^T$ *and* $\mathbb{M} \subset \mathbb{R}^P$ *is a compact set of the input space* $\mathbb{R}^P$*, where* $\mathcal{NL}$ *can be observed.*

If a multi–dimensional output is required, separate nonlinearities of this type will simply have to be stacked. Therefore, it will be sufficient now to concentrate on these multiple–input–single–output nonlinearities.

4.1.1 Concepts of Function Approximation

There are some basic concepts of function approximation within a compact set of a given input space and using a limited number of parameters (see figure 4.1):

- Global interpolation (or approximation) using finite function series
- Local interpolation between sample points (e.g. splines)
- Local approximation based on sample points

Well–known examples for the first concept are polynomial functions (like Taylor and Fourier series). Although this can be a very useful method, there is a main disadvantage in our case: As each term of the superimposed function components is allowed to contribute to the function output over the full range of its input space, no local (and therefore fast) adaptation is enabled.

The second approach is commonly used for look–up tables and provides completely local operation of all parameters, which we demanded just before. However, in many control applications, also extrapolation is required to some extent, which means that the adapted knowledge of the nonlinearity within a certain input region can be applied to another — yet "unlearned" — region. This is unfortunately impaired by strictly local operation of parameters.

In order to overcome this and meet both the option of fast local adaptation and the ability of extrapolation, the third concept is applied. Like approximation by a function series, it employs a weighted and superimposed sum of function components. However, this concept involves *localized basis functions*, which have a region of peak activity at their center and tend towards zero elsewhere. This will enable fast local adaptation and, as these functions overlap each other, extrapolation is also supported. This is why we regard the local and sample based concept as most suitable for nonlinear control [1]. In addition, the effect of noise in input data can be reduced by employing less parameters than the number of data sets available, which results in smoothing of the identified function.

Before proceeding to specific neural and fuzzy concepts we will first have to define suitable basis functions and to discuss how they support universal function approximation and parameter convergence.

4.1.2 Basis Functions for Function Approximation

We will define *localized basis functions*, which are suitable basis functions for nonlinear function approximation, as follows:

Definition 4.2 *A continuous bounded and non–negative function* $\mathcal{B} : \mathbb{R}^P \rightarrow \mathbb{R}_{0+}$ *is a* **localized basis function** *if it has a global maximum at* $\underline{x} = \underline{\chi}$ *and if, for all elements* x_p *of the* P*–dimensional vector* $\underline{x}$*, it meets*

$$\frac{\partial \mathcal{B}(\underline{x})}{\partial x_p} \quad \begin{cases} \geq 0 \\ \leq 0 \end{cases} \quad for \quad x_p \quad \begin{cases} < \chi_p \\ > \chi_p \end{cases}$$

and

$$\lim_{||\underline{x}|| \to \infty} \mathcal{B}(\underline{x}) = 0$$

This definition will allow uniform representation of different neural network and fuzzy approaches. Examples for localized basis functions are Gaussian radial basis functions, Parzen windows, and triangular windows. In [5] even rectangular windows are included. In figure 4.2 some symmetric candidates for these basis functions are shown.

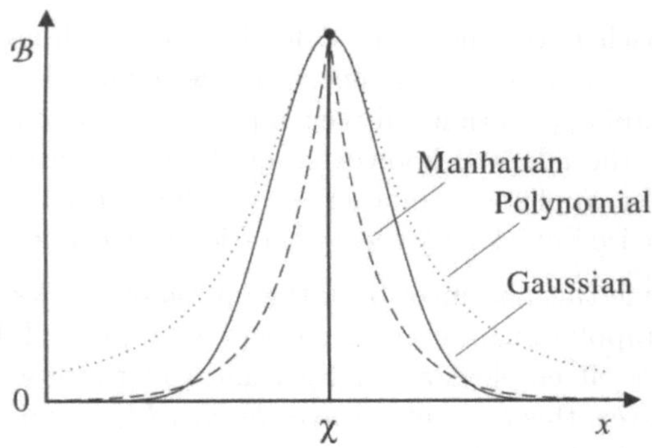

Figure 4.2: Examples of localized basis functions

4.1.3 Universal and Convergent Function Approximation

As mentioned above, a weighted sum of localized basis functions has shown good ability of function approximation. To identify a universal and convergent function approximator, two properties have to be verified: First, a *universal* function approximator has to be able to approximate any nonlinearity as defined above arbitrarily close by proper choice of design parameters and, second, the inherent approximation error between the nonlinearity and the approximator output has to *converge* to its global minimum.

As a mathematical proof of these demands is beyond the topic of this chapter, at least an intuitive approach to both will be given in the following.

Introducing an inherent approximation error $d(\underline{x})$, which corresponds to a non-ideal approximator, we will assume that for any nonlinearity $\mathcal{NL}(\underline{x})$ according to definition 4.1 in a compact set $\mathbb{M}$ of its input space a representation

$$\mathcal{NL}(\underline{x}) = \underline{\Theta}^T \underline{\mathcal{A}}(\underline{x}) + d(\underline{x}) \tag{4.1}$$

exists with the parameter vector of weights

$$\underline{\Theta}^T = [\vartheta_1 \; \vartheta_2 \; \dots \; \vartheta_N]^T$$

which correspond to the centers $\underline{\chi}_1, \underline{\chi}_2, \dots \underline{\chi}_N$ of activation functions with

$$\underline{\chi}_n^T = [\chi_{n,1} \; \chi_{n,2} \; \cdots \; \chi_{n,P}]^T$$

The activation function $\underline{\mathcal{A}}(\underline{x})$ is a vector of localized basis functions

$$\underline{\mathcal{A}}^T(\underline{x}) = [\mathcal{A}_1(\underline{x}) \; \mathcal{A}_2(\underline{x}) \; \dots \; \mathcal{A}_N(\underline{x})]^T$$

where all functions $\mathcal{A}_n(\underline{x})$ apply to definition 4.2. We will refer to $\mathcal{A}_n(\underline{x})$ as the *activation* of the n-th weight.

Based on localized basis functions, this representation supports local activity of all weights, i.e. each weight is mainly responsible for the approximator output at a specific input region. For infinitely small overlapping of individual activation functions, approximation changes to interpolation between samples. Thus, employing a sufficiently large number N of weights, any continuous function can be approximated arbitrarily close. For "partition of unity" radial basis functions such as the Gaussian, this has been proven in [6]. Hence, this concept provides universal function approximation.

Using the same concept to represent the estimate of a nonlinearity, an estimated nonlinearity $\widehat{\mathcal{NL}}$ can be introduced:[2]

$$\hat{y} = \widehat{\mathcal{NL}}(\underline{x}) = \underline{\hat{\Theta}}^T \underline{\mathcal{A}}(\underline{x}) \tag{4.2}$$

Provided a stable learning algorithm, convergence of the estimated function towards a given nonlinearity will be guaranteed if parameter convergence can be proved. To show this, we introduce an estimation error e, which is due to incomplete adaptation resulting in a difference between a given nonlinearity and its estimate:

$$e(\underline{x}) = \underline{\hat{\Theta}}^T \underline{\mathcal{A}}(\underline{x}) - \underline{\Theta}^T \underline{\mathcal{A}}(\underline{x})$$

Considering a state of optimal adaptation, the estimation error equals zero for any $\underline{x}$ in the compact set of the input space, where $\mathcal{NL}$ is to be identified. This can be expressed as:

$$e(\underline{x}) = \underline{\hat{\Theta}}^T \underline{\mathcal{A}}(\underline{x}) - \underline{\Theta}^T \underline{\mathcal{A}}(\underline{x}) = \left(\underline{\hat{\Theta}} - \underline{\Theta}\right)^T \underline{\mathcal{A}}(\underline{x}) = \underline{\Phi}^T \underline{\mathcal{A}}(\underline{x}) = 0$$

In this final state, the parameter error

$$\underline{\Phi} = \underline{\hat{\Theta}} - \underline{\Theta} \tag{4.3}$$

will be a constant vector, whereas the activation $\underline{\mathcal{A}}(\underline{x}) > \underline{0}$ must not be constant within a certain time period, which is given by the condition of *persistent excitation* in [11]. Hence, this inner product will equal zero if and only if either $\underline{\Phi} = \underline{0}$ or both vectors are orthogonal. If we assume that $\underline{\Phi} \neq \underline{0}$ then $\underline{\mathcal{A}}(\underline{x})$ may only represent a vector in a hyper–plane which is orthogonal to $\underline{\Phi}$.

For an ideal case, the contrary is simple to show: Employing non–overlapping basis functions, we take a set of N activation vectors located at the center of each activation function. This corresponds to a variation of the input vector all over the input space. Then each of these vectors equals a different unit vector, which combined are a basis of $\mathbb{R}^P$. Hence, by definition, these vectors are linearly independent and is $\underline{\Phi} = \underline{0}$, which in turn proves parameter convergence and consequently convergence of the estimated function.

---

[2] We will denote the estimate and all associated parameters using the ^ symbol. (Please note that the input vector $\underline{x}$ is identical for both the nonlinearity and its approximation.)

For Gaussian radial basis functions, uniqueness of functional representation has been proved in [9]. Together with stable adaptation, this proves convergence of the estimated function on a more general basis.

4.2 Neural Networks as Function Approximators

Various neural network and fuzzy approaches exist, which combine the advantages of the local approximation concept as outlined in the previous section. In the following, Radial Basis Function (RBF) networks with the subtypes of the General Regression Neural Network (GRNN) and the Distance Activation Neural Network (DANN) as well as fuzzy controllers will be discussed as representatives of most suitable approaches for adaptive control.

4.2.1 Radial Basis Function (RBF) Network

Radial Basis Function (RBF) networks represent a large family of neural networks for nonlinear function approximation. Compared to other approaches like multilayer perceptron networks or the network proposed by Kohonen, they are characterized by a constant correlation between the input space and their weights, i.e. each weight corresponds to a certain region of the input space. This provides meaningful interpretation of gained knowledge in a physical sense. Thus, applying definition 4.2, an RBF network can be represented as a weighted sum of basis functions resulting in an estimate

$$\widehat{y}(\underline{x}) = \widehat{\mathcal{NL}}(\underline{x}) = \sum_{n=1}^{N} \widehat{\vartheta}_n \mathcal{A}_n(\underline{x})$$

of the nonlinearity, which corresponds to equation (4.2). This not only enables local adaptation of weights but also provides extrapolation as far as $\underline{\mathcal{A}}(\underline{x}) > \underline{0}$ is ensured [1, 14].

A common choice of activation function is the Gaussian radial basis function

$$\mathcal{A}_n = \exp\left(-\frac{\mathcal{C}_n}{2\,\sigma^2}\right) \tag{4.4}$$

introducing a smoothing parameter σ, which provides different shapes. This representation is used in correspondence with the Gaussian standard deviation [15], where a function $\mathcal{C}_n$ correlated to the n–th weight is defined to be the squared length of the vector $\underline{x} - \underline{\chi}_n$ (see also figure 4.2):

$$\mathcal{C}_n := \|\underline{x} - \underline{\chi}_n\|^2 = (\underline{x} - \underline{\chi}_n)^T (\underline{x} - \underline{\chi}_n) = \sum_{p=1}^{P} (x_p - \chi_{n,p})^2 \tag{4.5}$$

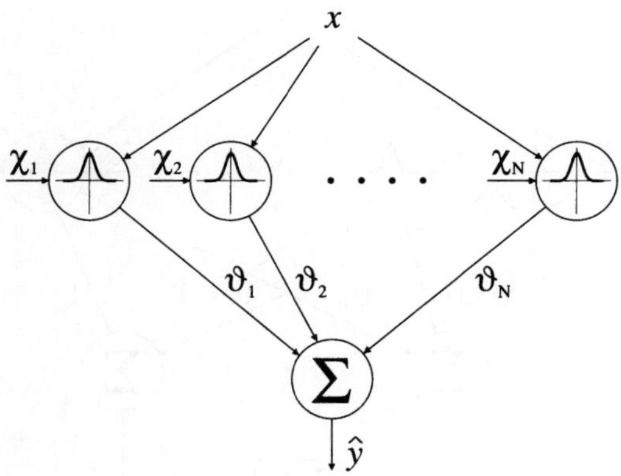

Figure 4.3: Topology of a radial basis function network

For this constellation, the topology of an RBF network can be outlined as given in figure 4.3. Alternatively, the Manhattan distance could be employed for $\mathcal{C}_n$:

$$\mathcal{C}_n := \sum_{p=1}^{P} |x_p - \chi_{n,p}| \tag{4.6}$$

4.2.2 General Regression Neural Network (GRNN)

The General Regression Neural Network (GRNN) has evolved from the RBF network to improve function approximation with a minimal number of weights. Given a finite number of samples of a continuous function, regression means the calculation of the most probable output for a specific input. Thus, the GRNN differs from the RBF representation by its normalized activation as shown in figure 4.4 [15]:

$$\mathcal{A}_n = \frac{\exp\left(-\dfrac{\mathcal{C}_n}{2\,\sigma^2}\right)}{\sum_{m=1}^{N} \exp\left(-\dfrac{\mathcal{C}_m}{2\,\sigma^2}\right)} \tag{4.7}$$

Hence is

$$\sum_{n=1}^{N} \mathcal{A}_n = 1$$

The normalized activation function of the GRNN yields a bounded range of output values of the network, which will be limited by its minimum and maximum weights, even if unlearned input vectors are presented. This leads to improved

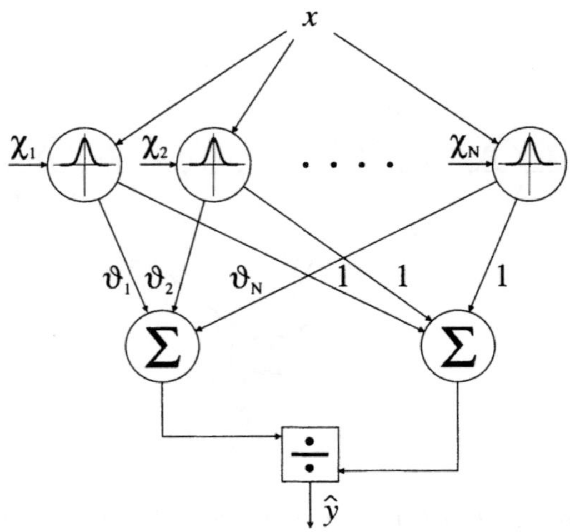

Figure 4.4: Topology of a general regression neural network

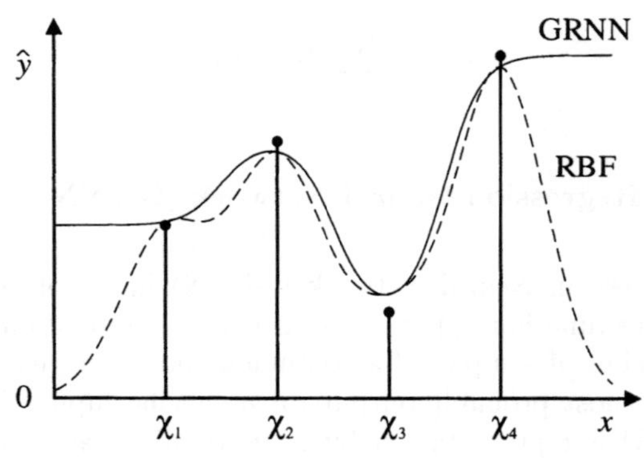

Figure 4.5: Improved interpolation and extrapolation of the GRNN compared to the RBF network

interpolation and extrapolation performance compared to a standard RBF network, where the output at unlearned regions and between the centers of weights tends towards zero. In contrast to that, extrapolation of a GRNN always maintains the value of the nearest weight (see figure 4.5).

As the number of neural network implementations on a microcontroller platform grows, and as very high update rates for high–performance networks on digital signal processors become indispensable, computational time has to be reduced. In the following, we will present some options to provide this.

4.2.2.1 GRNN at Multidimensional Input Space

Any GRNN with a dimension $P \geq 2$ of its input space will profit extraordinarily from the factorization of the exponential function. To apply this, two conditions have to be met: First, each activation function $\mathcal{A}_n$ must be identical to each other despite the shift and, second, all centers of the activation functions have to be arranged in a lattice that is defined by discretized coordinates of the input space. Then, instead of computing a function

$$\mathcal{C}_n = \sum_{p=1}^{P} c_{xp,n}$$

and subsequently the activation $\mathcal{A}_n$ for each weight, it is sufficient to calculate once the function components $c_{xp,n} := f(x_p - \chi_{n,p})$ for all "rows" and "columns" of the lattice (in the two–dimensional case) and to apply the exponential function to those only. The numerator of the activation function can then be determined by multiplication of the appropriate components for each dimension:

$$\begin{aligned}\mathcal{A}_n &= \exp\left(-\frac{\mathcal{C}_n}{2\sigma^2}\right)\\ &= \exp\left(-\frac{c_{x1,n} + c_{x2,n} + \cdots c_{xP,n}}{2\sigma^2}\right)\\ &= \exp\left(-\frac{c_{x1,n}}{2\sigma^2}\right)\exp\left(-\frac{c_{x2,n}}{2\sigma^2}\right)\cdots\exp\left(-\frac{c_{xP,n}}{2\sigma^2}\right)\end{aligned} \tag{4.8}$$

This is also applicable to the case of normalized activation as it can easily be shown for two dimensions with $X \cdot Y = N$:

$$\begin{aligned}\mathcal{A}_n &= \frac{\exp\left(-\frac{\mathcal{C}_n}{2\sigma^2}\right)}{\sum_{m=1}^{N}\exp\left(-\frac{\mathcal{C}_m}{2\,\sigma^2}\right)}\\ &= \frac{\exp\left(-\frac{c_{x1,n} + c_{x2,n}}{2\sigma^2}\right)}{\sum_{x=1}^{X}\sum_{y=1}^{Y}\exp\left(-\frac{c_{x1,x}}{2\,\sigma^2}\right)\exp\left(-\frac{c_{x2,y}}{2\,\sigma^2}\right)}\\ &= \frac{\exp\left(-\frac{c_{x1,n}}{2\sigma^2}\right)}{\sum_{x=1}^{X}\exp\left(-\frac{c_{x1,x}}{2\,\sigma^2}\right)} \cdot \frac{\exp\left(-\frac{c_{x2,n}}{2\sigma^2}\right)}{\sum_{y=1}^{Y}\exp\left(-\frac{c_{x2,y}}{2\,\sigma^2}\right)}\end{aligned}$$

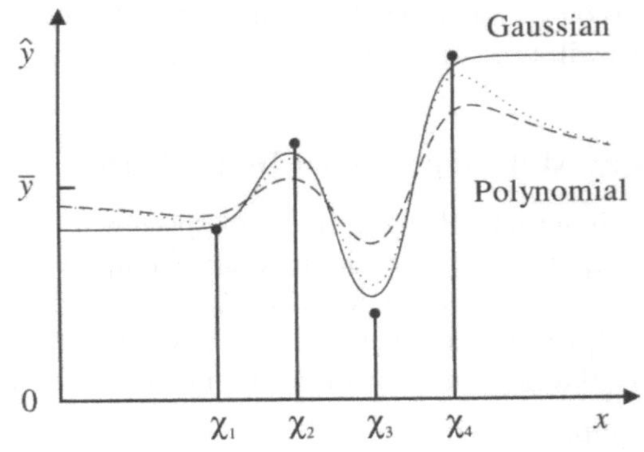

Figure 4.6: Extrapolation of the GRNN using Gaussian activation compared to polynomial activation of the DANN for two different smoothing factors σ

4.2.2.2 Polynomial Activation (DANN)

In the case of a one–dimensional GRNN or if the previous suggestion cannot be applied, implementation of a simplified activation function promises reduced computational effort. Thus, the exponential function could be replaced e.g. by a polynomial one as suggested for the Distance Activation Neural Network (DANN) in [10]:

$$\mathcal{A}_n = \frac{\left(1 + \dfrac{\mathcal{C}_n}{2\,\sigma^2}\right)^{-1}}{\sum\limits_{m=1}^{N} \left(1 + \dfrac{\mathcal{C}_m}{2\,\sigma^2}\right)^{-1}} \tag{4.9}$$

As the polynomial function declines much slower than the exponential function, local activity of weights will be supported less, which in turn leads to a larger number of parameters to obtain the same accuracy. Moreover, extrapolation always tends towards the mean value $\bar{y}$ of all weights, which ensures the boundedness of output but usually does not maintain the value of the nearest weight as shown in figure 4.6. Nevertheless, this can be an appealing alternative [4, 10].

4.2.2.3 Restricted Update Area

As most software implementations of neural networks have to cope with limited numerical resolution, there is no point in updating weights that do not exceed a minimum activation. Hence, according to numerical accuracy and to the quality of function approximation, an area centered at the current input vector can be determined outside of which no computation is necessary. A reasonable choice for this area is e.g. a hypercube $|x_p - \chi_{n,p}| \leq \Delta x_{min}$, which will also cooperate very well with the factorization of the exponential function as suggested in section 4.2.2.1 [4].

4.2.3 Other Neural Network Approaches

There are also further neural approximators that might be of interest for some applications, so we shall here give a brief overview.

Kohonen Neural Network: Kohonen's self–organizing neural network resembles the way regions of the cerebral cortex correspond to parts of the human body. In its original form, it consists of a single layer of neurons. To each neuron an output weight and a vector, which defines the center of the neuron's activation in the input space, are assigned. Initially, these vectors are spread randomly all over the input space. Adaptation then will not only tune the weights but will also shift the centers of the neurons towards regions of most frequent input vectors. This feature of self–organization provides efficient placement of neurons but, unfortunately, does not guarantee reasonable learning results under all circumstances. Besides high computational effort, this is why we cannot recommend this neural network for adaptive control tasks [14].

Multilayer Perceptron (MLP): This is perhaps the most common family of feedforward neural networks, which consist of multiple layers of *perceptron* called neurons. The input layer represents the components of the input vector and, for each perceptron of the subsequent layers, a weighted sum of the output of the preceding layer is computed. The output of this perceptron is then given by applying a monotonous activation function to this sum. For classification tasks, binary activation functions are used, whereas differentiable sigmoidal activation functions are preferred for control applications. Moreover, the latter enable the famous *Backpropagation* algorithm to adapt the weights of the different layers (also including the "hidden" ones) due to their influence on the network output. For this, the gradient of the approximation error is calculated to minimize this error. However, this algorithm cannot guarantee that the global minimum of the approximation error will be reached, which, in contrast, is the case for RBF networks [13].

Cerebellar Model Articulation Controller (CMAC): This neural network provides discrete output as its localized basis functions are rectangular windows. The memory space required is significantly reduced by Hash–Coding, which is a smart mapping of the input space to corresponding parameters. Because of its discrete output, this network is not suitable for continuous control applications as considered in this context [2].

4.3 Neuro–Fuzzy Systems as Function Approximators

Another control concept, which has received major attention in the recent years, is fuzzy control. Despite the different background of fuzzy logic and neural networks, both are able to perform the same basic task of nonlinear function representation. Like neural networks, fuzzy controllers employ a set of parameters, which determine the nonlinearity $\mathcal{NL}$. For certain classes of fuzzy controllers, it can mathematically be proven that they are able to approximate any smooth nonlinear function like a GRNN does [12, 17]. Moreover, it can be shown that fuzzy controllers and neural networks have a very similar mathematical structure and thus, learning algorithms of the neural network domain can be used to tune the parameters of a fuzzy controller. The result is a hybrid neuro–fuzzy system which combines the learning capability of neural networks and the linguistic knowledge representation of fuzzy controllers. The control law is then no longer a "nonlinear black box" to the user, but is accessible in the form of transparent IF–THEN rules. This is especially useful if linguistic knowledge about the plant is available for controller initialization.

In this section, we will first describe the motivation for neuro–fuzzy control from the fuzzy logic point of view. Afterwards, we will discuss the structural similarities between fuzzy controllers and neural networks and present an example of a neuro–fuzzy system, which is very similar to the GRNN in the previous section.

4.3.1 Principles of Neuro–Fuzzy Systems

Systems which combine fuzzy logic and neural networks are referred to as *neuro–fuzzy* systems, or equivalently as fuzzy neural networks. Various methods for designing such systems have been proposed in literature [3, 7, 8, 18]. Most of them are *hybrid* systems, i.e. they do not consist of distinct fuzzy and neural network parts, but they are fuzzy controllers and neural networks at the same time. Hybrid neuro–fuzzy systems are based on the fact that fuzzy controllers and neural networks own a very similar structure. The algorithm of a fuzzy controller can be represented as a network of several layers of interconnected parallel processing elements. In this structure, the parameters of the fuzzy controller (e.g. positions and widths of the fuzzy sets) are represented as multiplicative weights associated with connections and can thus be trained by learning algorithms from neural network theory.

As an example, we will consider the well–known *singleton defuzzification method* [16]

$$u = \frac{\sum_{n=1}^{N} \mu_n \widehat{\vartheta}_n}{\sum_{m=1}^{N} \mu_m} \tag{4.10}$$

where u is the crisp controller output[3], N is the number of rules, μ_n is the firing strength of rule n, and $\widehat{\vartheta}_n$ is the position of the maximum of the consequent membership function for rule n. This formula can be represented as a "layer" of two parallel processing elements, which calculate the numerator and denominator of equation (4.10)[4] (see figure 4.7). From the neural network point of view, these elements are standard McCulloch–Pitts–neurons with a linear activation function. Their inputs are the firing strengths μ_n, which are computed in the previous layer. The positions $\widehat{\vartheta}_n$ of the output sets correspond to the synaptic weights of the numerator neuron. (The synaptic weights of the denominator neuron are fixed at unity.)

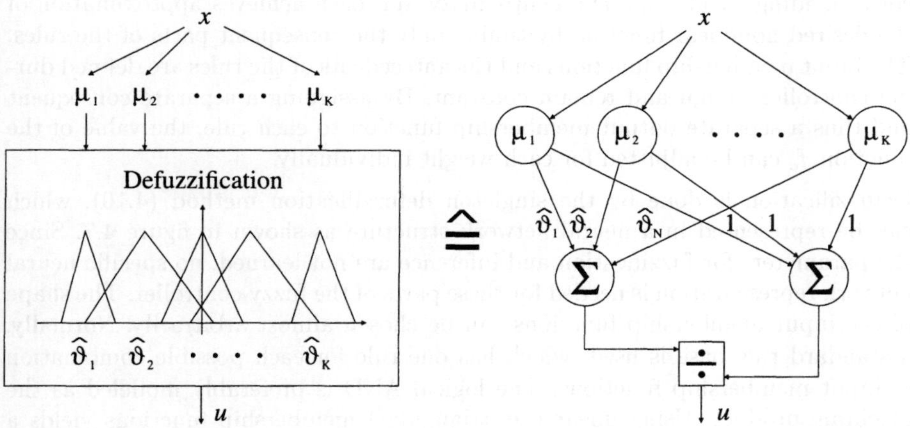

Figure 4.7: Neural network representation of the singleton defuzzification method

By representing the remaining parts of the fuzzy control algorithm — fuzzification and inference — in a similar structure, a fuzzy neural network is created, which is functionally equivalent to a fuzzy controller. The synaptic weights in this network can be trained using learning algorithms from the neural network domain. The result is a fuzzy controller, which can automatically adjust its parameters, e.g. the positions of the output fuzzy sets, in order to fulfill a specified control task. On the other hand, any fuzzy controller that adjusts its parameters by some adaptive mechanism can be viewed as a neural network.

---

[3] We will use u instead of $\widehat{y}$ here because the controller output is generally referred to as u in control literature.

[4] Note the close relationship to the GRNN described in section 4.2.2. The firing strengths μ_n correspond to the Gaussian basis function.

4.3.2 Neuro–Fuzzy Example: Tuning of Output Fuzzy Sets

Since there are different types of fuzzy controllers (in terms of shapes of membership functions or of logical operators) and different network structures, there is no unique method to represent the individual processing steps of a fuzzy controller in a neural network structure. We will now discuss one of the various possible approaches in detail, which is used in the control concept of *stable adaptive fuzzy control* [18].

Basically, the control law $u = f_c(\underline{x})$ of a fuzzy controller can be regarded as interpolation between certain sample points, where each rule specifies a sample point: the IF–part specifies an activation function μ_n and the THEN–part specifies the corresponding weight $\widehat{\vartheta}_n$. The neuro–fuzzy approach achieves approximation of the desired nonlinear function by tuning only the consequent parts of the rules. The input membership functions and the antecedents of the rules are defined during controller design and remain constant. By assigning a separate consequent and thus a separate output membership function to each rule, the value of the function f_c can be adjusted for each weight individually.

Defuzzification is done by the singleton defuzzification method (4.10), which can be represented in a neural network structure as shown in figure 4.7. Since the parameters for fuzzification and inference are not learned, no specific neural network representation is needed for these parts of the fuzzy controller. The shape of the input membership functions can be chosen almost arbitrarily. Normally, a standard rule base is used, which has one rule for each possible combination of input membership functions. The logical AND is preferably modeled as the algebraic product. Using linear (i.e. triangular) membership functions yields a multilinear approximation behaviour (linear variation in any single input leads to linear variation of the output).

If all adjustable parameters $\widehat{\vartheta}_n$ of the fuzzy controller (which are the positions of the output singletons) are combined to the parameter vector

$$\underline{\widehat{\Theta}}^T = \left[\widehat{\vartheta}_1 \; \widehat{\vartheta}_2 \; \ldots \; \widehat{\vartheta}_N\right]^T$$

the control law of the fuzzy controller can be written as

$$u = \underline{\widehat{\Theta}}^T \underline{\mathcal{A}}(\underline{x}) \tag{4.11}$$

where $\underline{\mathcal{A}}(\underline{x})$ contains the normalized firing strengths of the rules

$$\mathcal{A}_n(\underline{x}) = \frac{\mu_n(\underline{x})}{\sum_{m=1}^{N} \mu_m(\underline{x})}$$

the so–called *fuzzy basis functions*. Thus, the mathematical structure is the same as that of the GRNN. Hence, the GRNN and this type of adaptive fuzzy controller are functionally equivalent and can be freely interchanged. The same

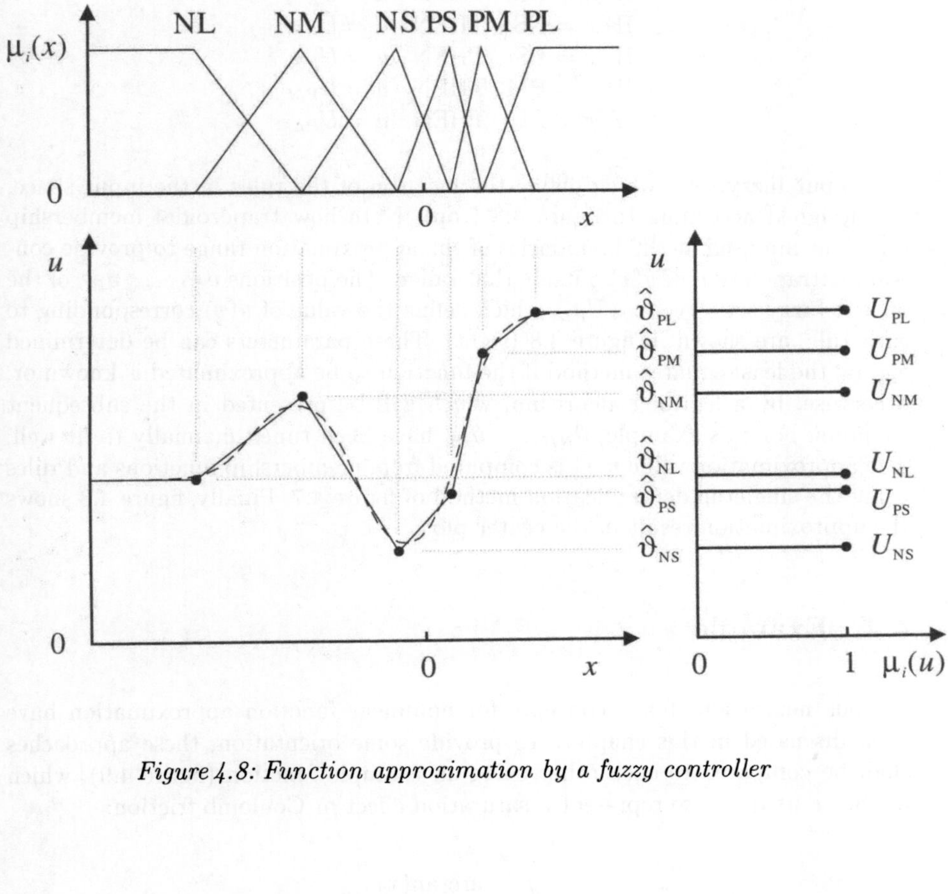

Figure 4.8: Function approximation by a fuzzy controller

learning algorithms can be applied for tuning the fuzzy controller parameters as for the GRNN.

An example for function approximation by a fuzzy controller is shown in figure 4.8, where the function to be approximated is similar to that of the previous sections. Approximation is done by a fuzzy controller with six sample points, i.e. six rules (for negative and positive small, medium and large values):

$$\begin{array}{lll} \text{IF } x = \text{NL} & \text{THEN} & u = U_{NL} \\ \text{IF } x = \text{NM} & \text{THEN} & u = U_{NM} \\ \text{IF } x = \text{NS} & \text{THEN} & u = U_{NS} \\ \text{IF } x = \text{PS} & \text{THEN} & u = U_{PS} \\ \text{IF } x = \text{PM} & \text{THEN} & u = U_{PM} \\ \text{IF } x = \text{PL} & \text{THEN} & u = U_{PL} \end{array}$$

The input fuzzy sets, which define the location of the rules in the input space, are designed according to figure 4.8 (top). Note how trapezoidal membership functions are used at the boundaries of the approximation range to provide constant extrapolation of $u(x)$ outside that range. The positions $\widehat{\vartheta}_{NL} \ \ldots \ \widehat{\vartheta}_{PL}$ of the output fuzzy sets $U_{NL} \ \ldots \ U_{PL}$, which define the value of $u(\underline{x})$ corresponding to each rule, are shown in figure 4.8 (right). These parameters can be determined e.g. by the least squares method if the function to be approximated is known or, otherwise, by a learning algorithm, which will be presented in the subsequent chapters. For this example, $\widehat{\vartheta}_{NL} \ \ldots \ \widehat{\vartheta}_{PL}$ have been tuned manually to fit well. The approximation value $u(\underline{x})$ is computed from membership functions and rules using the singleton defuzzification method of figure 4.7. Finally, figure 4.8 shows the approximation result in the center plot.

4.4 Example

Various neural and fuzzy concepts for nonlinear function approximation have been discussed in this chapter. To provide some orientation, these approaches shall be compared to each other using an example function (figure 4.9), which might be used e.g. to represent a saturation effect or Coulomb friction:

$$y = \arctan(x)$$

This function will be approximated employing 12 weights, which are equally spaced in the case of neural networks and are optimally placed in the case of the fuzzy approach.

In the following figures, approximation results are shown after optimal adaptation. In addition, the influence of the smoothing parameter and the training time required will be discussed. The plots also give the position of all weights as $\circ$ or $\times$ corresponding to different smoothing parameters.

The RBF–network (figure 4.10, solid line and $\circ$) provides sufficient accuracy near the center of the function but works far worse at the periphery. This may be enhanced by increasing the smoothing parameter σ, which at the same time deteriorates the approximation at the center (see dashed line and $\times$).

The GRNN (figure 4.11, solid line and $\circ$) shows outstanding overall performance for medium smoothing $\sigma = \sqrt{2}/N$. Please note, that the scaling of the error differs from that of the previous plot. Less smoothing makes the approximation to become step–shaped, whereas larger values of σ, impair the adaptation of steep slopes (see dashed line and $\times$).

The result of the DANN (figure 4.12) is similar to that of the GRNN, though, it cannot achieve the same accuracy. As the activation function is very wide spread over the input space, variation of the smoothing parameter has a minor effect on the quality of approximation and the DANN will also need more training cycles than the GRNN for the same reason. However, this will be compensated by minimized computational effort for the DANN.

Finally, the fuzzy controller employs 12 rules with optimal choice of the input fuzzy sets. As it can be seen in figure 4.13, the rules concentrate on areas of high curvature. Although the fuzzy controller employs linear interpolation, its approximation accuracy almost comes up to that of the GRNN. However, in most applications, the input fuzzy sets can only be chosen semi–optimal, which requires additional rules to obtain comparable approximation performance. This will be compensated by a rather short training time as all weights (rules) operate locally.

In general, for all these approaches, a trade–off has to be accepted between high accuracy of final approximation on the one hand, which will be achieved by a large

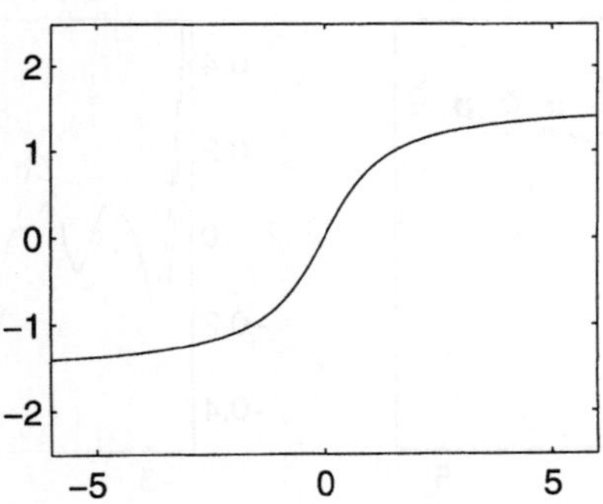

Figure 4.9: Example function $y = \arctan(x)$

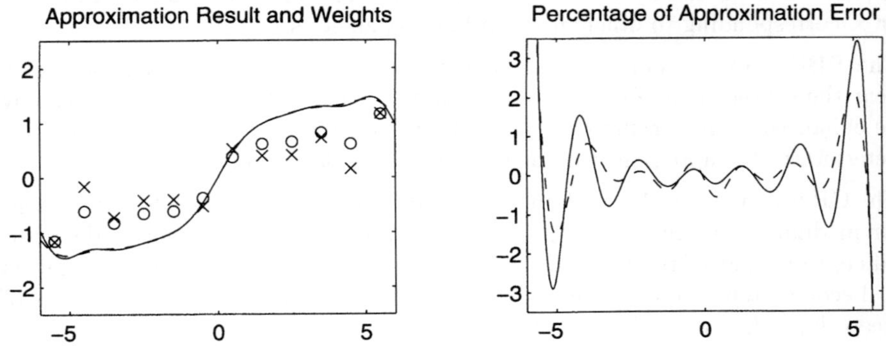

Figure 4.10: Approximation results of the RBF

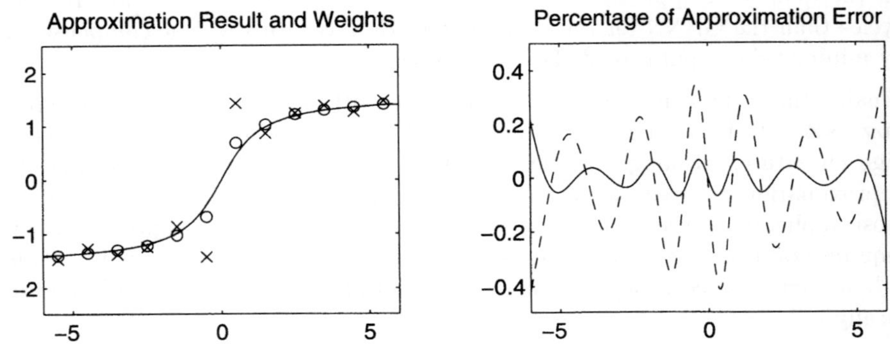

Figure 4.11: Approximation results of the GRNN

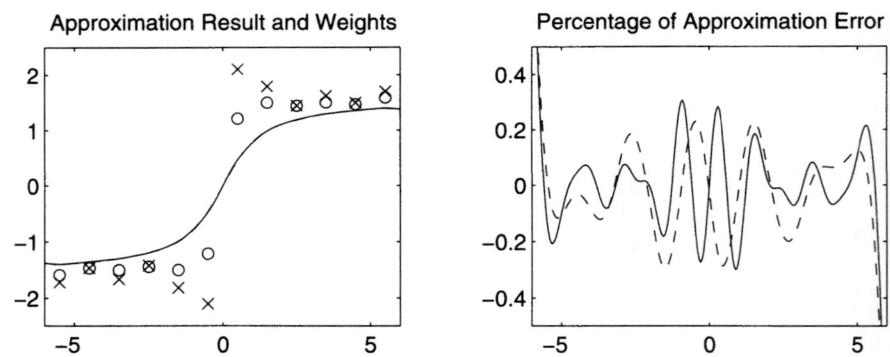

Figure 4.12: Approximation results of the DANN

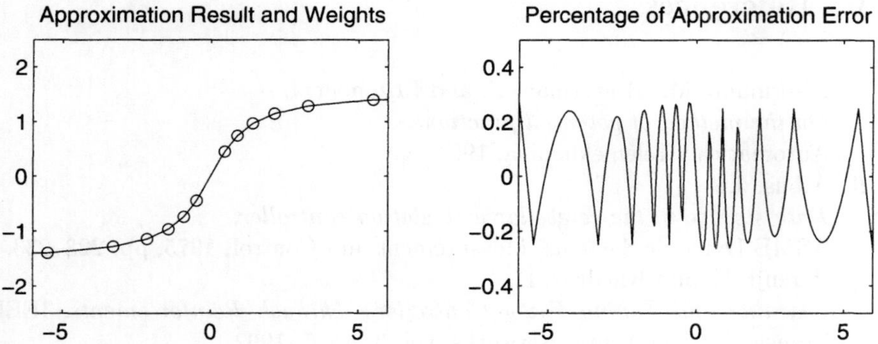

Figure 4.13: Approximation result of the fuzzy controller

number of weights and by very smooth activation functions, and fast adaptation on the other hand, which will be supported by few and local operating weights.

4.5 Conclusion

For nonlinear control, nonlinear function approximators are essential. Various neural networks and fuzzy control approaches have been discussed. To qualify for adaptive control, a function approximator has to support fast adaptation by mainly local operation of its weights and it should also provide reasonable extrapolation at unlearned input data. That is why, in the following chapters, we will prefer the General Regression Neural Network (GRNN) and a fuzzy controller using the singleton defuzzification method for theory and applications in the field of adaptive control.

4.6 References

[1] Aisermann, M., Braverman, E. and Rozonoer, L.:
On the method of potential functions.
Automatica i Telemekhanika, 1964.

[2] Albus, J.:
Data storage in the cerebellar articulation controller.
ASME Dynamic Systems, Measurement and Control, 1975, pp. 228–233.

[3] Berenji, H. and Khedkar, P.:
Learning and Tuning Fuzzy Controllers Through Reinforcements. IEEE Transactions on Neural Networks, vol. 3, no. 5, 1992

[4] Beuschel, M. and Lenz, U.:
Optimierte Implementation Neuronaler Netze in der Sprache C.
Internal Report, Technical University of Munich, 1995.

[5] Brause, R.:
Neuronale Netze — Eine Einführung in die Neuroinformatik.
B.G. Teubner, Stuttgart, 1995.

[6] Hakala, J., Koslowski, C. and Eckmiller, R.:
'Partition of Unity' RBF Networks are Universal Function Approximators.
ICANN, Sorrento, 1994.

[7] Jang, Jyh–Shing R.:
Self-Learning Fuzzy Controllers Based on Temporal Back Propagation.
IEEE Transactions on Neural Networks, vol. 3, no. 5, 1992.

[8] Jang, Jyh–Shing R.:
ANFIS: Adaptive–Network–Based Fuzzy Inference System.
IEEE Transactions on Systems, Man, and Cybernetics, vol. 23, no. 3, 1993.

[9] Kůrková, V. and Neruda, R.:
Uniqueness of Functional Representation by Gaussian Basis Function Networks.
ICANN, Sorrento, 1994.

[10] Lenz, U. and Schröder, D.:
Local Identification using Artificial Neural Networks.
Proc. Ninth Yale Workshop on Adaptive and Learning Systems, 1996, pp 83–88.

[11] Narendra, K.S. and Annaswamy, A.M.:
Stable Adaptive Systems.
Prentice Hall International Inc., 1989.

[12] Nguyen, H.T. and Kreinovich, V.:
On Approximation of Controls by Fuzzy Systems.
Proc. 5th IFSA World Congr. (Seoul), 1993, pp. 1414–1417.

[13] Rosenblatt, F.:
The perceptron: a probabilistic model for information storage and organization in the brain.
Psychological Review, Vol. 65, 1958, pp. 386–408.

[14] Schäffner, C.:
Analyse und Synthese neuronaler Regelungsverfahren.
Dissertation, TU München, 1995.

[15] Specht, D.F.:
A General Regression Neural Network.
IEEE Transactions on Neural Networks, Vol. 2, No. 6, 1991.

[16] Sugeno, M.:
Fuzzy Control: Principles, Practice and Perspectives.
IEEE Int. Conf. on Fuzzy Systems, 1992.

[17] Wang, Li–Xin.:
Fuzzy Systems Are Universal Approximators.
Proc. IEEE Int. Conf. Fuzzy Systems (San Diego), 1992, pp. 1163–1170.

[18] Wang, Li–Xin.:
Stable Adaptive Fuzzy Control of Nonlinear Systems.
IEEE Transactions on Fuzzy Systems, Vol. 1, 1993.

5 Systematic Intelligent Observer Design for Plants Characterized by an Isolated Nonlinearity

Ulrich Lenz

The starting point for control design is a theoretical analysis of the plant to be controlled resulting in a mathematical model. This model can either be linear or nonlinear. In the nonlinear case, a common method for controller design is to linearize the model at certain points of operation and then to use linear control theory.

But a large class of real plants being the object for controller design has a nonlinear behaviour in a global view, where the model is not linearizable without losing the real plant's behaviour. Figure 5.1 represents such a nonlinear plant, of which the position control is complicated by effects caused by nonlinearities such as friction, backlash or air resistance. In most cases such nonlinear effects handicap the application of the common linear control and observer theory. We

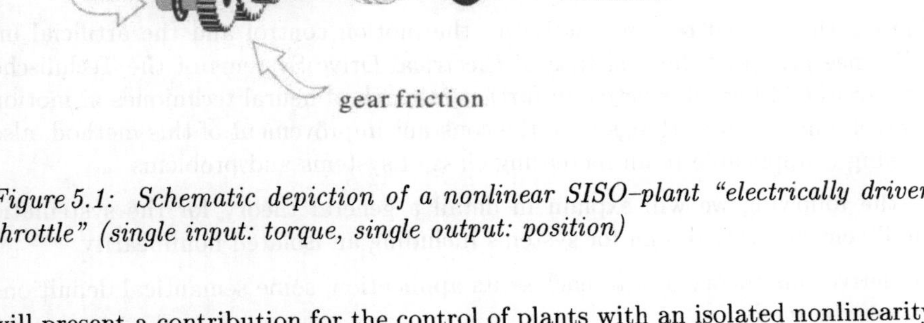

Figure 5.1: Schematic depiction of a nonlinear SISO–plant "electrically driven throttle" (single input: torque, single output: position)

will present a contribution for the control of plants with an isolated nonlinearity giving means for identifying the nonlinearity and observing the plant's states. In this chapter, an approach will be introduced which guarantees stable learning also in closed–loop applications. This approach provides some advantages for the control of nonlinear plants, since it leads to the identification of the nonlinearity and renders the nonlinear plant's states observable. It is our objective to ad-

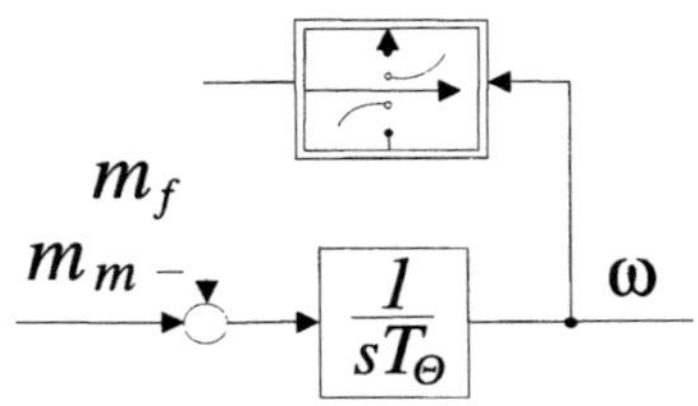

Figure 5.2: Friction depending on speed as an example for an isolated nonlinearity in a feedback path

vance and generalize approaches first developed by *C. Schäffner* in [21] and first proposed in [24]. Some approaches in his work considered the identification of speed–dependent friction, and the proposed method is suitable for plants showing a structure similar to figure 5.2. For this method the motor torque as input and the plant's speed as its output must be accessible for measurement.

In principle, nonlinearities exist in several ways within a real plant, e.g. in front of or between linear parts of a complex structure, or in the feedback path to a linear transfer function $F(s)$ (figure 5.3). In this chapter we will present a theory for the systematic intelligent observer design as the next step to [21], motivated by the success in validation results achieved by *Th. Frenz* [3]. The theory for observer design is applicable on plants in which static nonlinearities appear in several structural locations as displayed in figure 5.3.

Among the quoted research activities the motion control and the artificial intelligence group at the Institute of Electrical Drive Systems of the Technische Universität München is active in further research of neural techniques in motion control, among other things with the constant improvement of this method, also making it applicable to an increasing class of systems and problems.

In the following we will explain in detail a general theory for the systematic intelligent observer design for systems including an isolated nonlinearity.

To derive this theory and to enclose its application, some semantical definitions have to be made:

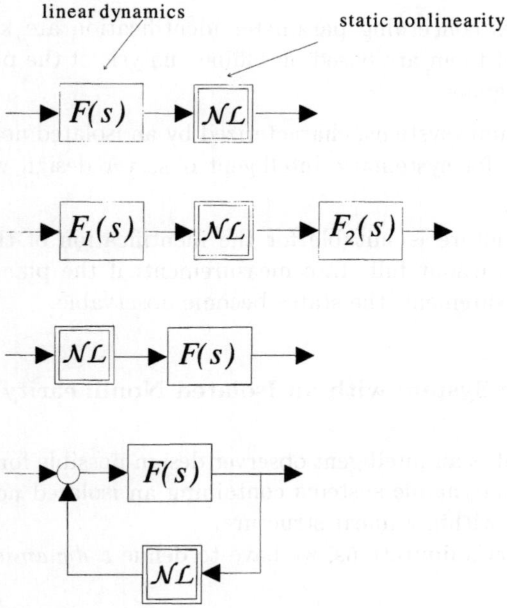

Figure 5.3: Possible structural locations of a nonlinearity within linear parts of the plant

5.1 Definitions: Dynamic Systems Containing an Isolated Nonlinearity

Within the theory of linear dynamic systems the systematic filter design for observing a linear time–invariant dynamic system is described in detail in literature and state of the art in control engineering. These filtering or observer approaches of e.g. Kalman and Luenberger can be found in [7, 8, 9, 28, 29, 14, 15]. Besides these linear theories some approaches do exist concerning the observer design for nonlinear dynamic systems [18, 2, 10, 16].

Until now, a theory for intelligent observer design for dynamic systems containing an isolated static nonlinearity does not exist.

Definition 5.1 Intelligent observer *An intelligent observer is an adaptive observer, able to adapt unknown dependencies within the observed plant online and to store these dependencies as knowledge. The adaptation is executed with guaranteed stability and leads to a knowledge which is interpretable. The adaptation permits the nonlinear plant's state observation.*

Several approaches concerning parameter identification are known for specific structures, most of them are based on offline analysis of the plant's response to some excitation signals.

For a class of dynamic systems, characterized by an isolated nonlinearity, we will derive this theory for systematic intelligent observer design with stable online adaptation.

This observer structure is suitable for the identification of the unknown nonlinearity with or without full state measurement; if the plant's states are not accessible for measurement, the states become observable.

5.1.1 Dynamic System with an Isolated Nonlinearity

This approach makes an intelligent observer design possible for a class of nonlinear time–invariant dynamic systems containing an isolated nonlinearity at any arbitrary location within a linear structure.
To mark the theory's limitations, we have to define a *dynamic system with isolated nonlinearity.*

Definition 5.2 Dynamic system with isolated nonlinearity *A dynamic system is called a dynamic system with isolated nonlinearity, if it is in a state space description expressible as*

$$\dot{\underline{x}} = A\underline{x} + \underline{b}u + \underline{\mathcal{NL}}(\underline{x}, u) \text{ and } y = \underline{c}^T\underline{x} + d\,u \qquad (\text{SISO} - \text{system}) \quad (5.1)$$

or as

$$\dot{\underline{x}} = A\underline{x}+B\underline{u}+\underline{\mathcal{NL}}(\underline{x}, \underline{u}) \text{ and } \underline{y} = C\underline{x} + D\,\underline{u} \qquad (\text{MIMO} - \text{system}) \quad (5.2)$$

where $\underline{\mathcal{NL}}(\underline{x}, u)$ *or* $\underline{\mathcal{NL}}(\underline{x}, \underline{u})$ *represents an isolated nonlinearity.*

For the observer design, the matrices A, B or $\underline{b}$, C or $\underline{c}$ and D or d must be (quasi–) time–invariant and known. This means, we use a-priori knowledge of the linear part of the system, e.g. resulting from a physical model of the system.

For reasons of description we neglect the general MIMO–case; the theory is derived based on a SISO–plant, but is easily applicable in the MIMO–case.

As well as the linear part of the system we require the knowledge of the dependencies of the nonlinearity and its location, where it acts.

Definition 5.3 Isolated nonlinearity *An isolated nonlinearity* $\underline{\mathcal{NL}}(\underline{x}, u)$ *within a dynamic system is expressible as the product of a scalar nonlinear term* $\mathcal{NL}(\underline{x}, u)$ *with a known coupling vector* $\underline{e}_{NL}$ *of the system's dimension,*

$$\underline{\mathcal{NL}}(\underline{x}, u) = \underline{e}_{NL}\; \mathcal{NL}(\underline{x}, u) \qquad (5.3)$$

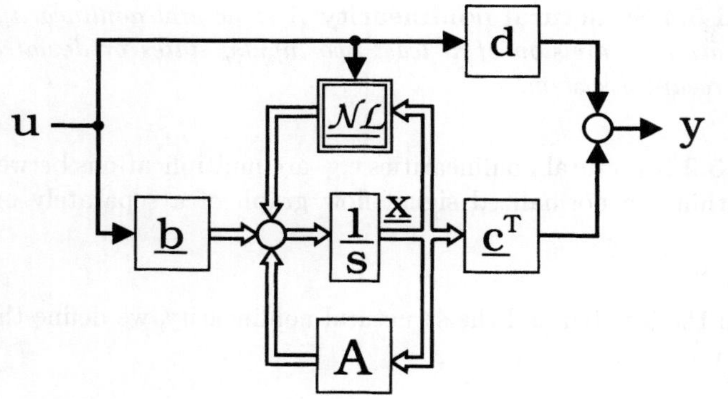

Figure 5.4: Signal flow chart of the state space representation of a SISO–system enclosing an isolated nonlinearity $\underline{\mathcal{NL}}(\underline{x}, u)$

where the nonlinear function differs from a simple multiplication or division of input values, states or deviated quantities. The dependencies of the nonlinearity are known.

This coupling vector $\underline{e}_{NL}$ describes the influence(s) of the nonlinearity on the system and must be known. In a signal–flow graph we represent an isolated

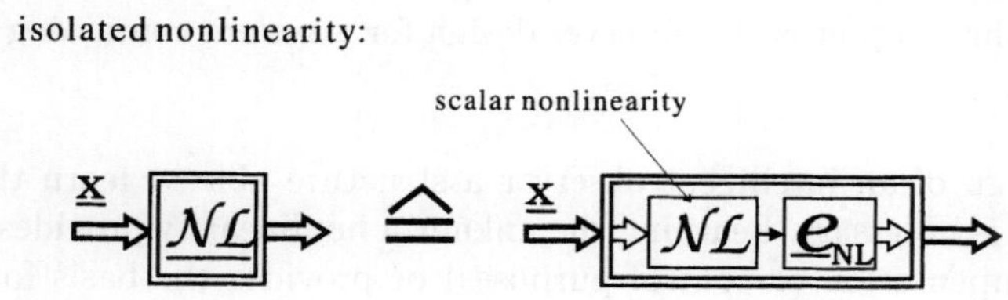

Figure 5.5: Isolated nonlinearity $\underline{\mathcal{NL}}(\underline{x})$ in state space signal flow graph representation

nonlinearity as depicted in figure 5.5.

Example 5.1 An isolated nonlinearity is the nonlinear relation between the excitation current I_E and the flux $\Psi(I_E)$ within the excitation circuit of e.g. a separately excited DC–machine [25, 26].

In contrast to an isolated nonlinearity, the structural nonlinearity occurs in nonlinear systems.

Definition 5.4 Structural nonlinearity *A structural nonlinearity consists of a multiplication or division of at least two inputs, states or deviated quantities within the regarded system.*

Example 5.2 Structural nonlinearities e.g. are multiplications between flux and current within the normalized signal flow graph of a separately excited DC–machine.

Apart from the isolated and the structural nonlinearity, we define the separable nonlinearity.

Definition 5.5 Separable nonlinearity *A separable nonlinearity can be represented by a vector* $\underline{\mathcal{NL}}_s(\underline{x}, u)$ *containing at least two different components unequal to zero. These components are scalar static nonlinearities over their input space* $\mathcal{NL}_{si}(\underline{x}, u)$

Example 5.3 A separable nonlinearity occurs due to friction on different parts (masses) or backlash in a mechanical multi–mass system.

A separable nonlinearity cannot be expressed as the product between a single scalar nonlinearity and a coupling vector, as it is possible for the representation of an isolated nonlinearity. In the following chapter 6 the identification of separable nonlinearities within the structure of a dynamic system is discussed in detail. The aim of this chapter is the observer design for a system containing an isolated nonlinearity.

For the design of an intelligent observer a structure able to learn the system's nonlinearity is necessary. Learning the unknown nonlinearity provides knowledge (e.g. for compensation or control purposes) or provides the basis for observing the states of such a nonlinear dynamic system. Both aspects might be useful for the controller design for such a dynamic system.

5.1.2 Approximation of a Static Nonlinearity

The observer theory to be presented here provides the means to identify an isolated static nonlinearity between the linear parts of a dynamic system in online–operation of the system.

Definition 5.6 Static nonlinearity *A static nonlinearity is defined by a limited real function of one or more real variables. This means*

$$\mathcal{NL}(\underline{x}, u) = f(\underline{x}, u) \tag{5.4}$$

with a limited real function f, of which the total system's state space is the definition area.
A static nonlinearity contains no internal states.

Typical static nonlinearities are e.g. friction, saturation effects, backlash, or the volumetric efficiency occuring with combustion engines. A static nonlinearity has no memory effects as e.g. hysteresis (see chapter 4).

5.2 Hybrid Notation of Signals in the Time and Frequency Domain

The deductions in the following sections are referred to signals in the time and frequency domain. Often a hybrid notation of signals in the time domain and transfer functions in frequency domain is useful due to a clear description.

In this hybrid notation

$$y(t) \;=\; H(s)\; x(t) \tag{5.5}$$

$y(t)$ is the output (in time domain) of a system with the transfer function $H(s)$ in frequency domain on the input $x(t)$ in time domain.

Using this notation we neglect all initial conditions.

5.3 Systematic Observer Design

A neural network is used within this observer structure to represent and identify the unknown real nonlinearity. The identification is based on an adaptation process which guarantees stability [21, 11].

The intelligent observer will identify the unknown nonlinearity and enable the observation of the system states.

This approach is suitable for systems fulfilling the following conditions:

5.3.1 Conditions

The nonlinearity's identification is possible based on the system's output, only if the nonlinearity's effect is visible in the system's output.

Definition 5.7 Visibility of a nonlinearity *A nonlinearity is visible at the output of a SISO–system, if the transfer function $H_s(s)$ from the nonlinearity's output to the plant's output*

$$H_s(s) = \underline{c}^T(sE - A)^{-1}\underline{e}_{NL} \tag{5.6}$$

is non–zero.

For the observer design, the linear part of the system must be (quasi–)time–invariant and known. This means, A, $\underline{b}$, $\underline{c}$ and d must be known as well as $\underline{e}_{NL}$ and the dependencies of the nonlinearity.

To derive an adaptation law for the neural net representing the nonlinearity, the error transfer function (defined later) must be asymptotically stable. This means, all poles of $H(s) = \underline{c}^T(sE - A + L)^{-1}\underline{e}_{NL}$ must be located in the negative half plane.

Under these stipulations for dynamic systems including an isolated nonlinearity, the design of an intelligent observer for identification and state observation is possible.

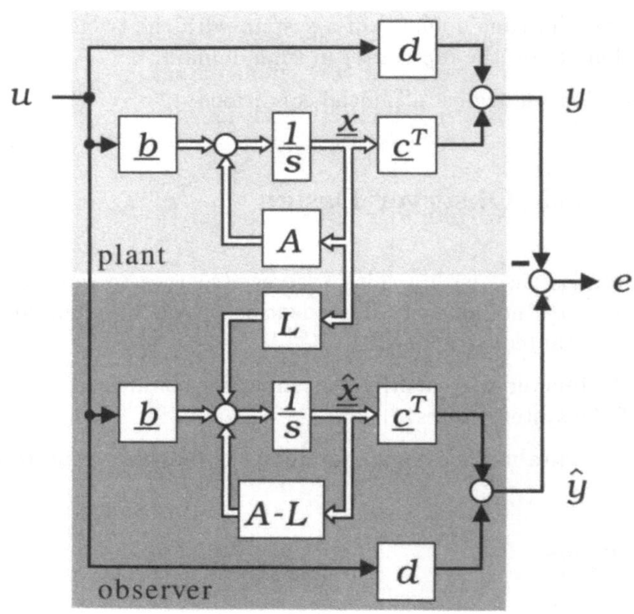

Figure 5.6: State observer for a linear dynamic system, applicable e.g. when noise occurs

5.3.2 Observer Design for Identification

Presenting the observer design with focus only on the identification leads to our global aim: simultaneous observation and identification. To begin with, we assume the states to be measurable and the plant to be observable in its linear part.

If noise occurs in a linear SISO–system, state observation can be realized using the structure depicted in figure 5.6. We will extend this structure for the identification of an isolated nonlinearity. This new approach will be presented for a SISO–system, but it is also easily extendable on the more general MIMO–case.

5.3.2.1 Observer Approach

The observer for the SISO–system with accessible states $\underline{x}$

$$\dot{\underline{x}} = A\underline{x} + \underline{b}u + \underline{e}_{NL}\mathcal{NL}(\underline{x}, u) \quad \text{and} \quad y = \underline{c}^T \underline{x} + d\,u \tag{5.7}$$

with A, $\underline{b}$, $\underline{c}$, d and $\underline{e}_{NL}$ known, can be applied as

$$\dot{\hat{\underline{x}}} = (A - L)\hat{\underline{x}} + \underline{b}u + L\underline{x} + \underline{e}_{NL}\hat{\mathcal{NL}}(\underline{x}, u) \quad \text{and} \quad \hat{y} = \underline{c}^T \hat{\underline{x}} + d\,u \tag{5.8}$$

leading to

$$e = \hat{y} - y \tag{5.9}$$

defined as the observer error.

> **With this approach we interpret the nonlinearity as an input to a linear system; this is important, since this way enables the use of linear methods for a nonlinear system (transfer functions, superposition ...).**

To represent the isolated nonlinearity, we extend figure 5.6 to the diagram presented in figure 5.7.

5.3.2.2 Dimensioning the Observer Feedback Matrix L

The observer feedback matrix L must be parametrized to guarantee the observer poles lying in the negative half plane and to ensure the settling of the observer states. Furthermore, it is appropriate to parametrize the matrix L according to known optimizing criterias, e.g. LQ–optimization or pole–placement.[1]

In principle, all linear theories remain valid, therefore we will not discuss this parametrization.

---

[1] If all states are measurable, a pole–zero–compensation in the error transfer function $H(s)$ seems to be appropriate for simplifying the learning rule ([3]).

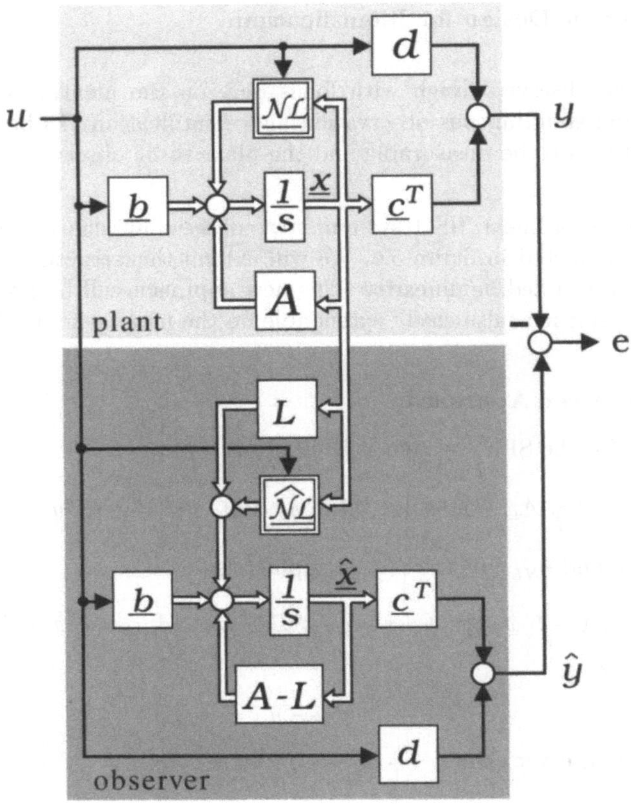

Figure 5.7: Intelligent observer for SISO–systems with isolated nonlinearity; applicable for identification and state observation (e.g. noise)

5.3.2.3 Specification of the Error Transfer Function $H(s)$

To derive the adaptation law for the neural network within the intelligent observer based on the error transfer function $H(s)$ we describe system and observer in the frequency domain.
We finally conclude for the error transfer function

$$
\begin{aligned}
e = H(s)(\hat{\mathcal{NL}} - \mathcal{NL}) &= \\
&= \underline{c}^T (sE - A + L)^{-1} \underline{e}_{NL} \left(\hat{\mathcal{NL}} - \mathcal{NL}\right) \qquad (5.10)
\end{aligned}
$$

The calculation of this transfer function $H(s)$ in the SISO case is executed in convenience using the approach

$$H(s) = \underline{c}^T(sE - A + L)^{-1}\underline{e}_{NL} = \frac{det\begin{bmatrix} sE - A + L & \vdots & -\underline{e}_{NL} \\ \cdots & \cdots & \cdots \\ \underline{c}^T & \vdots & 0 \end{bmatrix}}{det(sE - A + L)} \qquad (5.11)$$

to describe the observer's transfer characteristic of the nonlinearity's output to the observer's output $\hat{y}$.

5.3.2.4 Deriving a Stable Adaptation Law Using Known Error Models

A static nonlinearity can be represented with arbitrary exactness by a specific RBF neural network as

$$\mathcal{NL} = \underline{\Theta}^T \underline{\mathcal{A}}(\underline{x}, u) + d_a \qquad (5.12)$$

with the inherent approximation error $d_a = d_a(\underline{x})$, which tends to zero with a rising number of neurons and is therefore negligible. Shall the real unkown nonlinearity $\mathcal{NL}$ within the system

$$\dot{\underline{x}} = A\underline{x} + \underline{b}u + \underline{e}_{NL}\mathcal{NL}(\underline{x}, u) \text{ and } y = \underline{c}^T\underline{x} + d\,u \qquad (5.13)$$

be defined as a neural network by the unknown weights vector $\underline{\Theta}$ according to equation (5.12), this nonlinearity is represented in

$$\dot{\hat{\underline{x}}} = (A - L)\hat{\underline{x}} + \underline{b}u + L\underline{x} + \underline{e}_{NL}\hat{\mathcal{NL}}(\underline{x}, u) \text{ and } \hat{y} = \underline{c}^T\hat{\underline{x}} + d\,u \qquad (5.14)$$

using an unknown weights vector $\hat{\underline{\Theta}}$ of the neural net within the observer structure. Defining the observer error as the difference between observer output and

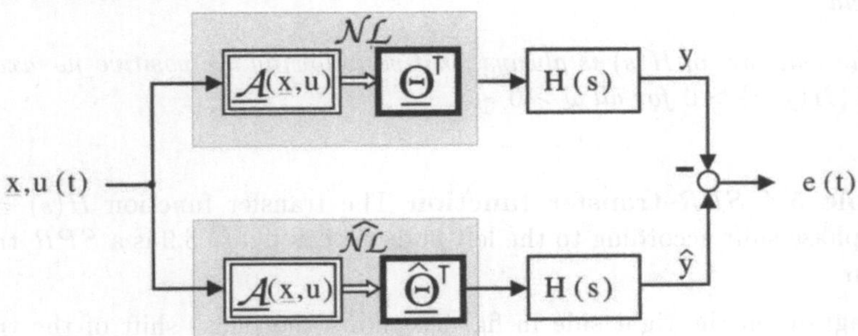

Figure 5.8: The observer structure leads to the observer error defined as the difference over $H(s)$ between the nonlinearity regarded as input of a linear plant (depending on $\underline{x}$ and u) and the neural net $\hat{\mathcal{NL}}$

measured system output and neglecting the subsiding effect of different initial conditions, leads to the error equation

$$e = H(s)(\hat{\mathcal{NL}} - \mathcal{NL}) = H(s)\left[\underline{\hat{\Theta}} - \underline{\Theta}\right]^T \underline{\mathcal{A}}(\underline{x}, u) \tag{5.15}$$

including the error transfer function (parametrized by L to be asymptotically stable)

$$H(s) = \underline{c}^T(sE - A + L)^{-1}\underline{e}_{NL} \tag{5.16}$$

Defining a parameter error vector $\underline{\Phi}$ as

$$\underline{\Phi} = \underline{\hat{\Theta}} - \underline{\Theta} \tag{5.17}$$

leads to an error equation

$$e = H(s)\underline{\Phi}^T \underline{\mathcal{A}}(\underline{x}, u) \tag{5.18}$$

known in the literature (e.g. [18, 27]). Figure 5.8 depicts equation (5.18). According to [3] with a fully measurable system state vector a skillful selection of L may lead in the error transfer function $H(s)$ to a pole–zero–compensation, resulting in a $H(s)$ fulfilling the SPR–condition. Doing so, reduces the computational efforts for the adaptation.

The transfer function $H(s)$ in equation (5.18) between the nonlinearity and the observer error normally is not a strictly positive real (SPR) transfer function with an absolute phase shift always less than 90°.

Definition 5.8 ***SPR*–transfer function** *A transfer function $H(s)$ is called a strictly positive real (SPR–) transfer function if and only if*

1. *$H(s)$ is asymptotically stable,*
   and

2. *the real part of $H(s)$ is always positive following the positive $j\omega$–axis*
   $\mathcal{R}\{H(j\omega)\} > 0$ *for all* $\omega \geq 0$

Example 5.4 ***SPR*–transfer function** The transfer function $H(s) = \frac{1}{1+5s}$ with a phase shift according to the left bode plot in figure 5.9 is a SPR–transfer function.

The diagram on the right side in fig. 5.9 shows the phase shift of the transfer function $H(s) = \frac{7+s}{7+2s+s^2}$ and we can conclude that this transfer function is not SPR, since the phase shift becomes less than -90° for some frequencies.

From this example we draw the conclusion, that from degree denominator = degree numerator + 1, the fulfilling of the SPR–condition does not follow.

In literature [27, 18] there exists for an error equation (5.18) characterized by a not–SPR–transfer function an adaptation law which guarantees stable adaptation. Together with this adaptation law

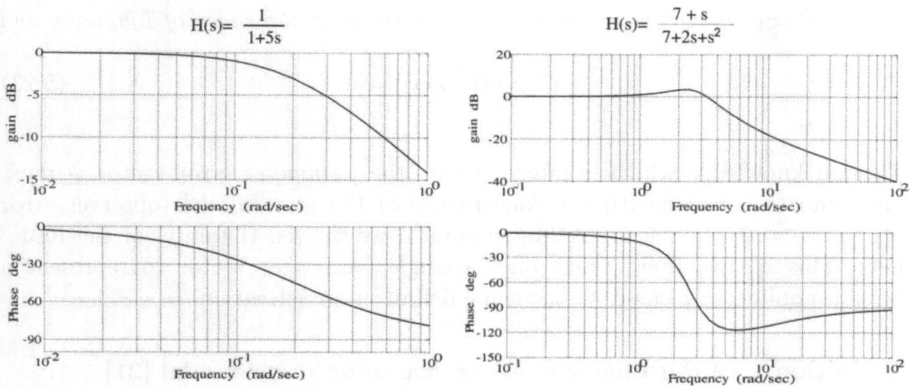

Figure 5.9: Bode plots for a SPR–transfer function (left side) and a not-SPR–transfer function (right side)

$$\dot{\underline{\Phi}} = \dot{\hat{\underline{\Theta}}} = -\eta\epsilon \underbrace{H(s)\underline{A}(\underline{x}, u)}_{delayed\,activation} \tag{5.19}$$

with the augmented error for adaptation

$$\epsilon = e + \hat{\underline{\Theta}}^T H(s)\underline{A}(\underline{x}, u) - H(s)\hat{\underline{\Theta}}^T \underline{A}(\underline{x}, u) \tag{5.20}$$

the error equation (5.18) forms a known error model. This interpretation of the error model corresponds to the "delayed activation method" [11, 12]. To learn the neurons' weights, we delay their activation until the weights error effects the system's output. The augmented error according to equation (5.20) is defined to compensate the dynamic effects due to the time–varying weights.

Based on Ljapunov's direct method, for this error model the convergence

$$\lim_{t\to\infty} e(t) = 0 \tag{5.21}$$

follows and due to the net's features from a *persistent excitation* by $\underline{A}(\underline{x}, u)$ the convergence

$$\lim_{t\to\infty} \hat{\underline{\Theta}}(t) = \underline{\Theta} \tag{5.22}$$

follows as well. This means the convergence of the neural net's weights by adaptation towards the real unknown parameter vector.

First, we will explain the mathematical proof of stability as it is presented in [18], then we will discuss the parameter convergence.

We define the following:

Definition 5.9 Convergence of an error model [21]
The (error) convergence of an error model is present, if the equilibrium point

$[e^*, \underline{\Phi}^{*T}]^T = \underline{0}$ *of the error model is globally stable in the sense of Ljapunov and if*

$$e^* = e(t \to \infty) = 0 \tag{5.23}$$

To gain knowledge which is interpretable, the parameter error vector $\underline{\Phi}$ must also tend to zero beneath the convergence of the error model (observer error $e(t \to \infty) = 0$). In the preceeding chapter 4 we showed the unity of the RBF–nets. This means, there exists only a single parameter vector to represent a certain nonlinearity based on the same definition of activation.

Definition 5.10 Parameter convergence of an error model [21]
Parameter convergence of an error model is present, if the equilibrium point $[e^*, \underline{\Phi}^{*T}]^T = \underline{0}$ *of the error model is globally stable in the sense of Ljapunov and if together with*

$$e^* = e(t \to \infty) = 0 \tag{5.24}$$

it follows

$$\underline{\Phi}^* = \underline{\Phi}(t \to \infty) = \underline{0} \tag{5.25}$$

This means that the parameter error vector is also zero, if the observer error has become zero. The adapted weights $\hat{\underline{\Theta}}$ then must be converged towards the real unknown weights $\underline{\Theta}$. From the unity of the net follows a knowledge which can be interpreted.

Against the industrial implementation of Artificial Intelligence methods in control algorithms there exists some disapproval due to safety reasons. In particular, online applications in a closed control loop demand a sophisticated safety examination. The most proposed neural approaches lack a proof of stability.

Therefore we list the proof of stability for the error model consisting of equations (5.18) and (5.19), which is used below:

Mathematical Proof of Stable Adaptation

The error model we use for adaptation in our intelligent observer is known in literature; we will present the mathematical proof of stability as it is depicted in [18].

This error model consists of the error equation

$$e = H(s)\underline{\Phi}^T \underline{\mathcal{A}}(\underline{x}, u) \; = H(s)\left[\hat{\underline{\Theta}} - \underline{\Theta}\right]^T \underline{\mathcal{A}}(\underline{x}, u) \tag{5.26}$$

and the adaptation law

$$\dot{\underline{\Phi}} = \dot{\hat{\underline{\Theta}}} = -\eta \epsilon H(s) \underline{\mathcal{A}}(\underline{x}, u) \tag{5.27}$$

with the augmented error for adaptation defined as

$$\epsilon = e + \hat{\underline{\Theta}}^T H(s)\underline{A}(\underline{x}, u) - H(s)\hat{\underline{\Theta}}^T \underline{A}(\underline{x}, u) \tag{5.28}$$

These three equations (5.26), (5.27) and (5.28) forming the error model are illustrated in the signal flow graph in figure 5.10. The mathematical proof seems

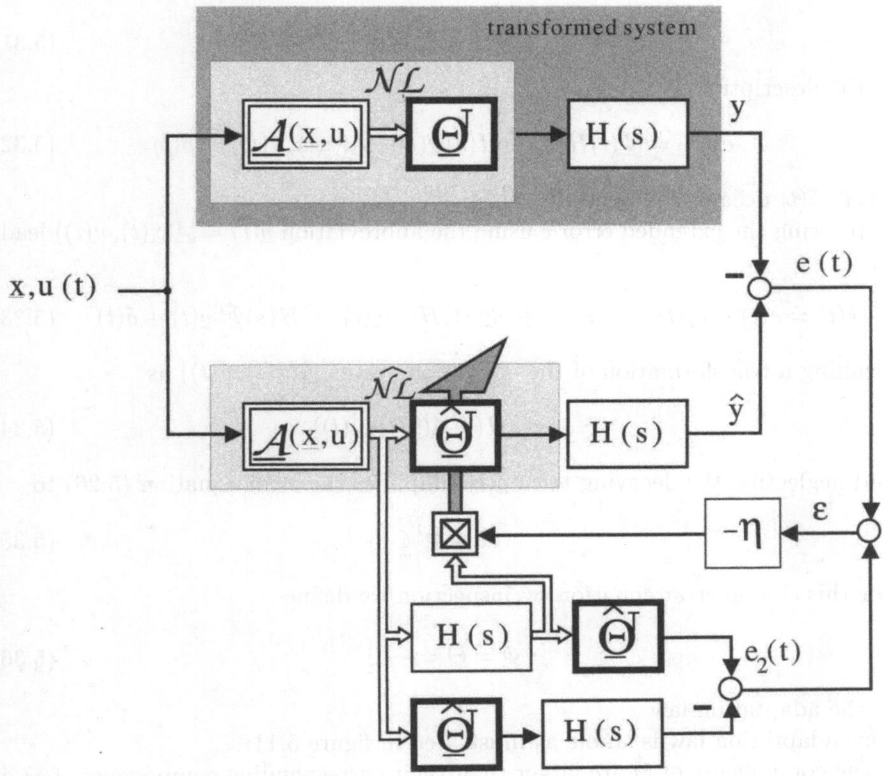

Figure 5.10: Signal flow graph of the error model with delayed activation; in the lower part, the extended error is depicted

to be easy, if we transform the input vector of the nonlinearity.

Transformation of the net input

We regard an extended observer error ϵ

$$\epsilon(t) = e(t) + e_2(t) \tag{5.29}$$

with

$$e_2(t) = \left[\underline{\hat{\Theta}}(t)H(s) - H(s)\underline{\hat{\Theta}}(t)\right]^T \underline{A}(\underline{x}(t), u(t)) \tag{5.30}$$

according to figure 5.10. Since $H(s)$ is an asymptotically stable transfer function, the error component $e_2(t)$ will tend to zero exponentially with time, when $\underline{\hat{\Theta}}(t)$ has reached a constant value $\underline{\Theta}$.
Therefore we are able to transform the equation using the parameter error defined as

$$\underline{\Phi}(t) = \underline{\hat{\Theta}} - \underline{\Theta} \tag{5.31}$$

to the description as

$$e_2(t) = \left[\underline{\Phi}(t)H(s) - H(s)\underline{\Phi}(t)\right]^T \underline{A}(\underline{x}(t), u(t)) + \delta(t) \tag{5.32}$$

where $\delta(t)$ decays exponentially with time.
Expressing the extended error ϵ using the abbreviation $\underline{v}(t) = \underline{A}(\underline{x}(t), u(t))$ leads to

$$\epsilon(t) = e(t) + e_2(t) = e(t) + \underline{\Phi}^T(t)H(s)\underline{v}(t) - H(s)\underline{\Phi}^T\underline{v}(t) + \delta(t) \tag{5.33}$$

Defining a transformation of the activation vector $\underline{A}(\underline{x}(t), u(t))$ as

$$\underline{\zeta} = H(s)\underline{A}(\underline{x}(t), u(t)) \tag{5.34}$$

and neglecting the decaying term $\underline{\delta}(t)$ simplifies the error equation (5.26) to

$$\epsilon = \underline{\Phi}^T\underline{\zeta} \tag{5.35}$$

For this simple error equation by inspection we define

$$\underline{\dot{\Phi}} = \underline{\dot{\hat{\Theta}}} = -\epsilon\underline{\zeta} \tag{5.36}$$

as the adaptation law.
This adaptation law is stable as illustrated in figure 5.11:
If the components of $\underline{\hat{\Theta}}$ are larger than their corresponding components of $\underline{\Theta}$, it results in a positive ϵ. So we must diminish the concerning components in the vector $\underline{\hat{\Theta}}$.

Ljaponov based stability proof

The clear mathematical proof of stability for this transformed error model consisting of equations (5.35) and (5.36) is easily carried out based on Ljapunov's second, direct method.
To apply this method, we put equation (5.35) in equation (5.36) and it follows

$$\underline{\dot{\Phi}} = -\underline{\zeta}\underline{\zeta}^T\underline{\Phi} \tag{5.37}$$

We suggest the scalar function

$$V(\underline{\Phi}) = \frac{\underline{\Phi}^T \underline{\Phi}}{2} \tag{5.38}$$

as a Ljapunov function candidate and calculate its time derivative along the trajectories defined by equation (5.37) resulting in

$$\dot{V}(\underline{\Phi}) = -\epsilon^2(t) \leq 0 \tag{5.39}$$

According to [18] we conclude:

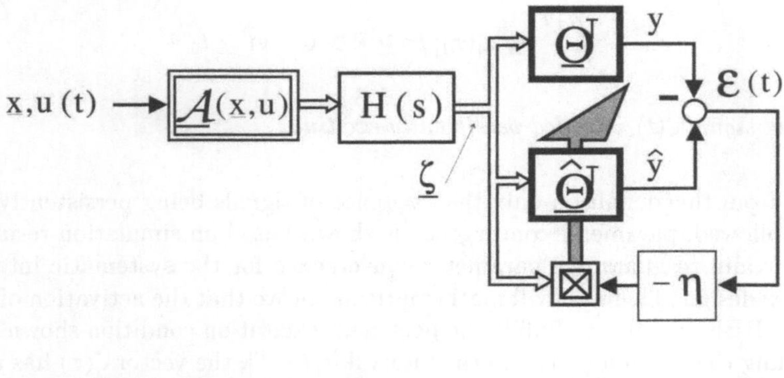

Figure 5.11: Signal flow graph of the transformed error model

- Since $V(\underline{\Phi})$ is a Ljapunov function, the parameter error vector $\underline{\Phi}$ and its time derivative remain bounded.
- If the nonlinearity's input $[\underline{x}(t), u(t)]^T$ and its time derivative are bounded, $\underline{\zeta}$ and its time derivative also remain bounded and it follows the convergence of the error ϵ to zero; $\hat{\underline{\Theta}}$ converges to a constant value.
- If the net's activation is persistently exciting, $\underline{\zeta}$ will also be, and we conclude besides the error convergence the parameter convergence.

$$\lim_{t \to \infty} \underline{\Phi}(t) = 0 \tag{5.40}$$

The isolated nonlinearity is interpretably identified only, if the net's activation fulfills the so–called *persistent excitiation* condition. We show the satisfaction of this condition in the following section for normalized RBF nets.

5.3.2.5 Reflections on Parameter Convergence

The problem of persistent excitation occurs always, if a single scalar error signal adapts several parameters. The statement of the problem is closely related to the existence of a unique solution of the error model related to a specific excitation.

To start with, we prove the fulfillment of the mathematical condition for persistent excitation based on the fact that the net's base functions are linearly independent and always greater than zero.
In [17, 1] a condition for *persistent excitation* of signals is given as cited in [18] and [27]:

Definition 5.11 Persistent excitation *If $\underline{\zeta}(t)$ is partially continuous and there exist positive constants T, t_0 and k such that for all unit vectors $\underline{w}$*

$$\int_t^{t+T} |\underline{w}^T \underline{\zeta}(\tau)| d\tau \geq k > 0 \quad \forall t \geq t_0 \tag{5.41}$$

then the signal $\underline{\zeta}(t)$ is called persistently exciting.

In [21] from this definition only the existence of signals being persistently exciting is followed, parameter convergence is shown based on simulation results.
It is our aim to guarantee parameter convergence for the systematic intelligent observer design. There we will mathematically prove that the activation of a normalized RBF–net always fulfills the persistent excitation condition shown above.
According to equation (5.41), in the interval $[t, t+T]$, the vector $\underline{\zeta}(\tau)$ has a finite projection along any unity vector $\underline{w}$ over a finite measurement of time starting with arbitrary $t \geq t_0$ [18]. Then parameter convergence follows.
This demand is satisfied, if **none** of the components of the vector $\underline{\zeta}(\tau)$ are identical to zero within the interval $[t, t+T]$, if the components of $\underline{\zeta}(\tau)$ are not constants in $[t, t+T]$, and if all components are linearly independent of each other.

Definition 5.12 Linear independent functions *A function f_u is called linearly independent of a set of n functions f_i, if f_u cannot be displayed as a linear combination of functions in the set f_i.*
This means,

$$f_u \not\equiv \sum_{i=1}^{n} k_i f_i \tag{5.42}$$

with constants k_i, if f_u is linearly independent of the functions f_i.

In our case, the vector $\underline{\zeta}(t)$ corresponds to the activation $\underline{A}(\underline{x}(t), u(t))$, delayed by a linear transfer function vector. The activation of a normalized RBF–net has the components

$$a_i = \frac{\frac{1}{1+\frac{d_i^2}{\sigma}}}{\sum_{i=1}^{q} \frac{1}{1+\frac{d_i^2}{\sigma}}} \tag{5.43}$$

for the DANN or

$$a_i = \frac{exp(-\frac{d_i^2}{2\sigma^2})}{\sum_{i=1}^{q} exp(-\frac{d_i^2}{2\sigma^2})} \tag{5.44}$$

for the GRNN with the distances d_i of the neurons i at their location $\underline{\chi}_i$ from the net input $\underline{x}$ defined by $d_i = || \underline{x} - \underline{\chi}_i ||$ for these nets. We have to examine the net's activation vector components considering their linear independency of each other according to definition 5.12. Both activation function equations (5.43) and (5.44) are nonlinearly independent in d_i. This means the linear independency of a single neuron's activation from the set of the other neurons' activations.
According to the equations (5.43) and (5.44) we conclude also that all components a_i for finite distances between the neurons and the net input $d_i = || \underline{x} - \underline{\chi}_i ||$ are always postive and non–zero.
Therefore the activation of a normalized RBF–net with always positive and linearly independent components is *persistently exciting.* This attribute is not affected by the delay of the single components caused by a linear dynamic transfer function.

By these deductions we have shown mathematically that the activation of the net we use (normalized RBF–net) is always *persitently exciting* and therefore, besides the error convergence parameter convergence will follow whilst learning. This deduction is based on the fact that each neuron is always (most of the time only marginally) activated. Considering the normalized RBF-net (GRNN or DANN) the neurons act mostly locally. However, due to the finite simulation step size, the convergence

$$\lim_{t \to \infty} \hat{\Theta}_i(t) = \Theta_i \tag{5.45}$$

of the adapted weights vector to the real unknown vector is only vaild for these components of which the receptive field has been activated whilst learning until the occurance of error convergence. Therefore we formulate the corollary:

Definition 5.13 Local parameter convergence for a normalized RBF–net *If there occurs no (delayed) observer error*

$$e = H(s)(\hat{\mathcal{NL}} - \mathcal{NL}) = H(s)\left[\underline{\hat{\Theta}} - \underline{\Theta}\right]^T \underline{A}(\underline{x}, u) \tag{5.46}$$

according to the systematic intelligent observer design while passing an area in the input space of the normalized RBF–net, we conclude that the weights $\hat{\Theta}_i$ *located within this area are convergent.*

Due to the variable smoothing factor σ a "passed area" is not definable in an exact mathematical sense; nevertheless, the meaning seems obvious.

Based on these deductions, the nonlinearity in equation (5.12) can be identified leading to a knowledge which is interpretable. Hence, this observer approach, depicted in detail, allows the identification of an isolated nonlinearity at any arbitrary position within a dynamic system with accessible states. In [32] an extension of this approach to two or more nonlinearities is presented.

5.3.2.6 Simplifying the Observer Design

Having access to the system's states for measurement gives the possibility to simplify the proposed observer approach: if the state is measurable on whose input the isolated nonlinearity acts, we will use this state to compare the system and the observer dynamic behaviour. The observer design simplifies to a single state to be observed.
In comparison to the full system approach presented earlier, the reduction always results in an error transfer function, which is strictly positive real (SPR). The error transfer function

$$H(s) = \underline{c}^T(sE - A + L)^{-1}\underline{e}_{NL} \tag{5.47}$$

will always be of the first order and fulfills the SPR–condition. Following [18] the adaptation law is simplified to

$$\dot{\underline{\hat{\Theta}}} = -\eta e \underline{A}(\underline{x}(t), u) \tag{5.48}$$

according to error model 3 in [18] and enables the weights adaptation with less computational efforts.

Yet, the aim of this deduction is an observer approach for systems with an isolated nonlinearity with a measurable in– and output, but not measurable internal states.

5.4 Intelligent Observer Design Following the Luenberger Approach

The basic idea to be presented here is applicable on several observer approaches; as an example we will base it on the Luenberger observer approach, which was published in 1963 [14, 15]. All expositions and deductions refer to a SISO–system (figure 5.12), but may be easily transfered to the more general MIMO case.

5.4.1 Prerequisites

In contrast to the deductions above, we assume the states of the regarded nonlinear dynamic system to be not measurable; only the system's in– and output are accessible for measurement. For the application of our observer we also have to demand observability:
The system is *observable in its linear part* and its linear components are known.

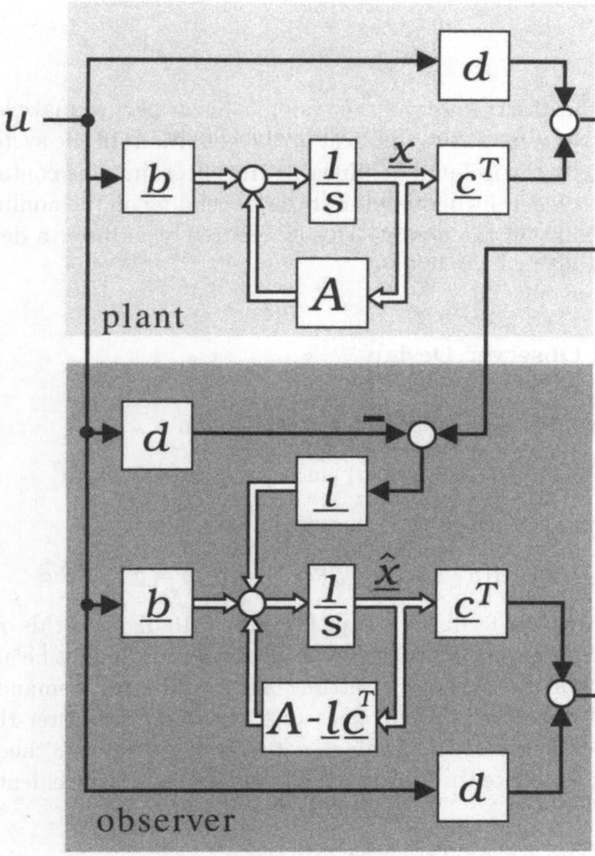

Figure 5.12: State observer for a linear SISO–system according to Luenberger, state reconstruction based on the system's output

Definition 5.14 Observable system *A dynamic system is completely observable, if each system state* $\underline{x}(t_0)$ *at the time* t_0 *is reconstructable knowing* $\underline{y}(t)$, $t_0 \le t \le t_1$, $t_1 < \infty$.

Definition 5.15 Observability in the linear part *A SISO–system with isolated nonlinearity of order n*

$$\dot{\underline{x}} = A\underline{x} + \underline{b}u + \underline{\mathcal{NL}}(\underline{x}, u) \text{ und } y = \underline{c}^T\underline{x} + du \tag{5.49}$$

is observable in its linear part, if the observability matrix Q_{BZ}

$$Q_{BZ} = \left[\underline{c} \;\; A^T\underline{c} \; ... \; (A^T)^{n-1}\underline{c}\right] \tag{5.50}$$

is regular, this means, if:

$$det(Q_{BZ}) \neq 0 \tag{5.51}$$

We have to demand the observability of the system's linear part to make sure the decaying of the observer error due to different initial conditions in the system and the observer. To identify the isolated nonlinearity, its effect must be contained in the system's output to generally observe the states depending on the nonlinearity. This means, the visibility of the nonlinearity is required according to definition 5.7 in the system's output.

5.4.2 Systematic Observer Design

The Luenberger observer's basic approach for the system

$$\dot{\underline{x}} = A\underline{x} + \underline{b}u + \mathcal{NL}(\underline{x}, u) \quad \text{and} \quad y = \underline{c}^T\underline{x} + du \tag{5.52}$$

with known A, $\underline{b}$, $\underline{c}$ and d reads in the same depiction

$$\dot{\hat{\underline{x}}} = (A - \underline{l}\,\underline{c}^T)\hat{\underline{x}} + \underline{b}u + \underline{l}(y - du) + \hat{\mathcal{NL}}(\hat{\underline{x}}, u) \quad \text{and} \quad \hat{y} = \underline{c}^T\hat{\underline{x}} + du \tag{5.53}$$

with the adjustable observer feedback vector $\underline{l}$. Figure 5.13 displays this observer approach in a signal flow graph. The observer feedback vector $\underline{l}$ might be adjusted to compromise between the observer's settling and its filtering demands; it is favourable to tune the observer's eigenvalues such that they are faster than the those of the corresponding system. Analogous to the known linear theory, the choice of $\underline{l}$ according to LQ–optimization or pole placement is convenient.

5.4.2.1 Deriving the Error Transfer Function

Neglecting the decaying effects of different system and observer initial conditions, we gain the error transfer function in analogy to equation (5.10) with the replacement $\underline{l}\,\underline{c}^T$ for the observer feedback matrix L as

$$\begin{aligned} e = H(s)(\hat{\mathcal{NL}} - \mathcal{NL}) &= \\ &= \underline{c}^T(sE - A + \underline{l}\,\underline{c}^T)^{-1}\underline{e}_{NL}\left(\hat{\mathcal{NL}} - \mathcal{NL}\right) \end{aligned} \tag{5.54}$$

To determine the transfer function $H(s)$ it is convenient to use the approach

$$H(s) = \underline{c}^T(sE - A + \underline{l}\,\underline{c}^T)^{-1}\underline{e}_{NL} = \frac{det\begin{bmatrix} sE - A + \underline{l}\,\underline{c}^T & \vdots & -\underline{e}_{NL} \\ \cdots\cdots & & \cdots \\ \underline{c}^T & \vdots & 0 \end{bmatrix}}{det(sE - A + \underline{l}\,\underline{c}^T)} \tag{5.55}$$

If the observer feedback vector $\underline{l}$ is adjustable without any restrictions, it makes sense to strive for a pole–zero–compensation in the transfer function $H(s)$.

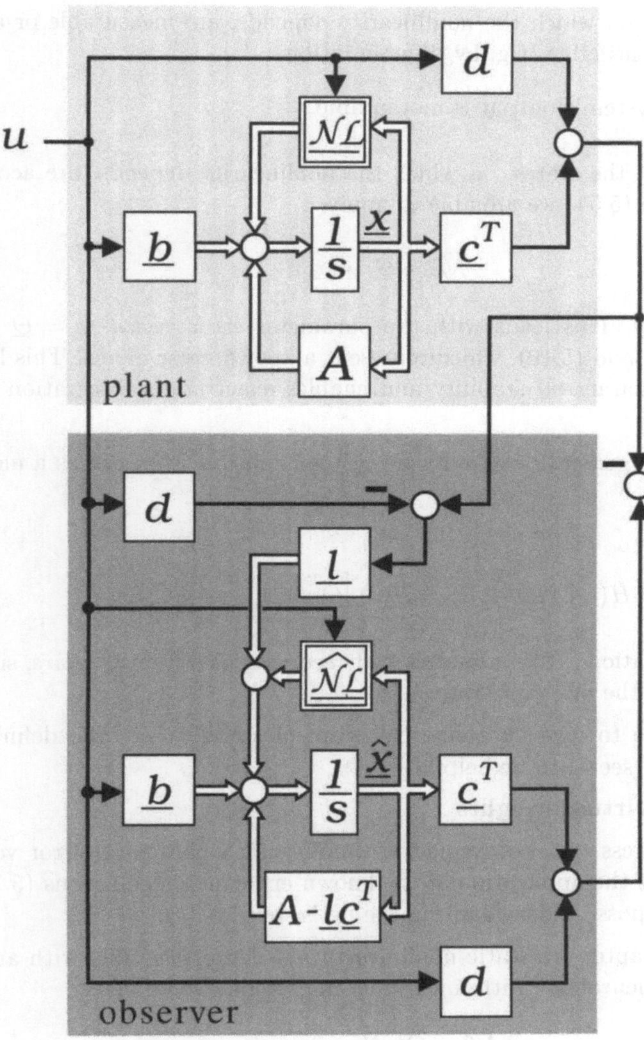

Figure 5.13: Intelligent observer for a dynamic SISO–system containing an isolated nonlinearity in the signal flow graph state space representation; application for state observation and identification of the nonlinearity based on the measured system output

5.4.2.2 Adaptation Law

To derive the net's adaptation law we have to distinguish between two cases:

1. the states, on which the nonlinearity depends, are measurable or at least exactly identifiable (e.g. by differentiation)

2. only the system's output is measurable

In the first case, the states, on which the nonlinearity depends, are accessible; thus in equation (5.54) we are able to apply

$$\hat{\mathcal{NL}} = \underline{\hat{\Theta}}^T \underline{\mathcal{A}}(\underline{x}, u) \tag{5.56}$$

in the same way. This leads with the parameter error vector $\underline{\Phi} = \underline{\hat{\Theta}} - \underline{\Theta}$ to equations (5.18) and (5.19), which represent a known error model. This leads to learning with guaranteed stability and enables exact state observation for the nonlinear system.

Yet, in the more general case, only the system's output is known as a measured signal:

Then we derive

$$e = H(s)(\hat{\mathcal{NL}} - \mathcal{NL}) = H(s)\left[\underline{\hat{\Theta}}^T \underline{\mathcal{A}}(\underline{\hat{x}}, u) - \underline{\Theta}^T \underline{\mathcal{A}}(\underline{x}, u)\right] \tag{5.57}$$

as the error equation. This case is named *recurrent observer structure*, since the net's inputs are the observer states.

We are not able to give an analytical proof of stability, yet the definition of 'virtual weights' seems to be helpful.

Definition of virtual weights

We have to express the error equation (5.57) with a parameter error vector in such a way, that the application of the known error model equations (5.18) and (5.19) becomes possible for adapting the net's weights.

According to chapter 4 a static nonlinearity $\mathcal{NL}$ is representable with arbitrary exactness by a neural net with linearly acting weights (RBF-net)

$$\mathcal{NL} = \underline{\Theta}^T \underline{\mathcal{A}}(\underline{x}, u) + d_a \tag{5.58}$$

with an inherent approximation error $d_a = d_a(\underline{x}, u)$. With an increasing number of neurons, this approximation error decreases and is negligible.

The real but unknown nonlinearity $\mathcal{NL}$ to be identified within the plant equation (5.52) shall be defined by the unknown weights vector $\underline{\Theta}$ of the net according to equation (5.58). Since the system's states are not measurable, the nonlinearity **cannot** be expressed within the observer equation (5.53) as a net whose activation is defined by the system's states, as we did before.

This means, due to the non–measurable states, the activation $\underline{\mathcal{A}}(\underline{x}, u)$ depending on the system's states cannot be calculated.

Yet, we develop

$$\mathcal{NL} = \underline{\Theta}^T \underline{\mathcal{A}}(\underline{x}, u) = \underline{\Theta}^T \underline{\mathcal{A}} = \underline{\check{\Theta}}^T \underline{\hat{\mathcal{A}}} = \underline{\check{\Theta}}^T \underline{\mathcal{A}}(\underline{\hat{x}}, u) \tag{5.59}$$

with for the moment undefined, but by

$$\underline{\Theta}^T \underline{\mathcal{A}} = \underline{\check{\Theta}}^T \underline{\hat{\mathcal{A}}} \tag{5.60}$$

determined virtual weights $\underline{\check{\Theta}}$. Matrix equation (5.60) is not solvable for the virtual weights; probable values are achieved based on an energetic inspection. The real nonlinearity $\mathcal{NL}$ together with the activation calculated on the basis of the observed states approximating weights are time–variant, since $\underline{\hat{x}}$ varies with time with respect to $\underline{x}$. But, due to the locally acting activation, we conclude

$$\underline{\hat{x}} \to \underline{x} \Rightarrow \underline{\check{\Theta}} \to \underline{\Theta} \tag{5.61}$$

the convergence of these virtual weights $\underline{\check{\Theta}}$ to the real weights, if the observer states are converging to the systems states, since

$$\underline{\hat{x}} \to \underline{x} \Rightarrow \underline{\hat{\mathcal{A}}} \to \underline{\mathcal{A}} \tag{5.62}$$

These virtual weights represent the real nonlinearity in the observer's state space; as a result of the learning, the difference between the system's and observer's states vanishes.
Based on the definition of these virtual weights we write

$$\begin{aligned} e = H(s)(\hat{\mathcal{NL}} - \mathcal{NL}) &= \\ &= \underline{c}^T (sE - A + \underline{l}\,\underline{c}^T)^{-1} \underline{e}_{NL} \left(\hat{\mathcal{NL}} - \mathcal{NL} \right) \\ &= \underline{c}^T (sE - A + \underline{l}\,\underline{c}^T)^{-1} \underline{e}_{NL} \left(\underline{\hat{\Theta}}^T \underline{\mathcal{A}}(\underline{\hat{x}}, u) - \underline{\check{\Theta}}^T \underline{\mathcal{A}}(\underline{\hat{x}}, u) \right) \\ &= \underline{c}^T (sE - A + \underline{l}\,\underline{c}^T)^{-1} \underline{e}_{NL} \left(\underline{\hat{\Theta}}^T - \underline{\check{\Theta}}^T \right) \underline{\mathcal{A}}(\underline{\hat{x}}, u) \\ &= \underline{c}^T (sE - A + \underline{l}\,\underline{c}^T)^{-1} \underline{e}_{NL} \underline{\Phi}^T \underline{\mathcal{A}}(\underline{\hat{x}}, u) \end{aligned} \tag{5.63}$$

with

$$\underline{\Phi} = \underline{\hat{\Theta}} - \underline{\check{\Theta}} \tag{5.64}$$

as the parameter error vector.
In analogy to the case with measurable states we develop

$$e = H(s) \underline{\Phi}^T \underline{\mathcal{A}}(\underline{\hat{x}}, u) \tag{5.65}$$

for this error equation from literature the adaptation law

$$\underline{\dot{\Phi}} = -\eta \epsilon H(s) \underline{\mathcal{A}}(\underline{\hat{x}}, u) \tag{5.66}$$

with the error for adaptation

$$\epsilon = e + \underline{\hat{\Theta}}^T H(s) \underline{A}(\underline{\hat{x}}, u) - H(s) \underline{\hat{\Theta}}^T \underline{A}(\underline{\hat{x}}, u) \tag{5.67}$$

with $\eta > 0$ is known, from which the convergence $\lim_{t \to \infty} e(t) = 0$ follows with global Ljapunov stability.
Since the virtual weights $\underline{\check{\Theta}}$ according equation (5.60) are time–variant and

$$\underline{\dot{\Phi}} = \underline{\dot{\hat{\Theta}}} \tag{5.68}$$

cannot be concluded, instead, using the adaptation law

$$\underline{\dot{\hat{\Theta}}} = -\eta \epsilon H(s) \underline{A}(\underline{\hat{x}}, u) \tag{5.69}$$

for the learning of the net's weights

$$\underline{\dot{\Phi}} = \underline{\dot{\hat{\Theta}}} - \underline{\dot{\check{\Theta}}} \tag{5.70}$$

we have to consider

$$\underline{\dot{\Phi}} = -\eta \epsilon H(s) \underline{A}(\underline{\hat{x}}, u) - \underline{\dot{\check{\Theta}}} \tag{5.71}$$

as the adaptation law. Formally speaking, this corresponds to the adaptation of a time–variant nonlinearity; this leads to the correct result, if the net is able to learn faster than the nonlinearity changes. According to equation (5.66) we conclude that the error convergence $\lim_{t \to \infty} e(t) = 0$ is given as long as the sign of $\underline{\dot{\Phi}}$ remains the same in spite of equation (5.70). This means, the term $\underline{\dot{\hat{\Theta}}}$ has to outweigh $\underline{\dot{\check{\Theta}}}$. The adaptation step size η has to be positive, this leads to the condition

$$0 < \eta_{min} < \eta < \eta_{max} \tag{5.72}$$

which has to be fulfilled to guarantee the learning of the time–variant virtual nonlinearity. The upper limit for the learning step size η_{max} is determined considerably by the discrete integration step size of the digital simulation. By choosing η we have to comply for equation (5.70) with $\underline{\dot{\hat{\Theta}}}$ outweighing $\underline{\dot{\check{\Theta}}}$.

An analytical way to determine η_{min} seems impossible, an appraisal could lead to probable values. Certainly, the shape of the real nonlinearity seems to effect this appraisal as well as the configuration of the neural net placed in the observer.

If the error transfer function

$$H(s) = \underline{c}^T (sE - A + \underline{l}\, \underline{c}^T)^{-1} \underline{e}_{NL} \tag{5.73}$$

fulfills the SPR–condition, for simplification the adaptation law

$$\underline{\dot{\hat{\Theta}}} = -\eta e \underline{A}(\underline{\hat{x}}, u) \tag{5.74}$$

according to error model 3 of [18] might be used for learning the weights.
Based on Ljapunov's stability theory (direct method), we conclude

$$\lim_{t\to\infty} e(t) = 0 \tag{5.75}$$

for these error models, if by choosing the learning step size η it is guaranteed that in

$$\dot{\underline{\Phi}} = \dot{\hat{\underline{\Theta}}} - \dot{\check{\underline{\Theta}}} \tag{5.76}$$

the term $\dot{\hat{\underline{\Theta}}}$ outweighs $\dot{\check{\underline{\Theta}}}$ in each element, leading to

$$\lim_{t\to\infty} (\hat{\underline{x}}(t) - \underline{x}(t)) = 0 \tag{5.77}$$

and to the convergence $\hat{\underline{A}} \rightarrow \underline{A}$.

Yet, in a practical application of this observer approach it is suitable to measure the states, on which the nonlinearity depends, as exactly as possible.

From the properties of the normalized RBF-net (GRNN or DANN) we conclude, due to the persistent excitation by $\underline{A}(\underline{x}, u)$, the convergence

$$\lim_{t\to\infty} \hat{\underline{\Theta}}(t) = \underline{\Theta} \tag{5.78}$$

of the net's weights to the unknown weights vector that represents the real nonlinearity. As stated earlier, due to the locally acting activiation of these nets, the convergence follows in the learned area. After the convergence of the net's parameter vector, there exists in the learned area a knowledge which is interpretable, and, in addition, the state observation of the nonlinear dynamic system has become possible.

5.5 Summary

In this chapter we introduced the systematic observer design for dynamic systems with an isolated nonlinearity. Under certain constraints, this intelligent observer is applicable to identify static nonlinearities in an interpretable way, thus enabling an observation of the system's states.

The basic idea in this approach is to define the nonlinearity's output as an input to a linear system, which can be examined with linear methods.

Whilst the nonlinearity is not yet exactly identified, this difference leads with the known linear system transfer behaviour to an observer error. Based on the Ljapunov theory, an adaptation of a specific type of neural network with linearly acting weights is executed, thus minimizing the observer error.

The next chapter 6 provides an extension of this theory on systems with separable nonlinearities. This chapter is followed by several examples of the theory's application on real plants coping with some occuring restrictions due to real implementation.

5.6 References

[1] Anderson, B. D. O.:
Exponential Stability of Linear Systems Arising from Adaptiv Identification.
IEEE Transactions on Automatic Control, 22–2, 1977.

[2] Engell, S.: (Hrsg.)
Entwurf nichtlinearer Regelungen.
R. Oldenbourg Verlag, München, 1995.

[3] Frenz, Th.:
Stabile neuronale online Identifikation und Kompensation statischer Nichtlinearitäten am Beispiel von Werkzeugmaschinenvorschubantrieben.
Dissertation, TU München, 1997.

[4] Gauß, C. F.:
Theory of Motion of the Heavenly Bodies.
New York, Dover, 1963.

[5] Hakala, J., Koslowski, C., Eckmiller, R.:
'Partition of Unity' RBF Networks are Universal Function Approximators.
Proceedings of the ICANN '94, Sorrento, Italy, Springer–Verlag, Berlin, Heidelberg, 1994, pp. 459–462.

[6] Hangl, F., Lenz, U., Schröder, D.:
Theorie des systematischen Entwurfs lernfähiger Beobachter für eine Klasse nichtlinearer Strecken.
GMA–Ausschuss 1.4 Theoretische Verfahren der Regelungstechnik, Interlaken, Switzerland, 28.9–1.10.1997.

[7] Kalman, R. E.:
Fundamental study of adaptive control systems.
Tech. Rept. ASD–TR–61, Vol. 1 (NASA N62–15355).

[8] Kalman, R. E.:
A new approach to linear filtering and prediction problems.
J. Basic Eng. 82D, 1960.

[9] Kalman, R. E., Bucy, R. S.
New results in linear filtering and prediction theory.
J. Basic Eng. 83D, 1961.

[10] Krebs, V.:
Nichtlineare Filterung.
Methoden der Regelungstechnik, R. Oldenbourg Verlag, München, 1980.

[11] Lenz, U., Schröder, D.:
Local Identification using Artificial Neural Networks.
Proc. Ninth Yale Workshop on Adaptive and Learning Systems, June 10-12, 1996, Yale University, CT, USA, pp 83–88.

[12] Lenz, U., Schröder, D.:
Identifikation isolierter Nichtlinearitäten mit Neuronalen Netzen.
GMA–Kongress Mess– und Automatisierungstechnik 96, Baden Baden, Germany, 9.–11. September 1996.

[13] Lenz, U., Schröder, D.:
Identifikation isolierter Nichtlinearitäten mit Neuronalen Netzen.
GMA–Ausschuss 1.4 Theoretische Verfahren der Regelungstechnik, Interlaken, Switzerland, 29.9–2.10.1996.

[14] Luenberger, D. G.:
Observing the state of linear systems.
IEEE Transactions on Military Electronics, 1964, pp. 74–80.

[15] Luenberger, D. G.:
Observers for multivariable systems.
IEEE Transactions on Automatic Control, pp. 190–197, 1966.

[16] Misawa, E.A., Hedrick, J.K.
Nonlinear observers — A state–of–the–art survey.
ASME Journal of Dynamic Systems, Measurement and Control 111, 1989, pp. 344-352.

[17] Morgan, A. P., Narandra, K. S.:
On the Uniform Asymptotic Stability of Certain linear Nonautonomous Differential Equations.
S.I.A.M. Journal of Control and Optimization, 15, 1977.

[18] Narendra, K., Annaswamy, A.M.:
Stable Adaptive Systems.
Prentice Hall, Englewood Cliffs, New Jersey, 1989.

[19] Narendra, K., Partasarathy, K.:
Identification and Control of Dynamical Systems Using Neural Networks.
IEEE Transactions on Neural Networks, Vol. 1, No. 1, March 1990, pp. 4–27.

[20] Parks, P.:
Ljapunov Redesign of Model Reference Adaptive Control Systems.
IEEE Transactions on Automatic Control, Vol. 11, 1966, pp. 362–367.

[21] Schäffner C.:
Analyse und Synthese neuronaler Regelungsverfahren.
Herbert Utz Verlag Wissenschaft, München, 1996.

[22] Schäffner, C., Schröder, D.:
Stable Nonlinear Observer Design with Neural Network.
IFAC–Workshop Motion Control, Munich, Germany, 1995.

[23] Schäffner, C., Schröder, D., Lenz, U.:
Application of Neural Networks to Motor Control.
International Power Electronics Conference, IPEC '95, Yokohama, Japan, Proc. Vol. 1, 1995, pp 46–51.

[24] Schroeder, D., Schaeffner, C., Lenz, U.:
Neural-Net Based Observers for Sensorless Drives.
20th International Conference on Industrial Electronics, Control and Instrumentation IECON '94, Bologna, Italy, Proceedings, Vol. 3, 1994, pp. 1599-1610.

[25] Schröder, D.:
Elektrische Antriebstechnik 1.
Springer–Verlag, Berlin, Heidelberg, 1994.

[26] Schröder, D.:
Elektrische Antriebstechnik 2, Regelung von Antrieben.
Springer–Verlag, Berlin, Heidelberg, 1995.

[27] Slotine, J., Li, W.:
Applied Nonlinear Control.
Prentice–Hall, Inc., Englewood Cliffs, NJ 07632, 1991.

[28] Sorenson, H. W.:
Kalman Filtering: Theorie and Application.
IEEE Press 1985.

[29] Sorenson, H. W.:
Kalman Filtering Techniques.
Advances in Control Systems Theory and Applications, Academic Press, 1966.

[30] Specht, D.:
A General Regression Neural Network.
IEEE Transactions on Neural Networks, Vol. 2, No. 6, November 1991, pp. 568–576.

[31] Straub, S., Schröder, D.:
Neuronalbasierter Beobachterentwurf für eine breite Klasse nichtlinearer Systeme.
Fachtagung Computational Intelligence, VDI Berichte 1381, 1998, pp. 361–380,

[32] Straub, S., Schröder, D.:
Identification of Nonlinear Dynamic Systems with Recurrent Neural Networks and Kalman Filter Methods.
ISCAS '96, Atlanta, USA. Proceedings, Vol. 3, 1996, pp. 341-345.

[33] Straub, S., Schröder, D.:
Neural Network based Identification Methods to solve Nonlinear Problems in Rolling Mill Subsystems.
2nd International Conference on Modelling of Metal Rolling Processes, London, GB, 1996, Proceedings pp. 366-377.

[34] Strobl, D., Lenz, U., Schröder, D.:
Systematic Design of Stable Neural Observers for a Class of Nonlinear Systems.
Proceedings of the 1997 IEEE International Conference on Control Applications, Hartford, Conneticut USA, October 5–7, 1997, pp. 377–382.

[35] Tzirkel–Hancock, E., Fallside, F.:
A Direct Control Method for a Class of Nonlinear Systems using Neural Networks.
Technical Report CUED/F–INFENG/TR.65, Cambridge University, Engineering Departement. Cambridge, England, 1991.

6 Identification of Separable Nonlinearities

Franz Hangl

The following considerations about the identification methods of so–called *separable nonlinearities* are based on [6]. In this chapter we will present an extension of the identification of isolated nonlinearities. In many applications it is necessary to identify multiple nonlinear influences simultaneously. A method for the identification of separable nonlinearities with guaranteed Ljapunov stability will be explained here.

Finally, we will give a short overview of possibilities how to implement a–priori knowledge of the nonlinearities especially to reduce the dimension of the used neural networks.

6.1 Plants with Separable Nonlinearities

Our identification method has been developed for plants with separable nonlinearities, which are very common in control systems. We will define them as follows.

Definition 6.1 *If a plant can mathematically be modeled as*

$$\underline{\dot{x}} = A\underline{x} + B\underline{u} + \underline{\mathcal{NL}}(\underline{u}, \underline{y}) \tag{6.1}$$

$$\underline{y} = C\underline{x} + D\underline{u} \tag{6.2}$$

$$\underline{\mathcal{NL}}(\underline{u}, \underline{y}) = \sum_{i=1}^{k} \underline{e}_{NLi} \mathcal{NL}_i \tag{6.3}$$

then it is called a plant with **separable nonlinearities**. *The nonlinear vector* $\underline{\mathcal{NL}}(\underline{u}, \underline{y})$*, consisting of* $1 \leq k \leq n$ *(n order of the system) scalar nonlinearities* $\mathcal{NL}_i = \underline{e}_{NLi}^T \underline{\mathcal{NL}}(\underline{u}, \underline{y})$*, represents the nonlinear part of the plant, where* $\underline{e}_{NLi}$ *are vectors of dimension n with one component not equal zero. The matrices A, B, C and D are constant and known.*

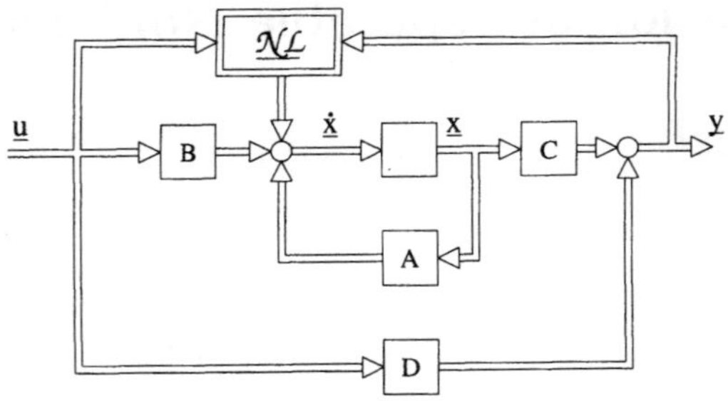

Figure 6.1: Plant with separable nonlinearities

In figure 6.1 a signal flow chart of such a plant is shown. It is the purpose of an identification method to guarantee stability and convergence of identification of the unknown components of the vector $\underline{\mathcal{NL}}$ using neural networks.

Basically our approach is a kind of disturbance observer, which enables us to estimate the system disturbance and to learn its dependency on the system output $\underline{y}$ and the input $\underline{u}$ at the same time. We approximate these relations with the General Regression Neural Networks according to chapter 4 and [11], because the defined extrapolation and interpolation behaviour of the GRNN is very attractive for adaptive control. Further advantages have already been presented in previous chapters.

6.2 Nonlinear Observer Approach

Based on the above definition of the plants under consideration, we will describe the observer design, but we have to distinguish between two situations. First we will consider plants with an accessible state vector $\underline{x}$. Then, the theory will be extended for plants, where the state vector is partially unknown. In both situations, we assume that the activations of the nonlinearities in the plant and thus of the neural networks in our observer are exactly known, which means that the nonlinear effects in the plant only depend on the system input $\underline{u}$ and the system output $\underline{y}$.

6.2.1 Identification with Accessible States

6.2.1.1 Adaptive Observer According to the Luenberger Observer

Using the mathematical representation of the plant

$$\dot{\underline{x}} = A\underline{x} + B\underline{u} + \underline{\mathcal{NL}}(\underline{u}, \underline{y}) \tag{6.4}$$

$$\underline{y} \;=\; C\underline{x} + D\underline{u} \tag{6.5}$$

we can determine, under the condition of an invertable output matrix C, the state vector $\underline{x}$ to

$$\underline{x} \;=\; C^{-1}\underline{y} - D\underline{u} \tag{6.6}$$

The accessible state vector $\underline{x}$ simplifies our observer design according to the linear observer theory of Luenberger. With a nonlinear observer like

$$\dot{\underline{\hat{x}}} \;=\; A\underline{x} + B\underline{u} + \widehat{\mathcal{NL}}(\underline{u}, \underline{y}) + L(\underline{\hat{x}} - \underline{x}) \tag{6.7}$$

$$\tag{6.8}$$

we obtain the time derivative of the observer error $\underline{e} = \underline{\hat{x}} - \underline{x}$

$$\dot{\underline{e}} \;=\; L\underline{e} + \widehat{\mathcal{NL}} - \mathcal{NL} \tag{6.9}$$

If the Luenberger matrix L is chosen diagonally with

$$L_{ii} \;<\; 0 \tag{6.10}$$

$$L_{ij} \;=\; 0,\; i \neq j, \tag{6.11}$$

$$\tag{6.12}$$

we have a decoupled estimation problem and so there is a direct correlation between each observer error e_i and each estimation error $\widehat{\mathcal{NL}}_i - \mathcal{NL}_i$ like

$$\dot{e}_i \;=\; L_{ii}e_i + \widehat{\mathcal{NL}}_i - \mathcal{NL}_i. \tag{6.13}$$

With the representation defined in previous chapters of nonlinear effects in the plant

$$\mathcal{NL}(\underline{u}, \underline{y}) \;=\; \sum_{i=1}^{n} \underline{e}_{NLi}\underline{\Theta}_i^T \underline{A}_i \tag{6.14}$$

$$\widehat{\mathcal{NL}}(\underline{u}, \underline{y}) \;=\; \sum_{i=1}^{n} \underline{e}_{NLi}\underline{\hat{\Theta}}_i^T \underline{A}_i \tag{6.15}$$

and its corresponding estimation in the adaptive observer, this yields a differential equation for each error e_i. Defining the n parameter– (estimation–) errors $\underline{\Phi}_i = \underline{\hat{\Theta}}_i - \underline{\Theta}_i$, it results in

$$\dot{e}_i = L_{ii}e_i + \underline{\Phi}_i\underline{A}_i(\underline{y}, \underline{u}) \tag{6.16}$$

To complete our observer design for plants with accessible states, it is necessary to create an adaptive law for $\underline{\hat{\Theta}}_i$, which guarantees stability and convergence of the adaption of the weights. Under the condition of a stable linear observer part, which is guaranteed through our choice of the Luenberger matrix L, the adaptive law for the weights

$$\dot{\underline{\hat{\Phi}}}_i = \dot{\underline{\hat{\Theta}}}_i = -e_i\eta_i\underline{A}_i(\underline{x}, \underline{y}) \tag{6.17}$$

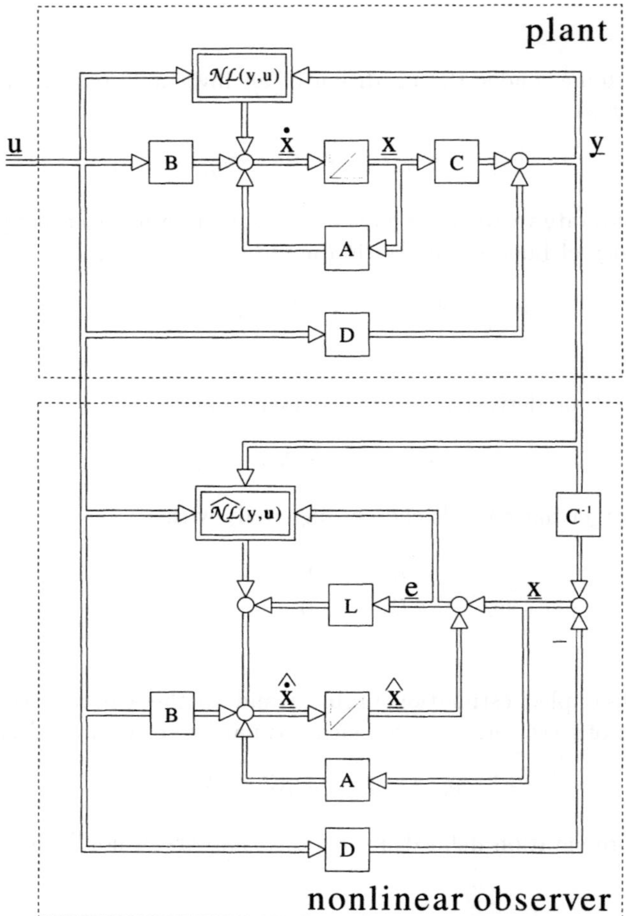

Figure 6.2: The nonlinear observer with accessible states

can be used according to previous chapters. With $\eta_i > 0$, an asymptotically stable and convergent adaptation of all weights-vectors $\underline{\widehat{\Theta}}_i$ is guaranteed. Therefore, with Ljapunov's theory, it is possible to ensure, that the observer error

$$\lim_{t\to\infty} \underline{e}(t) = \underline{0} \tag{6.18}$$

becomes zero. If the activations of the neural networks are persistently exciting it is also guaranteed that

$$\lim_{t\to\infty} \underline{\widehat{\Theta}}_i = \underline{\Theta}_i \tag{6.19}$$

that means the estimated weights tend towards their 'true' values in the plant.

In systems with accessible state vector $\underline{x}$, the identification of the unknown nonlinear part is not difficult. The problem can be reduced to a set of n observers with one nonlinear effect. The theory for that specific case has already been presented in the previous chapter 5. The GRNN, which we use for approximation, guarantees a stable and convergent identification of $\underline{\mathcal{NL}}$. In most applications the internal states of the plant are not directly accessible, hence, there are unknown states, which results in a more difficult identification problem. We have developed a method, which will be presented next to solve such an identification problem for a certain class of plants.

6.2.2 Identification of Nonlinearities in Plants with Unknown Internal States

In this section, our method of identification of separable nonlinearities in plants with unknown states will be presented. Any intelligent observer needs a minimum of information about the plant. For identifying k nonlinearities, it is necessary to measure k independent signals of the plant which has already been shown by P.C. Müller for linear disturbance observers in [4]. As a conclusion of our considerations, a theorem for identification is given at the end of this section.

6.2.2.1 Neural Observer Approach

Let us once more consider the plant described by the following equations.

$$\dot{\underline{x}} = A\underline{x} + B\underline{u} + \underline{\mathcal{NL}}(\underline{u}, \underline{y}) \tag{6.20}$$

$$\underline{y} = C\underline{x} + D\underline{u} \tag{6.21}$$

$\underline{x} \in \mathbb{R}^n$, $A \in \mathbb{R}^{n\times n}$, $C \in \mathbb{R}^{l\times k}$, $rank(C) = k$, $B \in \mathbb{R}^{n\times j}$, $D \in \mathbb{R}^{k\times j}$, with the vector

$$\underline{\mathcal{NL}}(\underline{u}, \underline{y}) = \sum_{i=1}^{i=k} \underline{e}_{NLi}\mathcal{NL}_i \tag{6.22}$$

consisting of k ($1 \leq k < n$) unknown components/nonlinearities. Each nonlinearity in the plant can be a function of the system input $\underline{u}$ and the system output $\underline{y}$, which are both known. We assume that the linear part of the system is observable. Corresponding to our considerations before, we design an observer according to the linear observer theory as follows. Using the system output $\underline{y}$, the observer can be described as

$$\dot{\hat{\underline{x}}} = A\hat{\underline{x}} + B\underline{u} + \sum_{i=1}^{i=k} \underline{e}_{NLi}\widehat{\mathcal{NL}}_i(\underline{u}, \underline{y}) + L(\hat{\underline{y}} - \underline{y}) \tag{6.23}$$

$$\hat{y} = C\hat{\underline{x}} + D\underline{u} \tag{6.24}$$

It is necessary to guarantee stability of the linear system part, which requests a so–called Luenberger matrix L where the poles of

$$det\,(sE - A - LC) = 0 \tag{6.25}$$

are located in the negative halfplane. As the nonlinearities can be considered as an unknown plant disturbance of the linear part, and it is guaranteed that starting values will decrease with time, we can describe the plant and the observer in the Laplace domain. This is not allowed if the matrices A, B, C and D are also nonlinear. The i–th output of the designed observer can thus be determined as follows

$$\widehat{y}_i = \underline{c}_i(sE - A - LC)^{-1}\left[B\underline{u} - LC\underline{x} + \sum_{i=1}^{i=k} \underline{e}_{NLi}\widehat{\mathcal{NL}}_i\right] + \underline{d}_i\underline{u} \tag{6.26}$$

where $\underline{c}_i$ represents the i-th row vector of the output matrix C. Similar to the observer output, we get for the i–th output of the plant:

$$y_i = \underline{c}_i(sE - A)^{-1}\left[B\underline{u} + \sum_{i=1}^{i=k} \underline{e}_{NLi}\mathcal{NL}_i\right] + \underline{d}_i\underline{u} \tag{6.27}$$

Hence, the i–th observer error becomes

$$\begin{aligned} e_i &= \widehat{y}_i - y_i \\ &= \underline{c}_i(sE - A - LC)^{-1}B\underline{u} \\ &\quad -\underline{c}_i(sE - A - LC)^{-1}LC(sE - A)^{-1}Bu \\ &\quad -\underline{c}_i(sE - A - LC)^{-1}LC(sE - A)^{-1}\sum_{i=1}^{i=k} \underline{e}_{NLi}\mathcal{NL}_i \\ &\quad +\underline{c}_i(sE - A - LC)^{-1}\sum_{i=1}^{i=k} \underline{e}_{NLi}\widehat{\mathcal{NL}}_i \\ &\quad -\underline{c}_i(sE - A)^{-1}B\underline{u} \\ &\quad -\underline{c}_i(sE - A)^{-1}\sum_{i=1}^{i=k} \underline{e}_{NLi}\mathcal{NL}_i \end{aligned} \tag{6.28}$$

Using the rules of linear algebra, equation (6.28) can be reduced to

$$e_i = \underline{c}_i(sE - A - LC)^{-1}\sum_{i=1}^{i=k} \underline{e}_{NLi}(\widehat{\mathcal{NL}}_i - \mathcal{NL}_i) \tag{6.29}$$

So the chosen observer design allows to create k observer errors, which are determined by

$$\underline{e} = C(sE - A - LC)^{-1}\sum_{i=1}^{i=k} \underline{e}_{NLi}(\widehat{\mathcal{NL}}_i - \mathcal{NL}_i) \tag{6.30}$$

The observer includes k differential error equations for the k unknown nonlinearities as shown above. Hence, it is obvious to define the dynamic error matrix H_F as

$$\underline{e} = H_F \begin{bmatrix} \widehat{\mathcal{NL}}_1 - \mathcal{NL}_1 \\ .. \\ .. \\ .. \\ \widehat{\mathcal{NL}}_k - \mathcal{NL}_k \end{bmatrix} \tag{6.31}$$

where H_F becomes to

$$H_F = \begin{bmatrix} \underline{c}_1(sE - A - LC)^{-1}\underline{e}_{NL1} & .... & \underline{c}_1(sE - A - LC)^{-1}\underline{e}_{NLk} \\ .... & & .... \\ .... & & .... \\ .... & & .... \\ \underline{c}_k(sE - A - LC)^{-1}\underline{e}_{NL1} & .... & \underline{c}_k(sE - A - LC)^{-1}\underline{e}_{NLk} \end{bmatrix} \tag{6.32}$$

This matrix describes the dynamic influence of the estimation errors of the individual nonlinearities in the plant to each observer error e_i. If the dynamic error matrix is diagonal we call it a noncoupled system. That is identical to k systems with isolated nonlinearities, which has already been shown in the previous chapter. In most applications, there will be a coupled system with a non-diagonal dynamic error matrix. Previously shown identification methods are not able to handle such systems, therefore it is necessary to transform the identification problem to a non-coupled system.

6.2.3 The Error Decoupling Filter

The result of the preceding section was a set of error equations with the error vector $\underline{e}$, which can be described as

$$\underline{e} = H_F \begin{bmatrix} \widehat{\mathcal{NL}}_1 - \mathcal{NL}_1 \\ \cdots \\ \cdots \\ \cdots \\ \widehat{\mathcal{NL}}_k - \mathcal{NL}_k \end{bmatrix} \tag{6.33}$$

To be successful with the following strategy, it is necessary to have an invertable matrix H_F. As mentioned at the beginning of section 6.2.2.1, we need k in the Laplace domain independent measurement signals of the plant each represented by its output vector $\underline{c}_i$ to ensure that. Under this condition it is possible to multiply equation (6.33) from the left side with the inverse dynamic error matrix, H_F^{-1}. This yields a transformed error vector $\underline{e}_{in}$ as follows:

$$\underline{e}_{in} = H_F^{-1}\underline{e} = H_F^{-1}H_F \begin{bmatrix} \widehat{\mathcal{NL}}_1 - \mathcal{NL}_1 \\ \cdots \\ \cdots \\ \cdots \\ \widehat{\mathcal{NL}}_k - \mathcal{NL}_k \end{bmatrix} \tag{6.34}$$

The matrix $E = H_F^{-1}H_F$ is of course diagonal. The multiplication has transformed the coupled system to a system of k non–coupled, in other words isolated, nonlinearities $\mathcal{NL}_i$ with their corresponding errors $e_{in,i}$. The filter H_F^{-1} will contain time derivatives in nearly all applications, which impairs the implementation of this filter on a real plant. Our aim is to design a decoupling filter which can be build up with integrating dynamic blocks only. Therefore, the realization becomes much easier and robust. This could be approached using a second diagonal filter

$$H_W = \begin{bmatrix} H_{11} & 0 & \cdots & 0 \\ 0 & H_{22} & \cdots & 0 \\ 0 & \cdots & \cdots & 0 \\ .. & \cdots & \cdots & .. \\ 0 & \cdots & 0 & H_{kk} \end{bmatrix} \tag{6.35}$$

which results in the complete decoupling filter to be

$$H_H = H_W H_F^{-1} \tag{6.36}$$

By choosing the elements of H_W in such a way, that all elements of H_H have a relative dynamic degree lower than zero, we can ensure that the complete decoupling filter consists only of linear gains and integrating blocks. Hence, the new filter has to be included in the above error equation which yields

$$\begin{aligned} \underline{e}_H &= H_H\underline{e} \qquad\qquad (6.37) \\ &= H_W H_F^{-1} H_F \begin{bmatrix} \widehat{\mathcal{NL}}_1 - \mathcal{NL}_1 \\ \cdots \\ \cdots \\ \cdots \\ \widehat{\mathcal{NL}}_k - \mathcal{NL}_k \end{bmatrix} \\ &= H_W \begin{bmatrix} \widehat{\mathcal{NL}}_1 - \mathcal{NL}_1 \\ \cdots \\ \cdots \\ \cdots \\ \widehat{\mathcal{NL}}_k - \mathcal{NL}_k \end{bmatrix} \end{aligned}$$

There is a direct correlation between the elements $e_{H,i}$ of the transformed error vector $\underline{e}_H$ and the corresponding estimation errors $\widehat{\mathcal{NL}}_i - \mathcal{NL}_i$. With the above decoupling filter consisting of integrating blocks and gains, the dynamic behaviour of the error equation can be optimized also, because the elements H_{ii} occur in the adaptive law. So it is possible to tune the adaptation performance of the neural networks by varying H_W.

6.2.3.1 The Adaptive Law

At last, an adaptive law for the weights vector of the individual neural networks has to be defined. As shown in the error equation (6.37), the transformed i−th observer error

$$e_{H,i} = H_{ii}(\widehat{\mathcal{NL}}_i - \mathcal{NL}_i) \tag{6.38}$$

can be written by using the i-th error vector of weights

$$\underline{\Phi}_i = \underline{\widehat{\Theta}}_i - \underline{\Theta}_i \tag{6.39}$$

as follows

$$e_{H,i} = H_{ii}\underline{\Phi}_i^T \underline{A}_i(\underline{x}, \underline{u}) \tag{6.40}$$

Creating an augmented error $\epsilon_{H,i}$, see [8],

$$\epsilon_{H,i} = e_{H,i} - H_{ii}\underline{\widehat{\Theta}}_i^T \underline{A}_i + \underline{\widehat{\Theta}}_i^T H_{ii}\underline{A}_i \tag{6.41}$$

a stable and convergent adaptive law for the individual vector of weights $\underline{\widehat{\Theta}}_i$ can be obtained. For a positive learning rate $\eta_i > 0$ the adaptive law

$$\dot{\underline{\Phi}}_i = \dot{\underline{\widehat{\Theta}}}_i = -\eta_i \epsilon_i H_{ii}\underline{A}_i \tag{6.42}$$

guarantees that both the transformed observer error

$$\lim_{t\to\infty} e_{H,i}(t) = 0 \tag{6.43}$$

and the actual observer error

$$\lim_{t\to\infty} \underline{e}(t) = \underline{0} \tag{6.44}$$

tend towards zero. If the activation of the neural network is persistently exciting, it is also guaranteed that the estimated weights

$$\lim_{t\to\infty} \underline{\widehat{\Theta}}_i = \underline{\Theta}_i \tag{6.45}$$

tend towards their 'true' values in the plant. The proof of stability is then identical to the case of an isolated nonlinearity.

It has been shown that it is possible to identify nonlinearities in plants under certain conditions, without the need to measure all states of the plant. The used observer approach like Luenberger provides a dynamic error matrix which represents the influence of estimation errors on each observer error. By decoupling these errors with a filter H_H, a stable adaptive law is obtained.

The idea of our approach has been to transform the observer errors into a form, which allows an easy application of established methods. As a conclusion we will present the theorem of guaranteed stable and convergent identification of separable nonlinearities.

Theorem 6.1 *If a nonlinear plant can be described in the form*

$$\begin{aligned}\underline{\dot{x}} &= A\underline{x} + B\underline{u} + \underline{\mathcal{NL}}\\ \underline{y} &= C\underline{x} + D\underline{u}\end{aligned}$$

with $\underline{x} \in \mathbb{R}^n$, $A \in \mathbb{R}^{n\times n}$, $C \in \mathbb{R}^{k\times n}$, $rank(C) = k$, $B \in \mathbb{R}^{n\times j}$, $D \in \mathbb{R}^{k\times j}$ *and the vector* $\underline{\mathcal{NL}}(\underline{u}, \underline{y})$

$$\underline{\mathcal{NL}} = \sum_{i=1}^{i=k} \underline{e}_{NLi} \mathcal{NL}_i$$

contains k $(1 < k \leq n)$ *unknown components, guaranteed stable and convergent identification of* $\underline{\mathcal{NL}}(\underline{u}, \underline{y})$ *is possible under the following conditions:*

- *The linear part of the system is observable.*
- *The Luenberger matrix is chosen in such a way that*

  $$det\,(sE - A - LC) = 0$$

  is stable.
- *The dynamic error matrix*

  $$H_F = \begin{bmatrix} \underline{c}_1(sE - A - LC)^{-1}\underline{e}_{NL1} & \ldots\ldots & \underline{c}_1(sE - A - LC)^{-1}\underline{e}_{NLk} \\ \ldots\ldots & & \ldots\ldots \\ \ldots\ldots & & \ldots\ldots \\ \ldots\ldots & & \ldots\ldots \\ \underline{c}_k(sE - A - LC)^{-1}\underline{e}_{NL1} & \ldots\ldots & \underline{c}_k(sE - A - LC)^{-1}\underline{e}_{NLk} \end{bmatrix}$$

  has to be invertable. The vectors $\underline{c}_1 ... \underline{c}_k$ *are the row vectors of the output matrix* C.

6.3 Implementation of A–Priori Knowledge

The identification of multiple nonlinearities requires substantial resources of computational power. Consequently it might be essential to reduce the complexity of the observers, especially the part containing the neural networks. The designer of an observer should therefore try to minimize the necessary input dimension of the neural network. In the following we will give a short overview of how to use a–priori knowledge for this aim. Special attention will also be drawn to the implementation of a–priori knowledge into the adaptive law.

Dealing with the identification of $\mathcal{NL} = \underline{\Theta}^T \underline{\mathcal{A}}$ by neural networks with $\widehat{\mathcal{NL}} = \underline{\widehat{\Theta}}^T \underline{\mathcal{A}}$, yields differential error equations such as

$$e = H(s)\underline{\Phi}^T \underline{\mathcal{A}} \tag{6.46}$$

with the transfer function $H(s)$. In this system we can divide each nonlinearity into a known and an unknown component, and, hence, define two types of a–priori knowledge as follows

$$\mathcal{NL} = \mathcal{NL}_p \mathcal{NL}' + \mathcal{NL}_s \tag{6.47}$$

These types are additive and multiplicative a–priori knowledge. A neural network should only learn the remaining unknown part $\mathcal{NL}'$. If there is no a–priori knowledge about $\mathcal{NL}$, the two parts become

$$\mathcal{NL}_s = 0 \tag{6.48}$$
$$\mathcal{NL}_p = 1 \tag{6.49}$$

The nonlinearity will be approximated by a GRNN in the well–known form $\widehat{\mathcal{NL}} = \underline{\widehat{\Theta}}^T \underline{\mathcal{A}}$, implementing a–priori knowledge

$$\widehat{\mathcal{NL}} = \mathcal{NL}_p \underline{\widehat{\Theta}}'^T \underline{\mathcal{A}}' + \mathcal{NL}_s \tag{6.50}$$

which will perhaps reduce the input dimension of $\mathcal{A}'$.

6.3.1 Additive A–Priori Knowledge

In this case we assume to have additive a–priori knowledge only, which simplifies equation (6.47) to

$$\mathcal{NL} = \mathcal{NL}' + \mathcal{NL}_s \tag{6.51}$$

and subsequently the error equation to

$$e = H(s)(\widehat{\mathcal{NL}}' + \mathcal{NL}_s - (\mathcal{NL}' + \mathcal{NL}_s)) \tag{6.52}$$

Hence, we obtain the observer error

$$e = H(s)\underline{\Phi}'^T \underline{\mathcal{A}}' \tag{6.53}$$

With the augmented error

$$\epsilon = e - H(s)\underline{\widehat{\Theta}}'^T \underline{\mathcal{A}}' + \underline{\widehat{\Theta}}'^T H(s)\underline{\mathcal{A}}' \tag{6.54}$$

and a positive learning rate $\eta > 0$, the stable adaptive law

$$\dot{\underline{\Phi}}' = \dot{\underline{\widehat{\Theta}}}' = -\eta \epsilon H(s)\underline{\mathcal{A}}' \tag{6.55}$$

can be used. The use of additive a–priori knowledge does not affect the adaptive law. Hence, the same observer design can be applied. Additive a–priori knowledge can either be added to the neural network output or built in by initializing the weight vector $\underline{\widehat{\Theta}}$. In the second case no reduction of the input dimension is possible, but the adaptation time can be reduced.

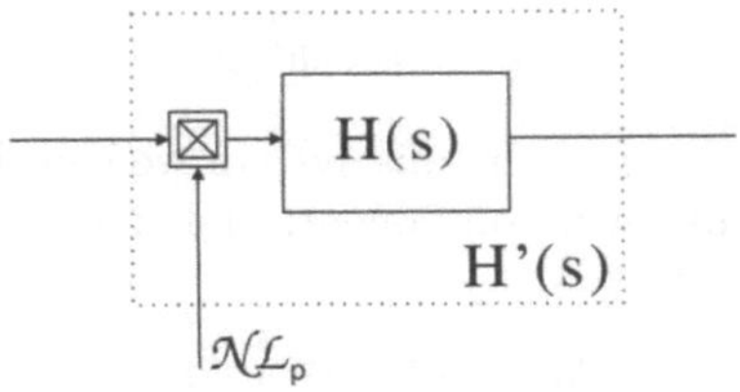

Figure 6.3: The time–variant dynamic $H'(s)$

6.3.2 Multiplicative A–Priori Knowledge

As shown above, implementing additive a–priori knowledge has no influence on the adaptive law, hence, we may neglect it while focusing on multiplicative a–priori knowledge. We will first concentrate on nonlinearities which can be represented as a product

$$\mathcal{NL} = \mathcal{NL}_p \mathcal{NL}' \tag{6.56}$$

of a known and an unknown component. To approximate $\mathcal{NL}$ in the form

$$\widehat{\mathcal{NL}} = \mathcal{NL}_p \widehat{\mathcal{NL}}' = \mathcal{NL}_p \underline{\widehat{\Theta}}'^T \underline{\mathcal{A}}' \tag{6.57}$$

Employing the error equation

$$e = H(s)\underline{\Phi}^T \underline{\mathcal{A}} \tag{6.58}$$

the observer error

$$e = H(s)\mathcal{NL}_p(\widehat{\mathcal{NL}}' - \mathcal{NL}') \tag{6.59}$$

changes to

$$e = H(s)\mathcal{NL}_p \underline{\Phi}'^T \underline{\mathcal{A}}' \tag{6.60}$$

It is only necessary to identify the unknown part using a neural network. Because of this, it is useful to define a new time–variant error transfer function $H'(s)$, see figure 6.3, as follows

$$H'(s) = H(s)\mathcal{NL}_p \tag{6.61}$$

With $H'(s)$ and the augmented error

$$\epsilon = e - H'(s)\underline{\widehat{\Theta}}'^T \underline{\mathcal{A}}' + \underline{\widehat{\Theta}}'^T H'(s)\underline{\mathcal{A}}' \tag{6.62}$$

the adaptive law

$$\dot{\underline{\Phi}}' = \dot{\underline{\widehat{\Theta}}}' = -\eta\epsilon H'(s)\underline{\mathcal{A}}' \tag{6.63}$$

with $\eta > 0$, guarantees stable and convergent identification of the unknown component $\mathcal{NL}'$.

We have seen that it is necessary to distinguish between two types of a–priori knowledge. If there is additive a–priori knowledge only, no change in the observer design is necessary. In the case of multiplicative a–priori knowledge, we defined a new time–variant error transfer function, which results in a modified adaptive law.

As mentioned above, the implementation of a–priori knowledge is a very effective way to reduce the input dimension of neural networks or at least to enhance the identification performance.

6.4 References

[1] Beuschel M., Hangl F.:
Kriterien zur Optimalen Auslegung Neuronaler Netze am Beispiel des GRNN.
Lehrstuhl für Elektrische Antriebstechnik, TU München, 1997.

[2] Beuschel M., Hangl F.:
Online–Invertierung Neuronaler Netze zur Kompensation von Nichtlinearitäten.
Lehrstuhl für Elektrische Antriebstechnik, TU München, 1997.

[3] Brause R.:
Neuronale Netze.
Teubner Verlag, Stuttgart.

[4] Engell S.:
Entwurf nichtlinearer Regelungen.
R. Oldenbourg Verlag, München.

[5] Föllinger O.:
Regelungstechnik.
Hüthig Verlag, 1991.

[6] Hangl F., Lenz U., Schröder D.:
Theorie des systematischen Entwurfs lernfähiger Beobachter für eine Klasse nichtlinearer Strecken.
Proceedings GMA Workshop Interlaken, 1997.

[7] Lenz U.:
Theorie des systematischen Entwurfs lernfähiger Beobachter für Strecken mit isolierter Nichtlinearität.
Lehrstuhl für Elektrische Antriebstechnik, TU München, 1997.

[8] Narendra K. S.:
Stable Adaptive Systems.
Prentice Hall, 1989.

[9] Schäffner C.:
Anwendung Neuronaler Netze in der Automatisierungstechnik.
Lehrstuhl für Elektrische Antriebstechnik, TU München, 1995.

[10] Schröder D.:
Elektrische Antriebe 2.
Springer–Verlag, Berlin, Heidelberg, 1995.

[11] Specht D. F.:
A General Regression Neural Network.
IEEE Transactions of Neural Networks, vol. 2, no. 6, 1991.

7 Identification and Compensation of Friction

Thomas Frenz

7.1 Introduction

In this chapter we are going to show an application example for the method of a systematic intelligent observer design presented in chapter 5. For a feed drive of a tool machine with unknown nonlinear friction a linear model is derived and a nonlinear observer is designed for the identification of the friction characteristic. The objective is to learn and compensate the sliding–friction of a feed drive of a lathe. Further information can be found in [1, 2, 3, 4, 5]. References on literature of the theoretical parts can be found in the chapters 3, 4 and 5

Figure 7.1: Experimental machine

The examined tool machine is a Gildemeister GD 200 as shown in figure 7.1. The feed drive consists of a servo drive, a toothed belt gear, a screw spindle, a ball–and–screw spindle drive and a slide; here a hydrodynamic sliding bearing is used.

The friction in feed drives originates mainly in the bearing assembly of the slide. It is advantageous for a well–damped mechanical system behaviour, but disadvantageous for an exact positioning and/or contouring control. Therefore we intend to learn the friction characteristic and compensate it using feedforward control. So we want to combine the advantage of a well damped mechanical system with a precise positioning and/or contouring control.

For the observer design first a linear model has to be derived. In this case, the feed drive is modeled as a linear two–mass system according to the theory of Lagrange. This results, including the differential equation of the servo drive, in a linear model of fifth order. The signal flow chart of the model is shown in figure 7.2.

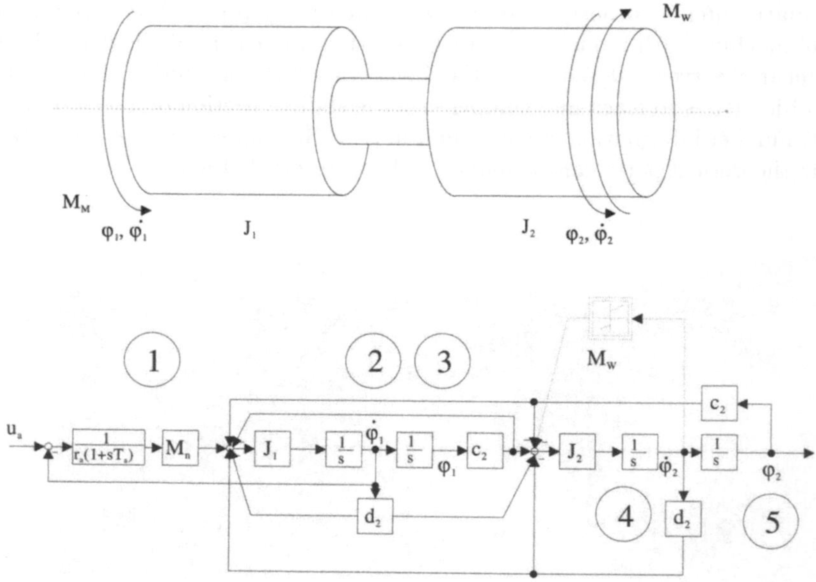

Figure 7.2: Two–mass model

Validations have shown, that with respect to its natural mechanical modes, the feed drive can be described by only two natural frequencies.

Figure 7.3 shows measured signals compared to simulated data of the linear model while exciting the speed controller of the system with sinusoidal excitation signal in the right column and a rectangular excitation signal in the left column. For

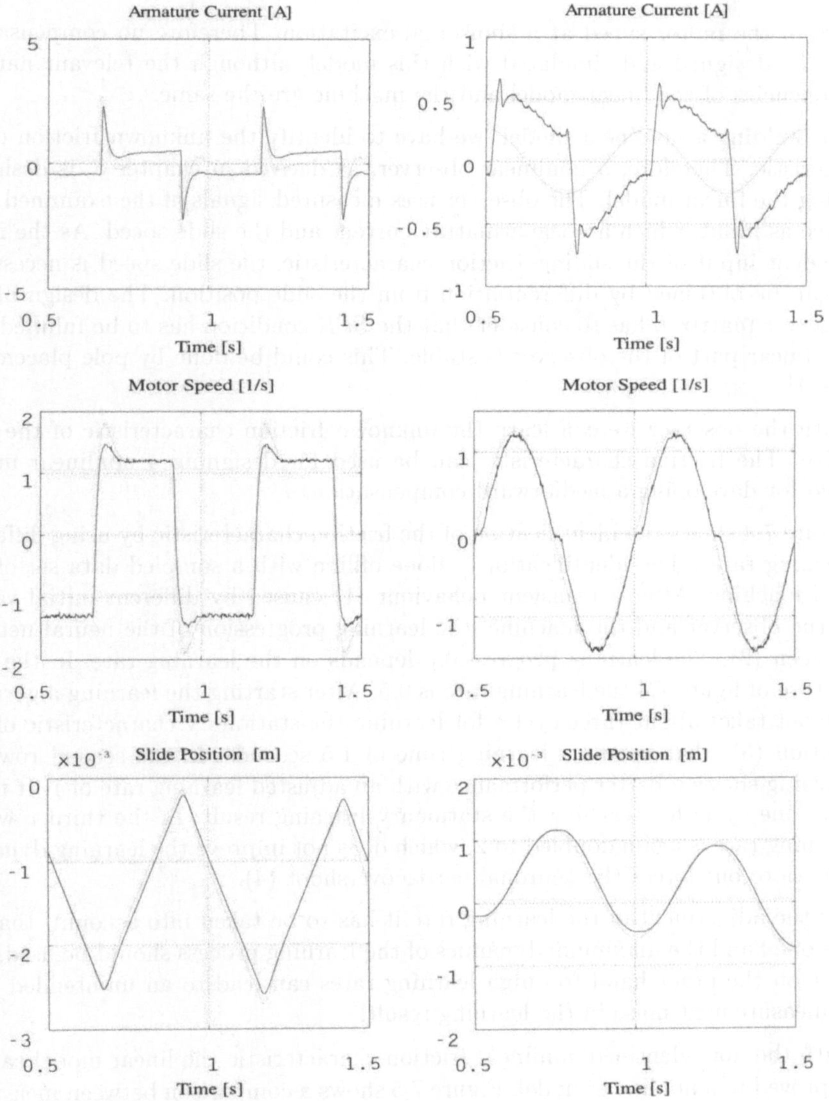

Figure 7.3: Simulated and measured data — linear model

the curves of the measured signals solid lines are used, the simulated data has dotted lines. It can easily be noticed, that significant errors remain between the simulated and the measured signals. The most important fact for compensation is the lack of the stick–slip–effect in the model, as can be seen at the change of sign of the motor speed at a sinusoidal excitation. Therefore no compensation can be designed and simulated with this model, although the relevant natural frequencies of the linear model and the machine are the same.

For building a nonlinear model, we have to identify the unknown friction characteristic. Therefore, a nonlinear observer, as derived in chapter 5, is designed using the linear model. The observer uses measured signals of the examined feed drive as input, which are the armature current and the slide speed. As the independent input of the sliding–friction characteristic, the slide speed is necessary, it can be obtained by differentiation from the slide position. The design of the observer matrix L has to consider that the *SPR*-condition has to be fulfilled and the linear part of the observer is stable. This could be done by pole placement, see [1].

With the observer we can learn the unknown friction characteristic of the feed drive. The friction characteristic can be used for designing a nonlinear model used for developing a feedforward compensation.

Figure 7.4 shows the identification of the friction characteristic by using different learning rates. The identification is done offline with a sampled data set of the tool machine. After a transient behaviour (1) caused by different initial values of the observer and the machine, the learning progression of the neural net can be seen (2). The learning progression depends on the learning rate. In the first picture of figure 7.4 the learning rate is 0.5. After starting the learning algorithm the net takes about three cycles for learning the stationary characteristic of the friction (3), that means a learning time of 1.5 seconds. In the second row the learning shows a better performance with an adjusted learning rate of 1. It takes only one cycle for reaching the stationary learning result. In the third row the learning rate is again doubled to 2, which does not improve the learning dynamic any more but forces the neuronal net to overshoot (4).

For the adjustment of the learning rate it has to be taken into account, that on the one hand the maximum dynamics of the learning process should be used, but that on the other hand too high learning rates can lead to an unintended gain of measurement noise in the learning result.

With the now identified nonlinear friction characteristic the linear model can be improved to a nonlinear model. Figure 7.5 shows a comparison between measured and simulated signals of the nonlinear model as in the example with the linear model before. It can be seen, that there are less errors between model and plant, and most important of all, the stick–slip–effect can be seen in the model. So this nonlinear model can be used for the design of a compensation strategy.

For compensation we want to use a feedforward structure. Theoretically, for a feedforward compensation two conditions have to be met. First, the disturbance value has to be known, and second, the inverse transfer function between the entering point of the disturbance and the set value has to be known. In reality the inverse transfer function is mostly known, but often it is characterized by

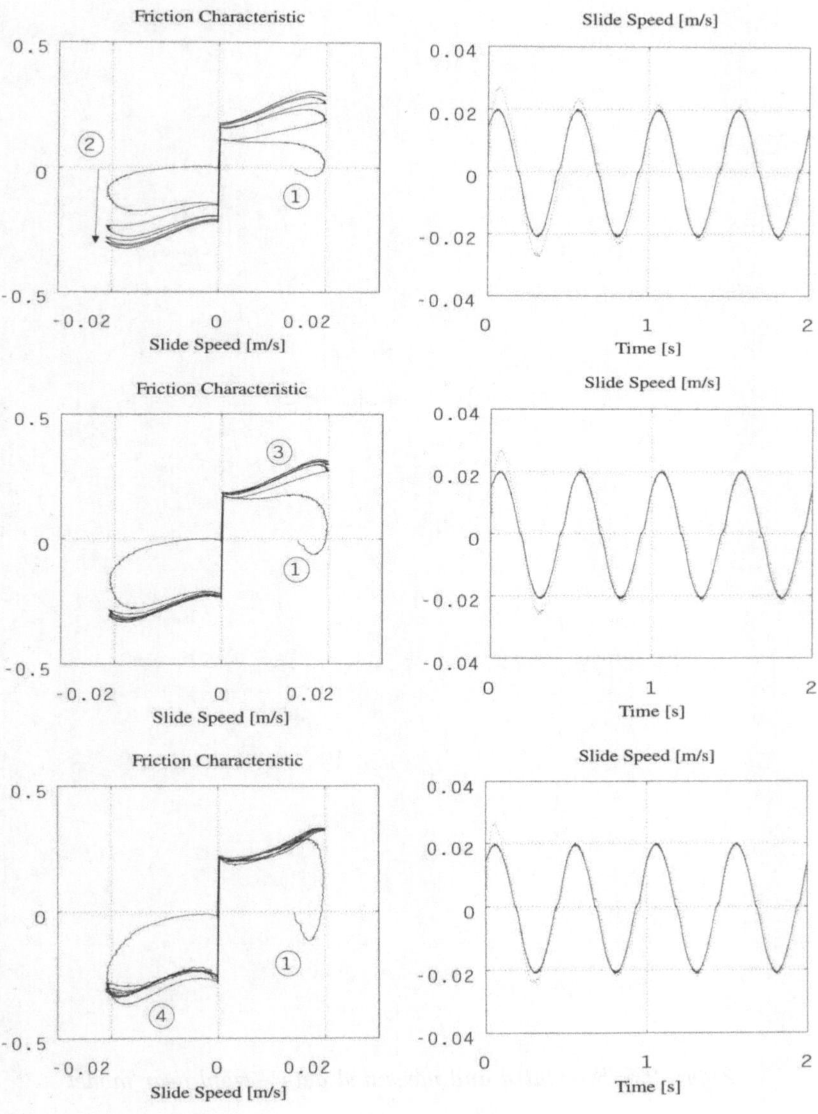

Figure 7.4: Identification of friction characteristic with different learning rates

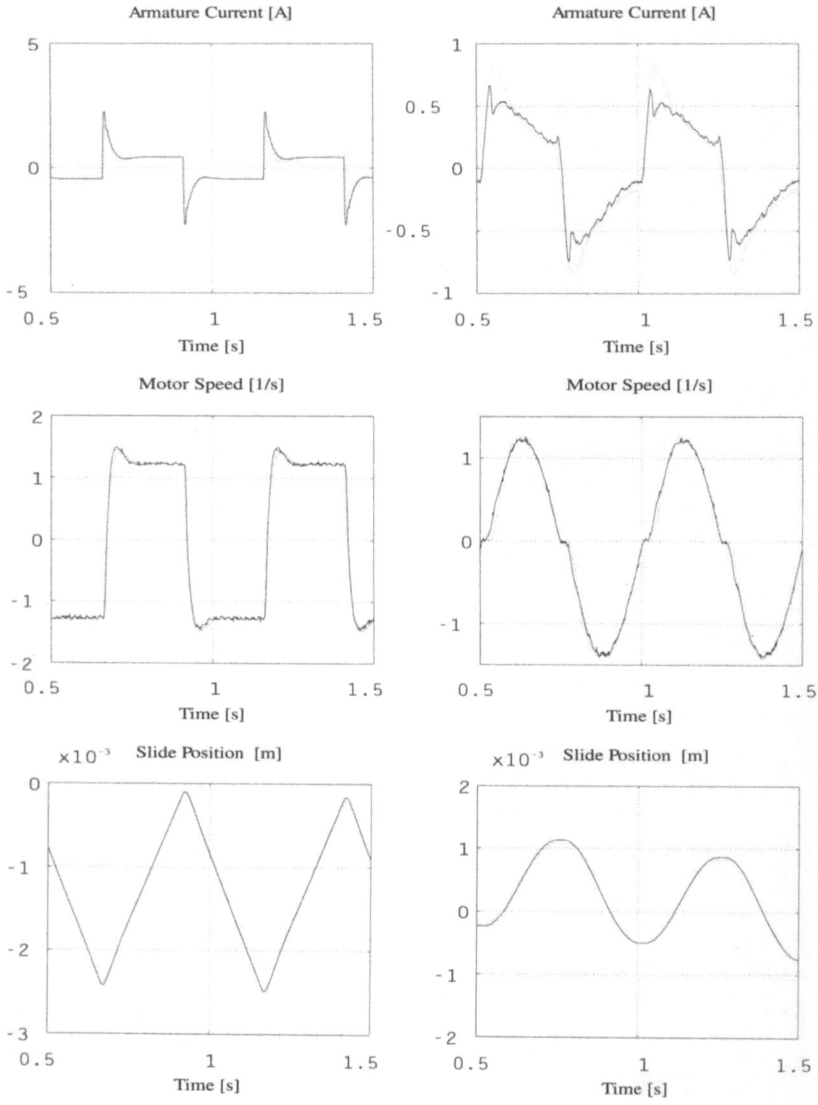

Figure 7.5: Simulated and measured data – nonlinear model

a multiple differential behaviour, which is technically not realizable. Also the disturbance is often unknown.

In this case, the disturbance can be recalculated by using the slide speed and a second neural network which uses the same set of parameters as the neural net in the observer. So the disturbance can be treated as known. The multiple differential behaviour is modeled by a PDT_1 element. Figure 7.6 shows the improved behaviour of the system for a such designed feedforward control. The signal flow chart of the feedforward controller is shown later. Now, this compensation is to be implemented on the application example for a validation of the simulated results.

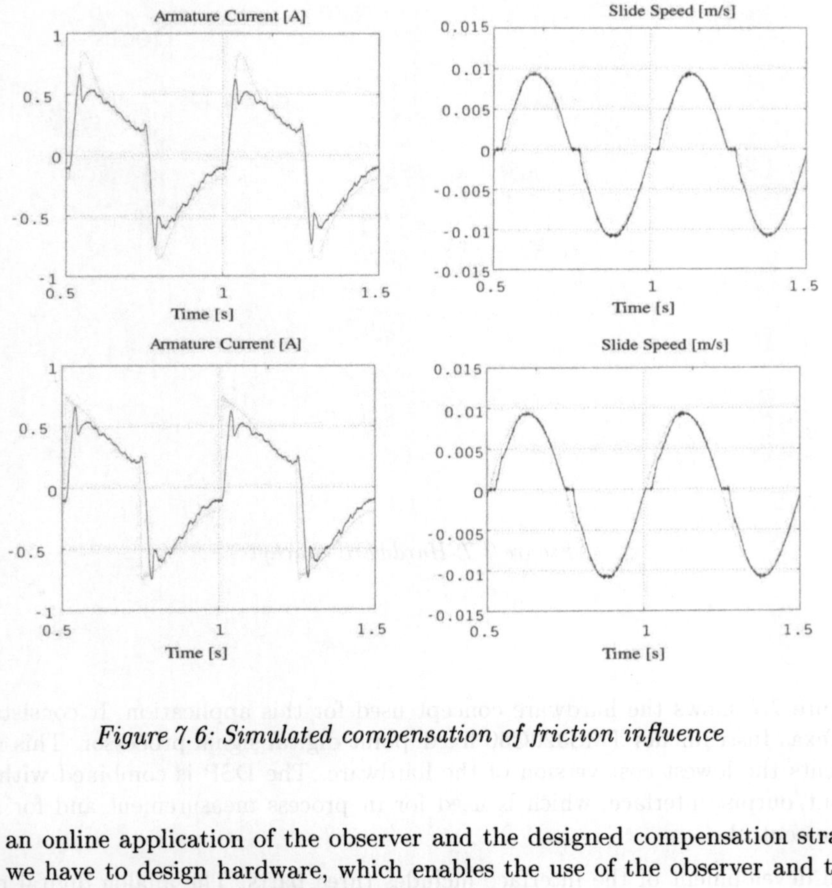

Figure 7.6: Simulated compensation of friction influence

For an online application of the observer and the designed compensation strategy we have to design hardware, which enables the use of the observer and the feedforward control with a small sampling time.

7.2 Design of Hardware

For a real time implementation of the Stable Local Neuroidentification, special hardware is necessary. The development of a DSP based system which fulfills the requirements of processing speed and input/output interface is discussed [6, 5].

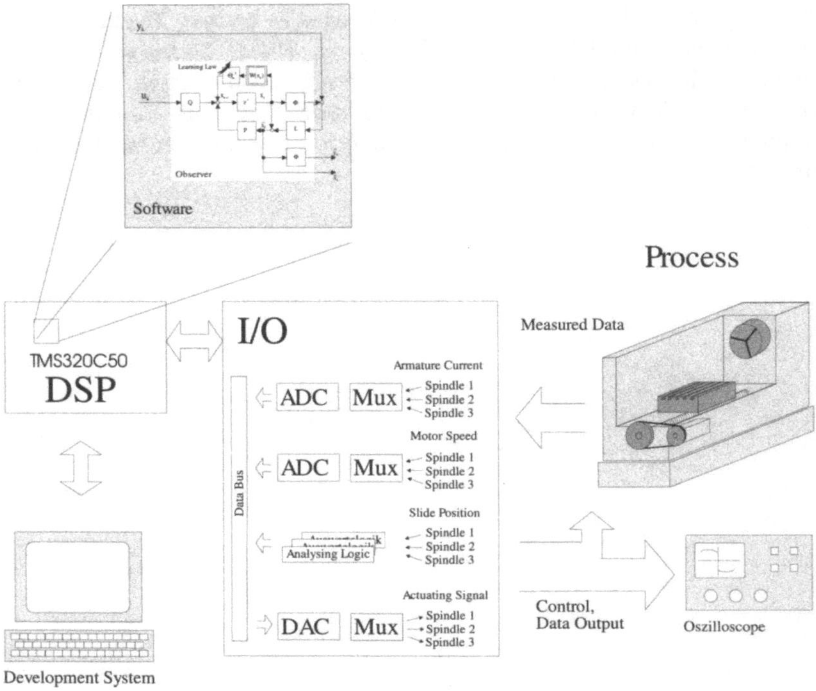

Figure 7.7: Hardware concept

Figure 7.7 shows the hardware concept used for this application. It consists of a Texas Instruments TMS320C50 fixed–point digital signal processor. This represents the lowest cost version of the hardware. The DSP is combined with an input/output interface, which is used for in–process measurement and for real time control.

The development of the interface includes three parts. The analog digital part is used to measure the analog signals of the examined machine tool. As analog digital converters three Analog Devices AD7876CN with 12 bit resolution and 10 μs sampling time are used with regard to a maximum of three examined axes. These ADCs, in combination with each a differential amplifier AMP03GP from Analog Devices and a four channel multiplexer ADG509 from Analog Devices enable the measurement of 12 analog input signals.

The digital analog part is used for the output of control signals and internal variables to be observed. Here the four channel DAC DAC4813 from Burr–Brown with 12 bit resolution and $10\mu s$ rise time has been chosen.

For catching the position and the speed signals a numerical counter has to be designed. In the case of machine tools the supplied position signals are often two 90° phase shifted rectangular signals. The examined feed drive has a resolution of $2\mu m$ for each axis. For the evaluation of this signal a one–step interpolation is used, which is not as exact as a two– or a four–step interpolation but very simple in hardware design. For each axis three 74F579 8 bit bidirectional counters are used.

This interface is connected with a Texas Instruments TMS320C50 Starter Kit, which presents a very cheap solution for embedding the digital signal processor in special hardware designs.

The software development ensues on a PC in the programming language C. Then a Texas Instruments C–Compiler generates an assembler-code, which can be downloaded via a RS232 interface to the DSP.

This hardware system is built up in order to control up to three spindles, but investigations have shown, that this is not possible in the desired sampling time. So for the use with more than one spindle a higher speed processor has to be used. A possible solution could be to fit the designed interface card to a TMS320C30, a floating point processor, which meanwhile is available also as a starter kit. In newer examinations the online identification is done with an Intel Pentium processor with 200 MHz, which leads to a further improvement of the time frame of the algorithm.

7.3 Implementation: Learning of Friction Characteristic

In this section the results of a validation investigation for the **Stable Local Neuroidentification** are shown. The figures are scanned plots of an oscilloscope and for a better presentation the marked areas are also presented zoomed.

Figure 7.8 shows the slide speed signal of the feed drive for a sinosoidal reference signal of the speed controller. As the slide speed could not be measured the signal shown in figure 7.8 had to be obtained by differentiation from the position signal.

The expected friction characteristic can be mainly divided into sliding friction, which depends on the velocity of the slide, and into static friction, which appears at zero velocity and when the velocity of the slide changes sign. The static friction is mainly responsible for problems in exact positioning and/or contouring control.

The influence of static friction is easy to detect in the marked areas. Here it causes the slide to stop until the PID speed controller compensates the friction influence and the slide breaks away. If the friction characteristic is known, it could be compensated much earlier.

Figure 7.9 shows the evaluated and the observed slide speed after learning the friction characteristic. The difference between both curves is very small, so it

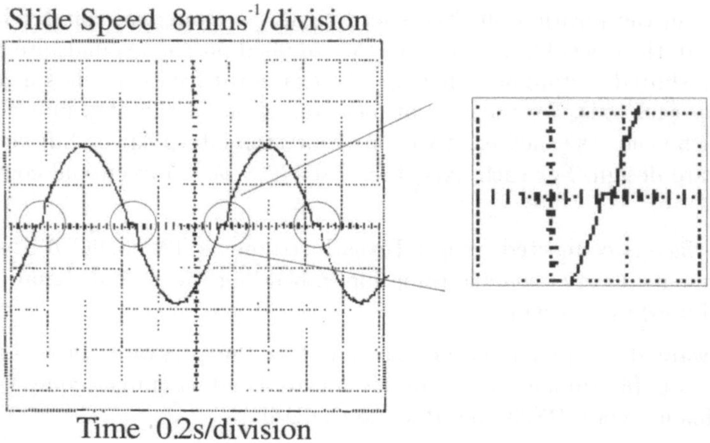

Figure 7.8: Evaluated slide speed

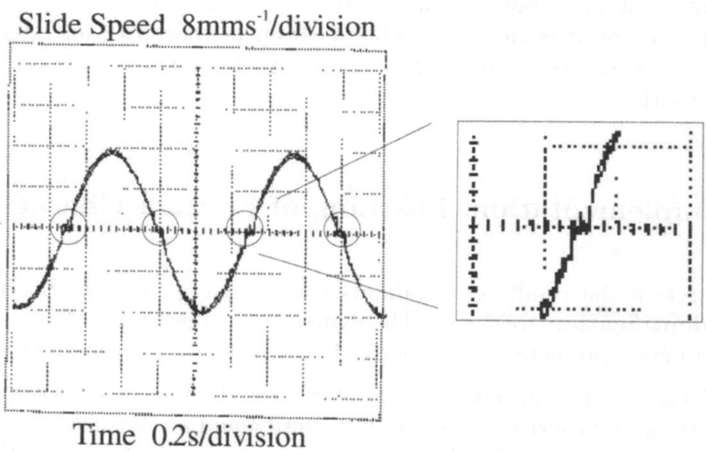

Figure 7.9: Evaluated and observed slide speed

cannot be seen clearly in this plot, compared to figure 7.10, in which the difference between both signals is plotted as an error signal. The learning is started at zero, and after ten seconds the error signal remains zero, so the friction characteristic has been learned. The speed of learning depends on the learning rate, which should be tuned so slowly, that stochastic disturbances do not influence the learning.

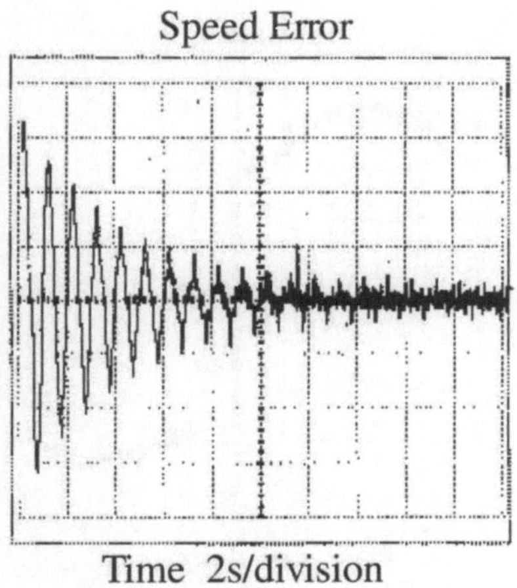

Figure 7.10: Error between evaluated and observed slide speed

The remaining error depends only on the inherent approximation error and on errors in the signal processing, because the slide speed evaluation needs more time than one sampling interval due to a very poor position signal. To decrease the error signal, a better position measuring device has to be installed and the neural network has to be enlarged. But here, this accuracy was accepted.

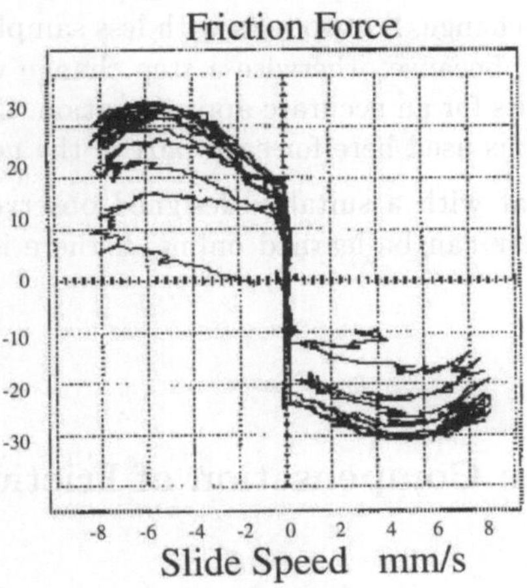

Figure 7.11: Learning process

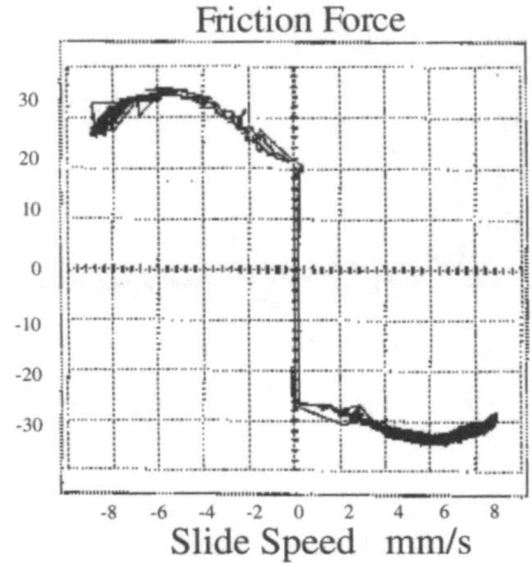

Figure 7.12: Learned friction characteristic

Figure 7.11 shows an example of the learning process. The friction force is shown as a function of the slide speed. For learning the sliding–friction characteristic it is not necessary to use any a-priori knowledge of the sliding–friction characteristic, but it is possible to incorporate a-priori knowledge. However, this is not necessary, because the learning is extremely fast and the possible advantage is very small.

Figure 7.12 shows the learned friction characteristic. Here a neural network is used, which is separated into two parts at zero speed, because it is known that there has to be a step change. So two nets with less sample points could be used for the approximation, because otherwise a step change would result in a large number of sample points for an accurate approximation. Over the effective range there are twenty neurons used here for each part of the net.

It has been shown that with a suitably designed observer the friction characteristic of the feed drive can be learned online, if there is a realtime operating capability.

7.4 Application: Compensation of Friction Influence

The learned friction characteristic is used to compensate the real friction. Therefore, the PDT_1 controller designed before is used for a feedforward control which applies a compensation signal to the PID speed controller. Figure 7.13 shows the signal flow chart of the observer and the feedforward compensation controller.

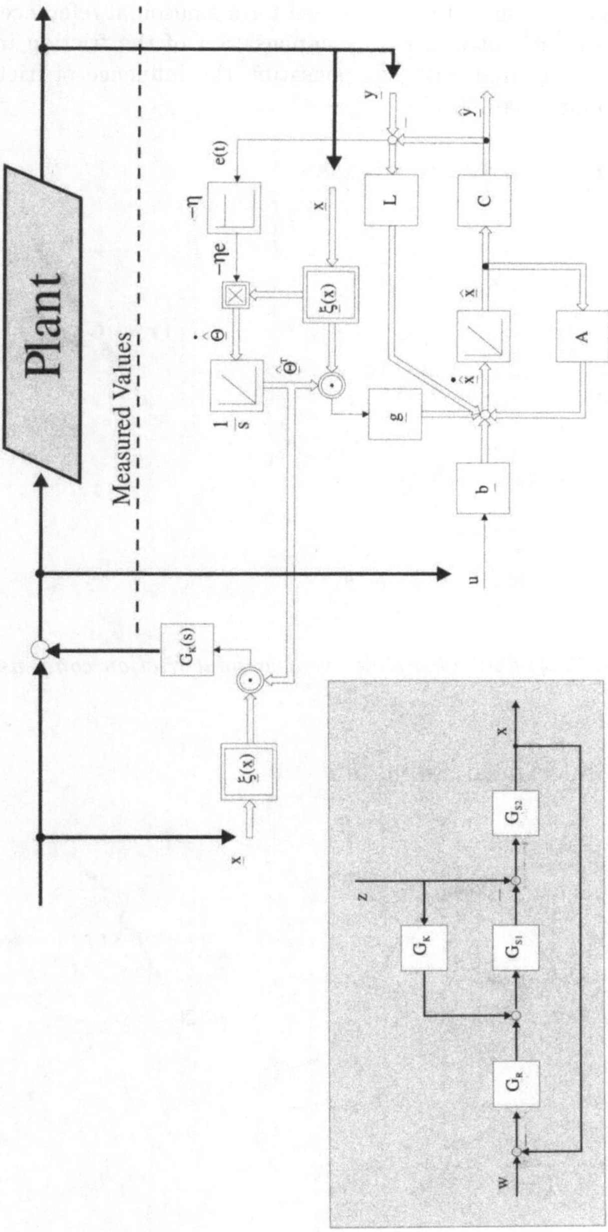

Figure 7.13: Signal flow chart of the feedforward compensation controller

Figure 7.14 and 7.15 show the slide speed for a sinusoidal reference signal of the speed controller without and with compensation of the friction influence. The marked areas show, that with compensation the influence of friction could be decreased by more than 90%.

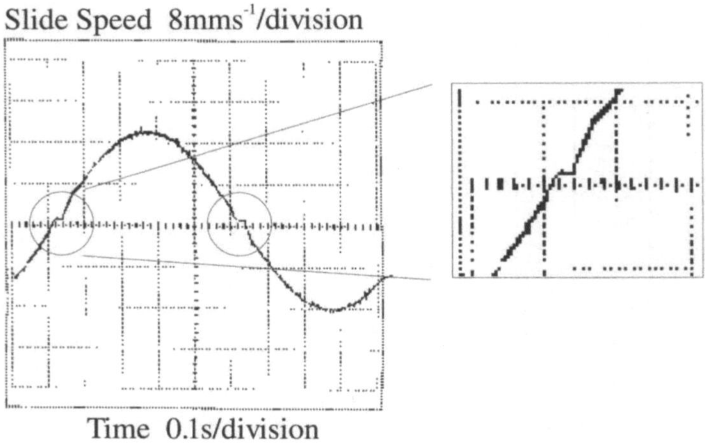

Figure 7.14: Evaluated slide speed without friction compensation

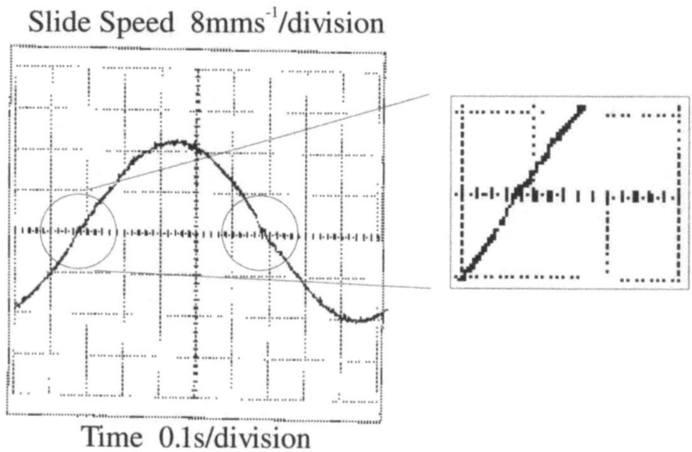

Figure 7.15: Evaluated slide speed with friction compensation

A better compensation could be achieved by reducing the sampling time, due to the learning and the compensation algorithm.

Figure 7.16 shows the applied compensation signal and the slide speed. Clearly the peak can be seen, which causes the motor to accelerate in order to break the slide away. In our example with the especially designed hardware based on a DSP the minimum sampling time was 3.5ms.

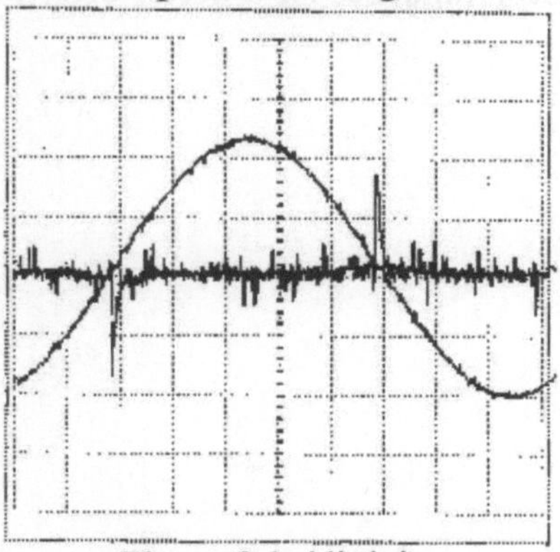

Figure 7.16: Signal for compensation and slide speed

In figure 7.15 it is shown that with the proposed method it is possible to compensate more than 90% of the nonlinear function shown in figure 7.12 without any previous knowledge. After a short period of time, the nonlinearity is learned and can be used for an exact compensation. Thus, in the simulated closed loop control the stick–slip–effect disappears. This is accomplished with a feedforward control with a PDT_1 controller.

7.5 Conclusion

The presented method shows in an application example the online learning of unknown nonlinearities without the exact knowledge of each system parameter. At the same time a compensation signal is generated for the compensation of the nonlinear function as far as it has been learned. With the progress in computing power today there is no longer need of special hardware. The task can be done with a standard PC and I/O cards for measurement.

The nonlinearity is learned very quickly, and most important of all, the learning is proven to be stable. The learned characteristics can be used for compensation, e. g. as shown for the friction of a feed drive in order to compensate for the stick–

slip–effect. Another possible application is the compensation of the eccentricity of re– and unwinder at rolling mills and other similar plants.

At the Institute of Electrical Drive Systems online examinations of systems of higher order, as there are e. g. radio telescopic aerials, have shown that with the method explained in chapter 5 static friction can be identified and used for the compensation of the friction influence for a better behaviour of the system.

7.6 References

[1] Frenz, Th.:
Stabile Neuronale Online Identifikation und Kompensation statischer Nichtlinearitäten am Beispiel von Werkzeugmaschinenvorschubantrieben.
Dissertation, TU München, 1998.

[2] Frenz, Th. and Schröder, D.:
Learning Unknown Nonlinearities Using a Discrete Observer in Combination with Neural Networks.
IEEE-IAS Annual Conference, Orlando, 1995, pp. 1800–1806.

[3] Frenz, Th. and Schröder, D.:
Nonlinear Modelling of a Feed Drive of a Lathe and Compensation of Friction Influence.
Proceedings of IFAC-Workshop Motion Control. München, 1995, pp. 331 – 338.

[4] Frenz, Th. und Schröder, D.:
Nichtlineare Modellbildung elektrischer Antriebsstränge.
VDI–Bericht 1220: Schwingungen in Antrieben. VDI–Verlag Düsseldorf, 1995, pp. 335 – 346.

[5] Frenz, Th. und Schröder, D.:
On Line Identification and Compensation of Friction Influence of Feed Drives of Machine Tools.
Proc. Seventh European Conference on Power Electronics and Applications (EPE), Trondheim, 1997

[6] Schönig, Ch.:
Implementierung eines nichtlinearen Beobachters mit DSP an einer Werkzeugmaschine
Diplomarbeit, Lehrstuhl für Elektrische Antriebstechnik, München, 1996.

8 Detection and Identification of Backlash

Dieter Strobl

8.1 Introduction

Backlash or gear play has been the subject of research to control engineers for many years. Its appearance is quite common in control systems with mechanical connections (e.g. gears). In contrast to other nonlinearities in electro–mechanical control systems, e.g. friction or eccentricities, backlash acts as a structure–switching nonlinearity. This means that during time periods when the two components of the backlash containing mechanical parts do not engage, the whole system is split into two decoupled subsystems. This causes special problems in the control of such systems. A model of backlash is shown in figure 8.1.

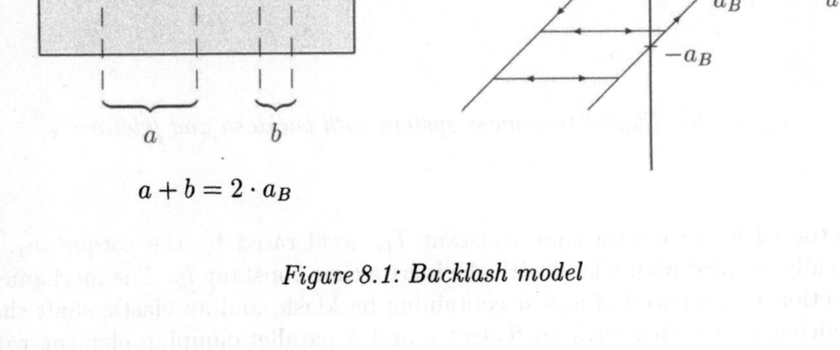

Figure 8.1: Backlash model

In this chapter we address the problem of identification of backlash. There are numerous contributions in literature that deal with the problem of controlling systems containing backlash. Tao and Kokotović [7, 8] propose an inverse backlash model to compensate the backlash nonlinearity. However, this concept assumes that backlash is located at either the input or the output of the system, and that the available control signal is able to provide both very high and stepwise amplitudes. In the control concepts in [1, 2, 4, 5] it must be known in advance, if a system contains backlash and if so, its magnitude.

Here we propose an observer structure which is able to determine if a system contains backlash, and, if it does, to learn the backlash width. The performance and stability of the approach can be proven mathematically. To identify the backlash characteristic, we use a neural network that is able to rebuild any continuous mathematical function.

8.2 Example System for Backlash Identification

8.2.1 Model of an Elastic Two–Mass System

We consider the following electro–mechanical drive train shown in figure 8.2.

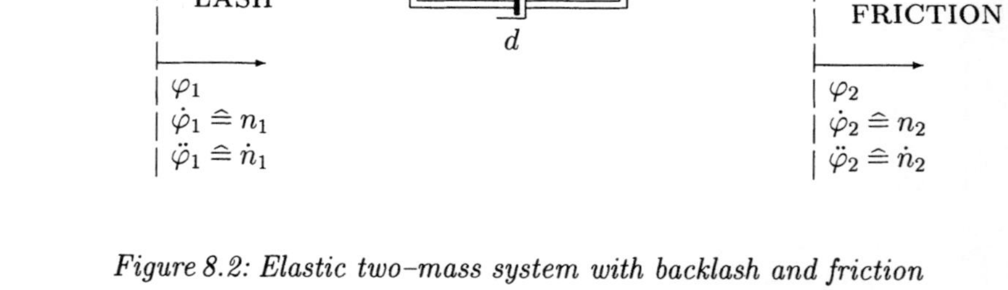

Figure 8.2: Elastic two–mass system with backlash and friction

A motor with the inertia time constant T_1, accelerated by the torque m_1, is elastically coupled with a load with the inertia time constant T_2. The mechanical connection is composed of a gear containing backlash, and an elastic shaft that is modeled as a spring with coefficient c and a parallel damping element with coefficient d. The motion of the load can be subdued to nonlinear friction. This elastic two–mass system is a common model for many mechanical systems (i.e. mechatronic systems).

The signal flow graph of the above system is shown in figure 8.3.

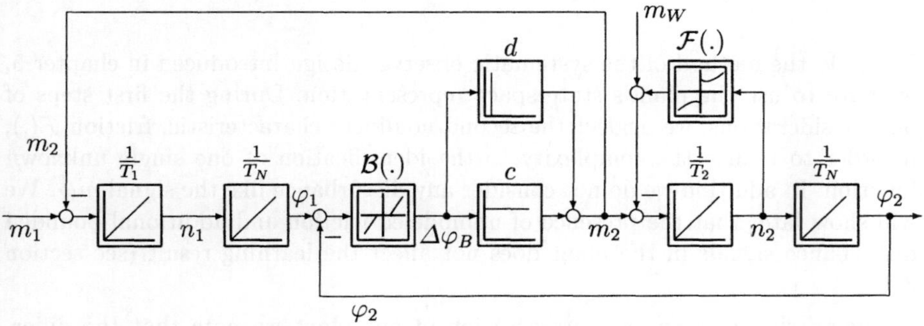

Figure 8.3: Normalized signal flow graph of the elastic two–mass system with backlash and friction

The blocks marked with $\mathcal{B}(.)$ and $\mathcal{F}(.)$ describe the nonlinearities of the plant, backlash and friction, respectively. The parameter T_N is the normalizing time constant, m_W is an additional disturbance torque. The signal flow graph is valid if we assume that the motor is so inert that variations in the position φ_1 occur slower than the twisting speed of the shaft. This means that the shaft does not transfer torque before the backlash comes into engagement.

Without restrictions in the following derivations we can set the gear transmission ratio $i = 1$.

8.2.2 Identifiability of the Backlash Characteristic

Using this common two–mass system description, it must be noted that the complicated mathematical characteristic of backlash, described in figure 8.1, now simplifies to a deadzone characteristic. This is because the two components of the part containing backlash only get into the state of disengagement, if the shaft is not twisted, i.e. if $|\varphi_1 - \varphi_2| < a_B$.

This feature is very important with respect to the identification with a neural network. We only use the neural network as part of our observer structure to learn an unknown static nonlinear function. Considering the characteristic in figure 8.1 we detect that this characteristic is contradictory to the mathematical definition of a function. Thus, such a backlash nonlinearity cannot be approximated by a neural network directly. However, in an elastic two–mass system of figure 8.3, backlash only appears as a deadzone which in fact is a mathematical function and thus can be learned by a neural network.

8.2.3 State Space Description of the Nonlinear System

To apply the method of the systematic observer design introduced in chapter 5, we have to use the plant's state space representation. During the first steps of our considerations, we neglect the second nonlinear characteristic, friction $\mathcal{F}(.)$, in order to reduce the complexity to the identification of one single unknown function. In addition we do not consider any disturbance like the signal m_W. We will show later that the presence of unmodeled friction and additional bounded disturbance signals in the plant does not affect the learning result (see section 8.4).

Deducing the state space representation of our plant we note that the differentiating block corresponding to the damping coefficient d produces a second nonlinear function: the time derivative of the backlash function $\dot{\mathcal{B}}$. In order to get a representation with only one unknown nonlinearity, we have to transform the signal flow graph into a representation without differentiation. This results in figure 8.4.

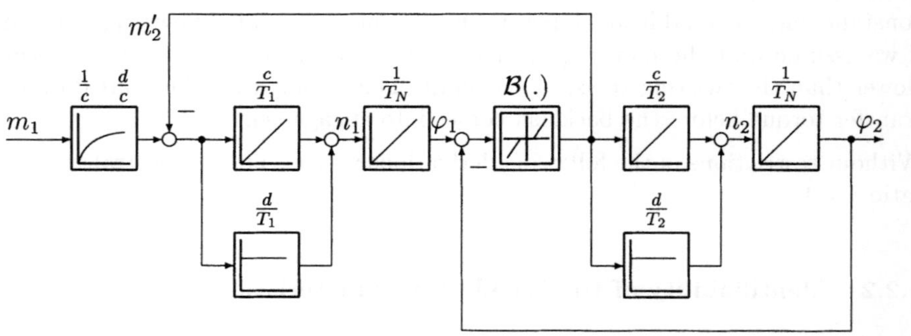

Figure 8.4: Transformed signal flow graph of the elastic two–mass system without differentiation

To finally get the state space equations of our system, we use the output signals of the integrators as state variables. The input signal is the filtered motor torque m_1:

$$u = \frac{\frac{1}{c}}{1 + s\frac{d}{c}} \cdot m_1 \tag{8.1}$$

We assume that both motor position φ_1 and load position φ_2 are available for measurement, so we have two sensed output signals:

$$y_1 = \varphi_1 \quad , \quad y_2 = \varphi_2 \tag{8.2}$$

With these definitions the following state space representation of the transformed nonlinear elastic two–mass system is achieved:

$$\begin{bmatrix} \dot{x}_1 \\ \dot{x}_2 \\ \dot{x}_3 \\ \dot{x}_4 \end{bmatrix} = \begin{bmatrix} 0 & 0 & 0 & 0 \\ \frac{1}{T_N} & 0 & 0 & 0 \\ 0 & 0 & 0 & 0 \\ 0 & 0 & \frac{1}{T_N} & 0 \end{bmatrix} \cdot \begin{bmatrix} x_1 \\ x_2 \\ x_3 \\ x_4 \end{bmatrix} + \begin{bmatrix} \frac{c}{T_1} \\ \frac{d}{T_1 T_N} \\ 0 \\ 0 \end{bmatrix} \cdot u + \begin{bmatrix} -\frac{c}{T_1} \\ -\frac{d}{T_1 T_N} \\ \frac{c}{T_2} \\ \frac{d}{T_2 T_N} \end{bmatrix} \cdot \mathcal{B}(\underline{x}) \tag{8.3}$$

$$\underline{\dot{x}} = A \cdot \underline{x} + \underline{b} \cdot u + \underline{e}_{NL} \cdot \mathcal{NL}$$

$$\begin{bmatrix} y_1 \\ y_2 \end{bmatrix} = \begin{bmatrix} 0 & 1 & 0 & 0 \\ 0 & 0 & 0 & 1 \end{bmatrix} \cdot \begin{bmatrix} x_1 \\ x_2 \\ x_3 \\ x_4 \end{bmatrix} \tag{8.4}$$

$$\underline{y} = C \cdot \underline{x}$$

with

$$\mathcal{B}(\underline{x}) = \mathcal{B}(\varphi_1 - \varphi_2) = \begin{cases} \varphi_1 - \varphi_2 + a_B & , \quad \varphi_1 - \varphi_2 \le -a_B \\ 0 & , \quad |\varphi_1 - \varphi_2| < a_B \\ \varphi_1 - \varphi_2 - a_B & , \quad \varphi_1 - \varphi_2 \ge a_B \end{cases} \tag{8.5}$$

Looking at these system equations with the system matrix A, we notice that the necessary assumption of observability for an observer design is not given here. The whole system is not observable in the presence of only one error signal. This always has to be considered, if the plant under investigation contains a nonlinearity in the forward path of the signal flow graph. As a consequence, the state variables left of the nonlinearity are not observable with an output signal lying at the right of it, and vice versa.

To overcome the impossibility of designing an observer for the whole system, in section 8.3 we therefore propose two different observers for the two sides left and right of the backlash block, which both lead to the stable neural identification of unknown backlash.

8.3 Identification of Backlash

In the previous section we derived that the systematic observer design in chapter 5 is not applicable to the identification of backlash, if we want to design an observer for the whole system. To overcome these difficulties we will reduce the observer design to subsystems. The assumptions in choosing a suitable subsystem are:

- known subsystem input signal,
- known activation of the nonlinearity,
- at least one measurable output signal.

The signal flow graph in figure 8.4 shows two possible subsystem observer structures:

- load–side observer with input signal φ_1 and output signal φ_2,
- motor–side observer with input signal m_1 and output signal φ_1.

Choosing these options we get two different concepts to identify backlash in an elastic two–mass system:

1. Load–Side Backlash Observer: LBO
2. Motor–Side Backlash Observer: MBO

The basic idea of these two methods is to transform again the representation of the plant, in order to meet the assumptions of the systematic observer design of chapter 5. We will make use of the condition that both motor and load position are available for measurement. But both concepts are applicable as well if motor and load speed would be measurable. We will describe the two observer structures in detail in sections 8.3.2 and 8.3.3.

8.3.1 Representation of Backlash for the Identification with a Neural Network

As already expressed in the headline of this chapter, our goal is to decide whether a given plant contains backlash, and if so, to determine its magnitude. We want the neural network to learn the backlash characteristic only if it is present in the plant.

However, if we consider the deadzone characteristic of backlash (see figure 8.5 a), and if we have in mind that in real drive trains the elasticity of the shaft is so small that the possible twisting angle is very small as well, we can state that the input signal to the backlash block $\Delta\varphi$ will nearly exclusively move between $-a_B$ and a_B. This means that the neural network can also only learn the backlash characteristic between $-a_B$ and a_B, which unfortunately has an output value of $\Delta\varphi_B = 0$ in this area.

On the other hand, if the plant contains no backlash, the corresponding block or characteristic would be described by $\Delta\varphi_B = \Delta\varphi$, which is unequal to zero in the region of interest.

As a consequence, the neural network would produce a function unequal to zero if there is no nonlinearity to learn, and it would produce a zero line if a backlash nonlinearity is present in the plant. To achieve that, learning only takes place if there is a nonlinearity to reproduce, we transform the backlash description from a model with deadzone into a model with a limiting function, as shown in figure 8.5 b).

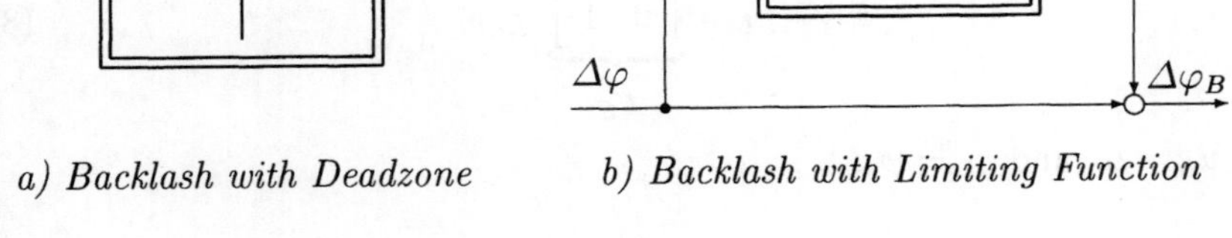

a) Backlash with Deadzone *b) Backlash with Limiting Function*

Figure 8.5: Backlash model in elastic two–mass systems

Therefore, we assume that there is no backlash, and learning will take place only if this assumption is wrong.

8.3.2 Load–Side Backlash Observer (LBO)

8.3.2.1 State Space Representation

In this section we will describe the design of a stable neural observer for backlash detection and identification based on the load side of the considered two–mass system. This subsystem is again shown in figure 8.6.

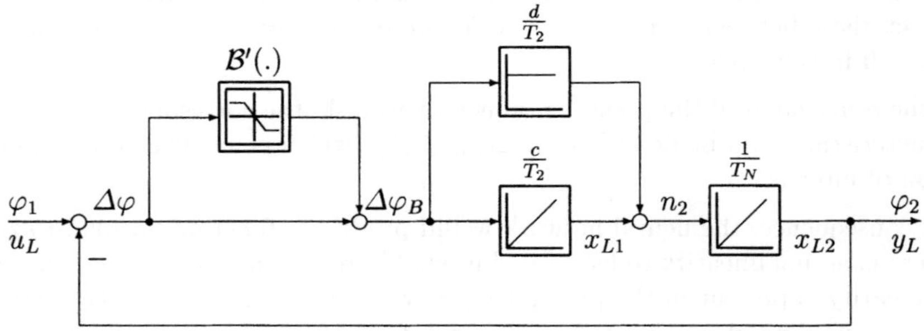

Figure 8.6: Load–side subsystem

The state space description of this subsystem is given by (8.6) and (8.7)

$$\underline{\dot{x}}_L = \underbrace{\begin{bmatrix} 0 & -\dfrac{c}{T_2} \\ \dfrac{1}{T_N} & -\dfrac{d}{T_2 T_N} \end{bmatrix}}_{A_L} \cdot \underline{x}_L + \underbrace{\begin{bmatrix} \dfrac{c}{T_2} \\ \dfrac{d}{T_2 T_N} \end{bmatrix}}_{\underline{b}_L} \cdot u_L + \underbrace{\begin{bmatrix} \dfrac{c}{T_2} \\ \dfrac{d}{T_2 T_N} \end{bmatrix}}_{\underline{e}_{B'L}} \cdot \mathcal{B}'(u_L - x_{L2}) \tag{8.6}$$

$$y_L = \underbrace{\begin{bmatrix} 0 & 1 \end{bmatrix}}_{\underline{c}_L^T} \cdot \underline{x}_L \tag{8.7}$$

with $u_L = \varphi_1$ and $y_L = \varphi_2$.

8.3.2.2 Observer Design and Error Model

For this system we design a Luenberger observer:

$$\dot{\hat{\underline{x}}}_L = A_L \cdot \hat{\underline{x}}_L + \underline{b}_L \cdot u_L + \underline{e}_{B'L} \cdot \hat{\mathcal{B}}' - \underline{l}_L \cdot e_L \tag{8.8}$$

$$\hat{y}_L = \underline{c}_L^T \cdot \hat{\underline{x}}_L \tag{8.9}$$

$$e_L = \hat{y}_L - y_L \tag{8.10}$$

with the observer vector $\underline{l}_L = \begin{bmatrix} l_{L1} & l_{L2} \end{bmatrix}^T$.

The estimation of the nonlinearity $\hat{\mathcal{B}}'$ is represented by the output signal of a neural network with trainable parameters $\underline{\hat{\Theta}}$:

$$\hat{\mathcal{B}}'(u_L - x_{L2}) = \underline{\hat{\Theta}}^T \cdot \underline{\mathcal{A}}(u_L - x_{L2}) \tag{8.11}$$

The observer matrix A_{OL} is given by

$$A_{OL} = A_L - \underline{l}_L \cdot \underline{c}_L^T = \begin{bmatrix} 0 & -\frac{c}{T_2} - l_{L1} \\ \frac{1}{T_N} & -\frac{d}{T_2 T_N} - l_{L2} \end{bmatrix} \tag{8.12}$$

The observer coefficients l_{L1} and l_{L2} have to be determined to build a stable and fast observer behaviour. The characteristic polynomial of matrix A_{OL} is

$$s^2 + s \cdot (\frac{d}{T_2 T_N} + l_{L2}) + \frac{1}{T_N}(\frac{c}{T_2} + l_{L1}) \overset{!}{=} s^2 + s \cdot 2d_0\omega_0 + \omega_0^2 \tag{8.13}$$

with d_0 = damping coefficient and ω_0 = natural frequency of the intrinsic observer behaviour. From the comparison of the polynomial coefficients we get:

$$\Longrightarrow \qquad l_{L1} = T_N \omega_0^2 - \frac{c}{T_2} \qquad , \qquad l_{L2} = 2d_0\omega_0 - \frac{d}{T_2 T_N} \tag{8.14}$$

This observer design leads to an error model with the following error equation:

$$e_L = H_L(s) \cdot (\hat{\mathcal{B}}' - \mathcal{B}') = H_L(s) \cdot \underline{\Phi}^T \cdot \underline{\mathcal{A}}(u_L - x_{L2}) \tag{8.15}$$

with the parameter error vector

$$\underline{\Phi} = \underline{\hat{\Theta}} - \underline{\Theta} \tag{8.16}$$

where $\underline{\Theta}$ is the real unknown parameter vector.

The error transfer function $H_L(s)$, which describes the behaviour of the subsystem between the output of the nonlinearity and the composition of the error signal, is calculated according to (8.17):

$$\begin{aligned} H_L(s) &= \underline{c}_L^T \cdot [sE - A_{OL}]^{-1} \cdot \underline{e}_{B'L} \\ &= \frac{s \cdot \frac{d}{T_2 T_N} + \frac{c}{T_2 T_N}}{s^2 + s \cdot (\frac{d}{T_2 T_N} + l_{L2}) + \frac{1}{T_N}(\frac{c}{T_2} + l_{L1})} \end{aligned} \tag{8.17}$$

As mentioned in chapter 5, there are various error models with corresponding adaptive laws. The adaptive law becomes quite simple if we can achieve that the error transfer function $H_L(s)$ is SPR (*strictly positive real*). If we remember that a PT_1–function always fulfills this condition, we can try to compensate the zero of the numerator polynomial with one zero of the denominator polynomial.

The zero of the numerator is given by

$$s_0 = -\frac{c}{d} \tag{8.18}$$

The zeroes of the denominator are

$$s_\infty = -\omega_0 \cdot (d_0 \pm \sqrt{d_0^2 - 1}) \tag{8.19}$$

In order to get real poles we set the damping rate $d_0 = 1$, which leads us to the following condition for the observer natural frequency:

$$\omega_0 = \frac{c}{d} \tag{8.20}$$

We finally get the SPR error transfer function

$$H_L(s) = \frac{\frac{d}{T_2 T_N}}{s + \omega_0} = \frac{\frac{d}{T_2 T_N}}{s + \frac{c}{d}} \tag{8.21}$$

As described in chapter 5, the following simple adaptive law corresponds to the error model with SPR error transfer function:

$$\underline{\dot{\Phi}} = \underline{\dot{\hat{\Theta}}} = -\eta \cdot e_L \cdot \underline{\mathcal{A}} \tag{8.22}$$

with the positive learning factor $\eta > 0$.

However, it must be taken into account that the chosen natural observer frequency $\omega_0 = \frac{c}{d}$ can reach very high values, because spring coefficient c can rise up to some powers of ten and the damping coefficient usually is very small (some percent). In time discrete realizations of the observer this can require a very small sampling time.

If it is not possible to achieve $H_L(s)$ as a SPR function, we must apply the following augmented error adaptation law (8.23) and (8.24), which corresponds to the *delayed activation method*, see chapter 5 for details.

$$\underline{\dot{\Phi}} = \underline{\dot{\hat{\Theta}}} = -\eta \cdot \epsilon_L \cdot H_L(s) \cdot \underline{\mathcal{A}} \tag{8.23}$$

$$\epsilon_L = e_L + \underline{\hat{\Theta}}^T \cdot H_L(s) \cdot \underline{\mathcal{A}} - H_L(s) \cdot \underline{\hat{\Theta}}^T \cdot \underline{\mathcal{A}} \tag{8.24}$$

With the stability theory of Ljapunov it can be proven [3] that for such an error model the error asymptotically tends to zero

$$\lim_{t \to \infty} e_L(t) = 0 \tag{8.25}$$

and, with the characteristics of the GRNN and persistent excitation with $\underline{A}(\underline{x}, u)$, that the parameters of the neural network converge to their real values:

$$\lim_{t\to\infty} \underline{\hat{\Theta}}(t) = \underline{\Theta} \tag{8.26}$$

The complete identification model with load–side observer and SPR error transfer function $H_L(s)$ is shown in figure 8.7.

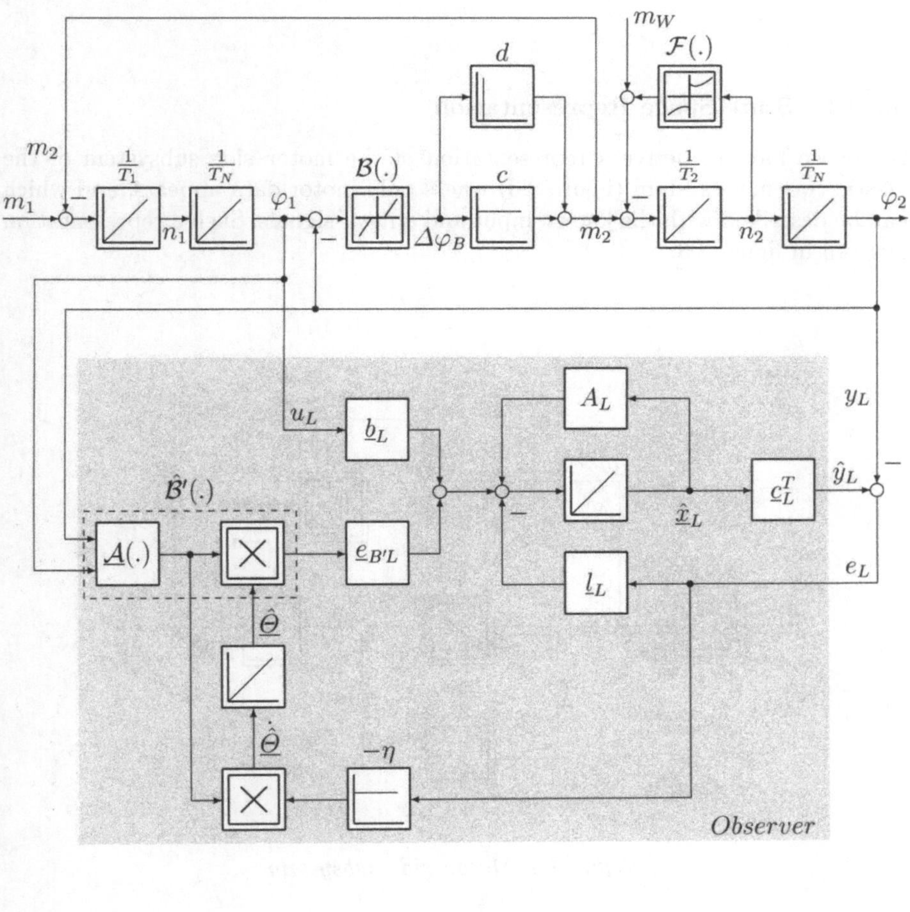

Figure 8.7: Complete signal flow graph of backlash identification with load–side observer

8.3.3 Motor–Side Backlash Observer (MBO)

With the backlash identification method LBO of section 8.3.2 a powerful tool is available for the detection of backlash in drive trains described as elastic two–mass systems. However, we must know in advance all data about the load side of the plant (inertia etc.). But in many cases it is much more likely that only the motor data is available (motor data sheet). In this case a similar method to detect and identify backlash would be helpful. This motor–side observer will be described in this section.

8.3.3.1 State Space Representation

Again we have to derive a representation of the motor–side subsystem of the elastic two–mass system (figure 8.3) where only motor data appears, and which can be described with the known input and output signals. Such a representation is given in figure 8.8.

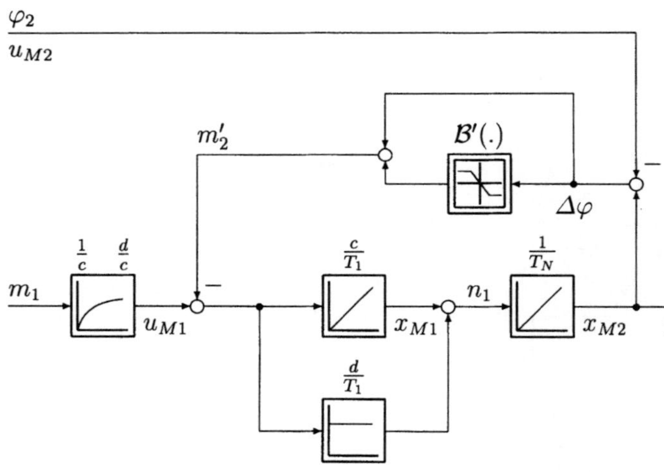

Figure 8.8: Motor–side subsystem

In contrast to the LBO concept, we have two input signals, $u_{M1} = \frac{\frac{1}{c}}{1+\frac{d}{c}s} \cdot m_1$ and $u_{M2} = \varphi_2$, and one output signal $y_M = \varphi_1$. The complete state space equations of this system are given in (8.27) and (8.28):

$$\dot{\underline{x}}_M = \underbrace{\begin{bmatrix} 0 & -\frac{c}{T_1} \\ \frac{1}{T_N} & -\frac{d}{T_1 T_N} \end{bmatrix}}_{A_M} \cdot \underline{x}_M + \underbrace{\begin{bmatrix} \frac{c}{T_1} & \frac{c}{T_1} \\ \frac{d}{T_1 T_N} & \frac{d}{T_1 T_N} \end{bmatrix}}_{B_M} \cdot \underline{u}_M + \underbrace{\begin{bmatrix} -\frac{c}{T_1} \\ -\frac{d}{T_1 T_N} \end{bmatrix}}_{\underline{e}_{B'M}} \cdot \mathcal{B}'(x_{M2} - u_{M2}) \quad (8.27)$$

$$y_M = \underbrace{\begin{bmatrix} 0 & 1 \end{bmatrix}}_{\underline{c}_M^T} \cdot \underline{x}_M \quad (8.28)$$

8.3.3.2 Observer Design and Error Model

For the motor–side subsystem (8.27) and (8.28) we again design a Luenberger observer:

$$\dot{\hat{\underline{x}}}_M = A_M \cdot \hat{\underline{x}}_M + B_M \cdot \underline{u}_M + \underline{e}_{B'M} \cdot \hat{\mathcal{B}}' - \underline{l}_M \cdot e_M \quad (8.29)$$

$$\hat{y}_M = \underline{c}_M^T \cdot \hat{\underline{x}}_M \quad (8.30)$$

$$e_M = \hat{y}_M - y_M \quad (8.31)$$

with the observer vector $\underline{l}_M = \begin{bmatrix} l_{M1} & l_{M2} \end{bmatrix}^T$.

We approximate the unknown nonlinearity with a trainable neural network:

$$\hat{\mathcal{B}}'(x_{M2} - u_{M2}) = \hat{\underline{\Theta}}^T \cdot \underline{\mathcal{A}}(x_{M2} - u_{M2}) \quad (8.32)$$

To determine the natural behaviour of the observer we have to calculate the observer matrix A_{OM}:

$$A_{OM} = A_M - \underline{l}_M \cdot \underline{c}_M^T = \begin{bmatrix} 0 & -\dfrac{c}{T_1} - l_{M1} \\ \dfrac{1}{T_N} & -\dfrac{1}{T_N} - l_{M2} \end{bmatrix} \quad (8.33)$$

From A_{OM} results the characteristic polynomial and the observer coefficients l_{M1} and l_{M2} can be determined:

$$s^2 + s \cdot \left(\frac{d}{T_1 T_N} + l_{M2} \right) + \frac{1}{T_N} \cdot \left(\frac{c}{T_1} + l_{M1} \right) \quad = \quad s^2 + s \cdot 2 d_0 \omega_0 + \omega_0^2 \quad (8.34)$$

$$\Longrightarrow \quad l_{M1} = T_N \omega_0^2 - \frac{c}{T_1} \quad , \quad l_{M2} = 2 D \omega_0 - \frac{d}{T_1 T_N} \quad (8.35)$$

Consequently, this design leads to the error model with the error equation already known from the LBO concept:

$$e_M = H_M(s) \cdot (\hat{\mathcal{B}}' - \mathcal{B}') = H_M(s) \cdot \underline{\Phi}^T \cdot \underline{\mathcal{A}}(x_{M2} - u_{M2}) \quad (8.36)$$

and the parameter error vector

$$\underline{\Phi} = \hat{\underline{\Theta}} - \underline{\Theta} \tag{8.37}$$

The error transfer function $H_M(s)$ in this case results in

$$\begin{aligned} H_M(s) &= \underline{c}_M^T \cdot (sE - A_{OM})^{-1} \cdot \underline{e}_{B'M} \\ &= \frac{-\frac{d}{T_1 T_N} \cdot (s + \frac{c}{d})}{s^2 + s \cdot (\frac{d}{T_1 T_N} + l_{M2}) + \frac{1}{T_N} \cdot (\frac{c}{T_1} + l_{M1})} \end{aligned} \tag{8.38}$$

Again it would be theoretically possible to transform $H_M(s)$ into a strictly positive real transfer function by choosing ω_0. However, for the reasons given in section 8.3.2.2 we abandon this possibility. This leads to the learning law given in equations (8.39) and (8.40) to form a globally stable error model:

$$\dot{\underline{\Phi}} = \dot{\hat{\underline{\Theta}}} = -\eta \cdot \epsilon_M \cdot H_M(s) \cdot \underline{A} \tag{8.39}$$

$$\epsilon_M = e_M + \hat{\underline{\Theta}}^T \cdot H_M(s) \cdot \underline{A} - H_M(s) \cdot \hat{\underline{\Theta}}^T \cdot \underline{A} \tag{8.40}$$

with $\eta > 0$.

The complete identification model with motor–side observer is again shown in figure 8.9.

8.4 Simulation Examples

After the description of the theoretical aspects of the two methods for the identification of backlash in the previous section, we now want to show the performance of our concepts with computer simulations.

In order to get a good identification result, we have to design the neural network appropriately. This means that the distance between two trainable weights must be so small that at least three neurons or weights lie in the area between $-\varphi_B$ and φ_B, which however are unknown. In most situations the network will be designed too large to avoid an input space with poorly distributed weights. After a first identification run the network can be adjusted properly.

The first simulation plot in figure 8.10 shows the learning result of a backlash characteristic with the load–side observer concept (LBO) in a quite stiff system, where the input signal $\Delta\varphi$ almost always stays between $-\varphi_B$ and φ_B. Outside of this area no learning takes place.

We can determine very clearly that the backlash width in the plant has the value of $2a_B = 0.02$.

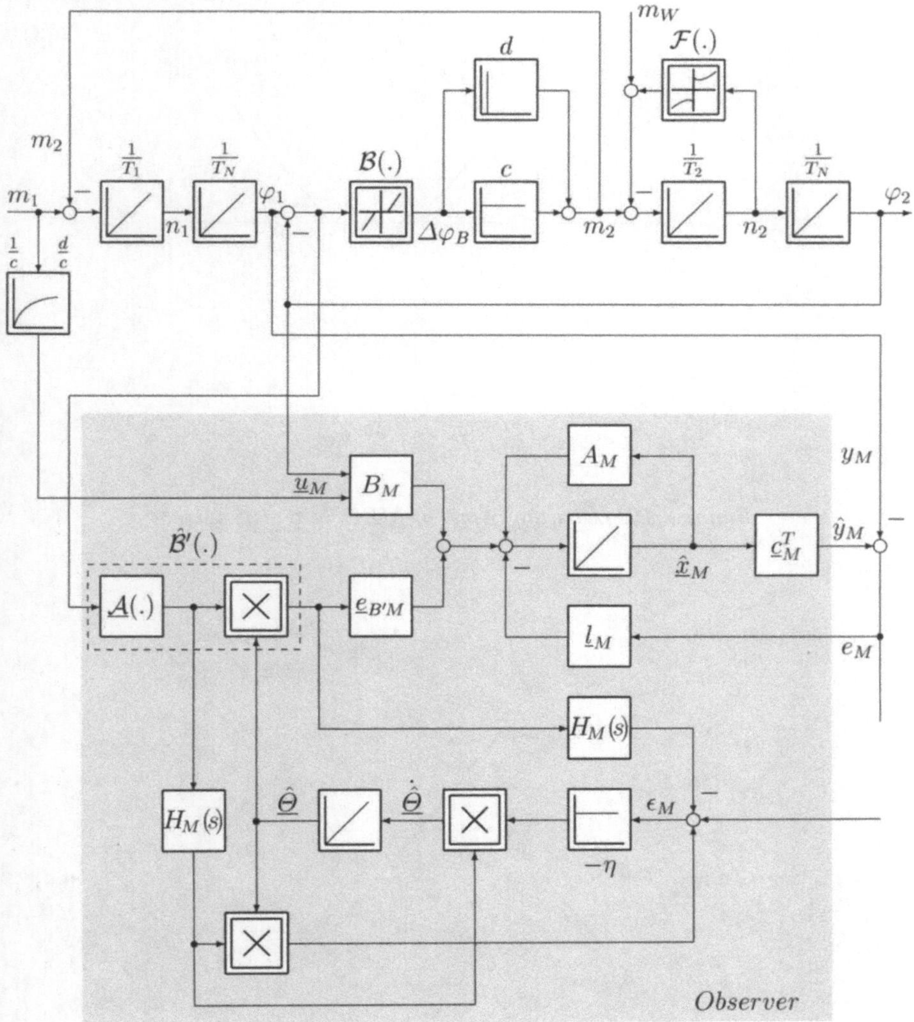

Figure 8.9: Complete signal flow graph of backlash identification with motor–side observer

The second simulation shown in figure 8.11 describes the results with the motor–side observer in a very flexible plant. Here $\Delta\varphi$ is much larger than in a stiff plant. Therefore, the neural network rebuilds the limiting characteristic of figure 8.5 b) in a wide range.

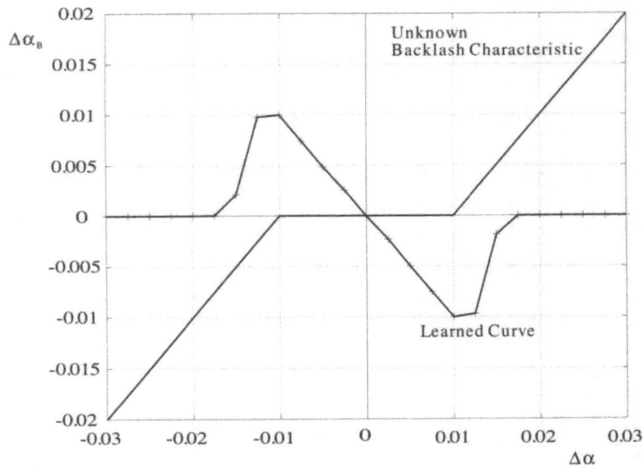

Figure 8.10: Learning result of MBO in a stiff system

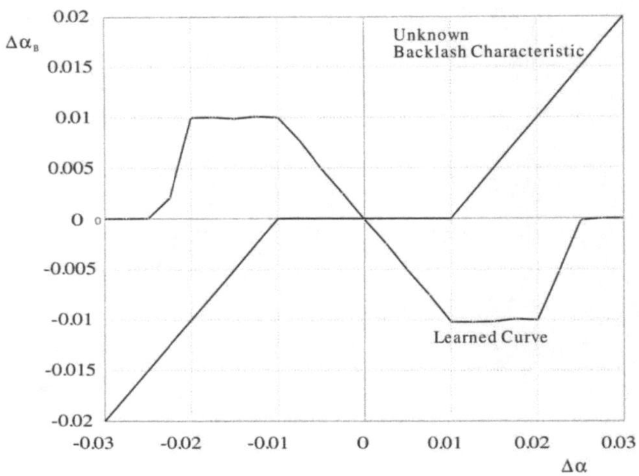

Figure 8.11: Learning result of LBO in a flexible plant

Here as well, we can determine the inherent backlash width very clearly.

Both simulations were carried out with and without bearing friction in the plant. There was no recognizable difference in the learning results, which means that the proposed method is robust to some extent against disturbances or other nonlinearities that were not modeled.

In the concept with motor–side observer we can go even further and state, that any additional nonlinearity between the load torque m_2 and the measured load angle φ_2 would have no impact on the backlash identification, provided that the system remains stable. This is a great advantage of this method and extremely extends the applicability of the method with the motor–side observer.

8.5 Experimental Validation

8.5.1 Experimental Set–Up and Parameters

To implement the two methods for the identification of backlash described in the previous sections (namely backlash identification with load–side observer and backlash identification with motor–side observer) in a real plant, we built the experimental set–up depicted in figure 8.12.

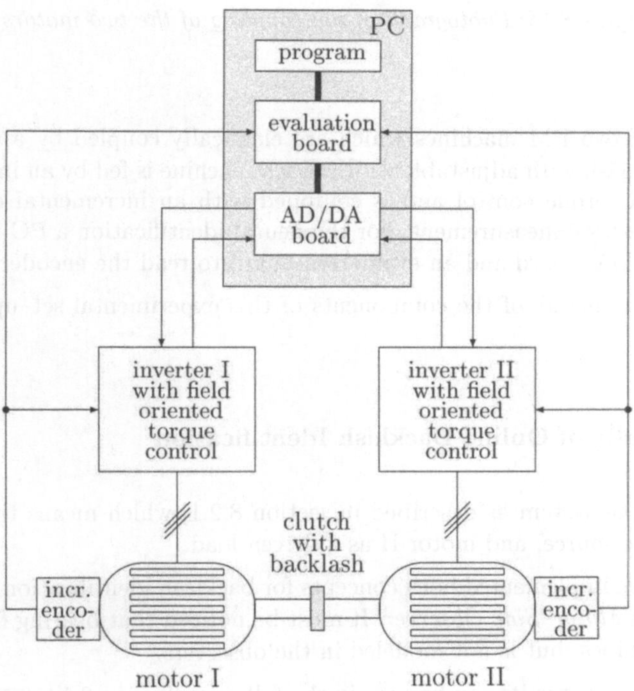

Figure 8.12: Experimental set–up for backlash identification

The mechanical set–up can be seen in the photograph in figure 8.13.

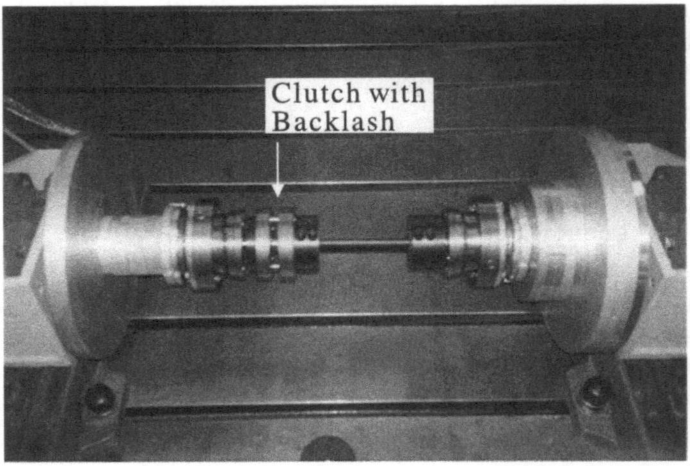

Figure 8.13: Photograph of the coupling of the two motors

It consists of two PM machines which are elastically coupled by a clutch that contains backlash with adjustable width. Each machine is fed by an inverter with field oriented torque control and is equipped with an incremental encoder for speed and position measurement. For the neural identification a PC is equipped with an AD/DA–board and an evaluation board to read the encoder signals.

The main parameters of the components of this experimental set–up are listed in table 8.1.

8.5.2 Results of Online Backlash Identification

We operate the system as described in section 8.2.1, which means that motor I acts as torque source, and motor II as a driven load.

On this PC we implemented both concepts for backlash identification, *Load–Side Observer* and *Motor–Side Observer*. It must be noticed that bearing friction acts on both machines, but is not modeled in the observers.

The identification results can be seen in the following figures 8.14, 8.15 and 8.16.

Figure 8.14 was achieved with an adjusted backlash width of $2a_B = \frac{0.027rad}{2\pi rad} = 0.0043 = \pm 0.00215$, in the figures 8.15 and 8.16 we used a backlash width of $2a_B = \frac{0.054rad}{2\pi rad} = 0.0086 = \pm 0.0043$. As it can be seen, the neural network

Motors:	
Type:	Synchronous Induction Motor with Permanent Magnet
Basic Speed:	$N_N = 2000\ min^{-1}$
Basic Torque:	$M_N = 23\ Nm$
Nominal Current:	$I_N = 10.9\ A$
Nominal Power:	$P_N = 4.8\ kW$
Moment of Inertia:	$J_I = 163.261 \cdot 10^{-3}\ kgm^2$
	$J_{II} = 330.129 \cdot 10^{-3}\ kgm^2$
Shaft and Clutch:	
Spring Constant:	$c = 783\ \frac{Nm}{rad}$
Damping Coefficient:	$d = 0.03$ (normalized)
Backlash Width:	$2a_B = 0\ rad\ /\ 0.027\ rad\ /\ 0.054\ rad$
PC:	
Processor:	Pentium
Clock Speed:	$166\ MHz$
RAM:	$40\ MB$
Harddisk:	$850\ MB$
Encoder:	
Resolution:	$2048\ \frac{Inkr.}{360^\circ}$, $1024 \cdot 2048\ \frac{Inkr.}{360^\circ}$ with Interpolation
Max. Frequency:	$400\ kHz$

Table 8.1: Parameters of the experimental set–up for online backlash identification

approximates the present backlash characteristic very well and exact. The figures 8.15 and 8.16 show the online learning procedure in the input space. Both concepts (LBO and MBO) were tested and both achieved similar identification results.

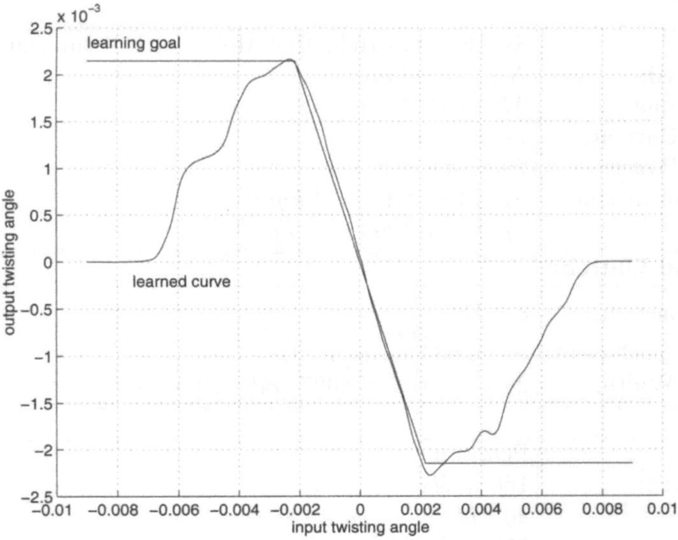

Figure 8.14: Learning result at the real system

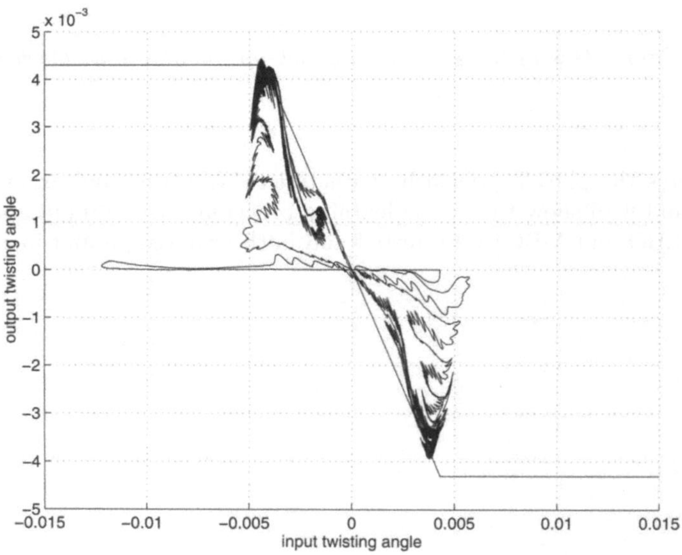

Figure 8.15: Learning procedure at the real system (current output of the neural network)

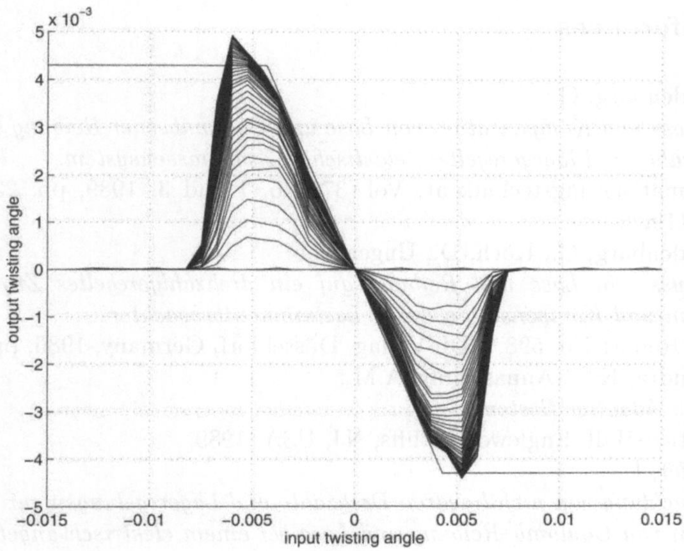

Figure 8.16: Learning procedure at the real system (evaluation of the neural network over the whole input space)

8.6 Conclusion

In this chapter we derived two different methods to detect and identify backlash in electro–mechanical systems which can be described as two–mass systems. The control engineer can use an observer structure with a trainable neural network for the better known subsystem, motor side or load side. Both concepts form a globally stable error model [3].

The knowledge after the learning phase can be used to compensate the backlash influence on the control behaviour of the system. One possible way of compensation is described in [4].

8.7 References

[1] Brandenburg, G.:
Einfluss und Kompensation von Lose und Coulombscher Reibung bei einem drehzahl– und lagegeregelten, elastischen Zweimassensystem.
Automatisierungstechnik at, Vol. 37, No. 1 and 3, 1989, pp. 23–31 and 111–119.

[2] Brandenburg, G., Koch, D., Unger, H.:
Einfluss von Lose und Reibung auf ein drehzahlgeregeltes Zweimassensystem und Kompensation durch Lastmomentbeobachter.
VDI–Report No. 598, VDI–Verlag, Düsseldorf, Germany, 1986, pp. 53–68.

[3] Narendra, K.S., Annaswamy, A.M.:
Stable Adaptive Systems.
Prentice–Hall, Englewood Cliffs, NJ, USA, 1989.

[4] Schäfer, U.:
Entwicklung von nichtlinearen Drehzahl– und Lageregelungen zur Kompensation von Coulomb–Reibung und Lose bei einem elektrisch angetriebenen, elastischen Zweimassensystem.
Dissertation, TU München, Germany, 1991.

[5] Schäfer, U., Brandenburg, G.:
State Position Control for Elastic Pointing and Tracking Systems with Gear Play and Coulomb Friction.
Proceedings of the 4th EPE Congress, Florence, Italy, 1991.

[6] Schäffner, C.:
Analyse und Synthese neuronaler Regelungsverfahren.
Dissertation, TU München, Germany, 1996.

[7] Tao, G., Kokotović, P.V.:
Adaptive Control of Systems with Unknown Output Backlash.
IEEE Transactions on Automatic Control, Vol. 40, No. 2, 1995, pp. 326–330.

[8] Tao, G., Kokotović, P.V.:
Continuous–Time Adaptive Control of Systems with Unknown Backlash.
IEEE Transactions on Automatic Control, Vol. 40, No. 6, 1995, pp. 1083–1087.

[9] Tustin, A.:
The Effects of Backlash and of Speed–Dependent Friction on the Stability of Closed–Cycle Control Systems.
Journal of IEE, No. 94, 1947, pp. 143–151.

9 Nonlinear Observer Structures for the Identification of Isolated Nonlinearities in Rolling Mills

Stephan Straub

9.1 Introduction

As it was shown in the last chapters, neural networks are useful tools to solve nonlinear problems concerning the control of dynamic systems. The method of the identification of systems with isolated nonlinearities (chapters 5, 7 and 8) can be used to get information about unknown nonlinearities and non–measurable states of the system (observer function) in a stable manner. For plants like rolling mills, the guarantee of stability in the operating area is an indispensable feature. In this chapter two possible nonlinear problems in rolling mills are described and solved with neural network based identification methods. It can be seen how the extracted knowledge of the nonlinear system is used to get better control results. The problems which are described are the identification of winder eccentricities and the identification of the nonlinear roll bite behaviour. In the last section experimental results are shown.

9.2 Neural Networks in Rolling Mills

9.2.1 Plant Description

Figure 9.1 shows the simplified structure of a one stand cold rolling mill. The material is transported from the unwinder to a roll bite, where the thickness of the material (in this case steel) is reduced to a specified thickness. The material is then wound up and used for further processing. In industrial production the tolerance limits of steel, used by car manufacturers, are very low (below one per cent of the reference value). Because of that, more and more effort concerning the design of new control methods has to be made to meet the demands of the

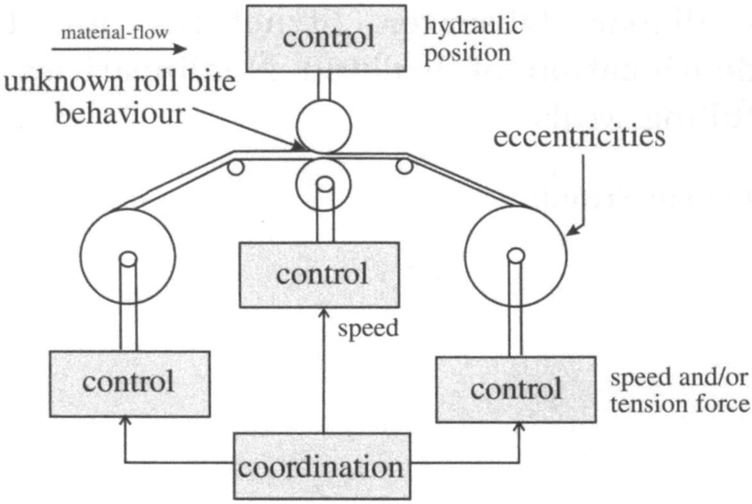

Figure 9.1: Simplified structure of a one stand cold rolling mill

industry. Rolling mills are composed of different subsystems which are coupled by the material. This fact increases the difficulty to control the subsystems. In this context two possible applications of the method described in chapter 5 are shown. The general indirect control concept can be seen in figure 9.2. Based on the learned (identified) knowledge of the plant, where a neural observer is adapted in such a manner that the observer error tends to zero, the controller compensates for the negative effects of nonlinearities or disturbances which are not exactly known. As it can be seen later, the compensation is really dependent on the application itself. In opposite to the indirect control method presented here, a global (direct) method is presented in the chapters 10 and 11.

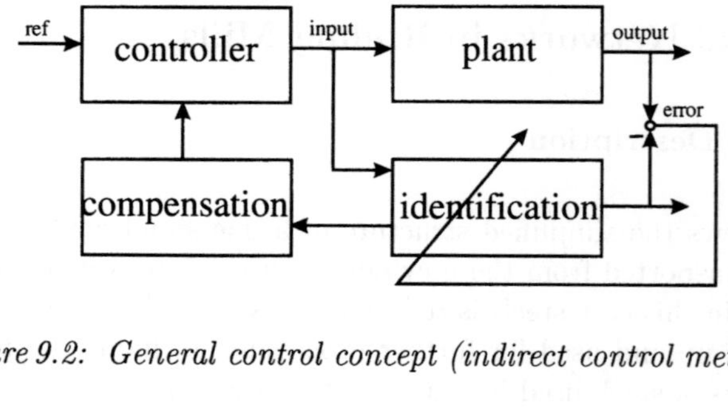

Figure 9.2: General control concept (indirect control method)

9.2.2 Compensation of Winder Eccentricities

In real plants the non–circularity of the winders can lead to undesired oscillations of the tension force, which lead to problems in controlling the output thickness of the material behind the roll bite. Such eccentricities can be caused for example by clamping the steel. This fact leads to an elevation at the clamping point which can also be seen, when further material is wound up (figure 9.3). The aim in this context is to identify the radius with respect to the angle position online. Based on this knowledge, a compensation algorithm can be implemented to damp the oscillations. Figure 9.3 shows a possible form of the eccentricity for one rotation in case of the rewinder where an additional increase of the mean value of the radius can be seen. This means, that there is a time–variant disturbance which has to be identified. It must be noted that this is only a possible form of eccentricity which is used for simulation in this case.

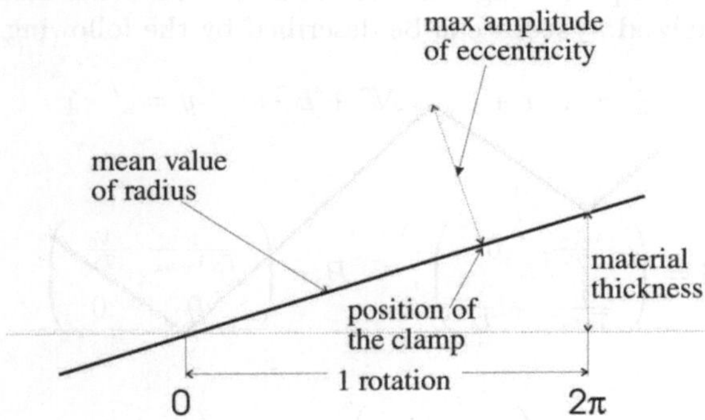

Figure 9.3: Possible form of eccentricity

Figure 9.4 shows the plant–observer structure of the whole system based on the method of identification of isolated nonlinearities. The drive system is assumed to be a two–mass system. This kind of system and the resulting problems from the control point of view are described in detail in chapter 2. In the chapters 7 and 8 applications concerning friction and backlash identification in two–mass systems are discussed.

It is assumed that the motor speed N_M and the (tension) force F_{12f} are measurable. The inputs to the whole system are the reference value of the motor torque M_{ref} and the material velocity V_2 at a fixed point between the winder system and the first processing station. $\hat{N}_W$ is the observed load speed and $\hat{\varphi}$ the (observed) angle position of the winder. In this case the adaptation of the neural network is done via the error between the filtered and observed force (F_{12f} and $\hat{F}_{12f}$). Another possible approach would be the adaptation via the error of the motor speed. In this case the two nonlinearities, friction and eccentricity, both affecting the dynamic behaviour of the drive system, have to be separated. The advantage of this method would be that no information about the velocity V_2 is needed.

Generally the force observer is a nonlinear system of the first order, which is described by the transient reaction of tension ϵ_{01} and tension ϵ_{12}. ϵ_{01} is the tension of the stored material in the winder and ϵ_{12} between the winder system and the first processing station. F_{12} can be computed by dividing ϵ_{12} by the material constant ϵ_n. The dynamic behaviour is based on the mass flow continuum of the material which is described by the following equation:

$$\frac{d}{dt}\left(\frac{L_N}{1+\epsilon_{12}}\right) = \frac{V_W}{1+\epsilon_{01}} - \frac{V_2}{1+\epsilon_{12}} \tag{9.1}$$

V_W is the winder velocity (multiplication of load speed N_W and radius), V_{max} the maximum speed of the material and L_N the constant length of the material between winder and processing station. With an additional measurement filter the whole **linearized system** can be described by the following equations:

$$\dot{\underline{x}} = A \cdot \underline{x} + \underline{e}_{NL} \cdot \mathcal{NL} + B \cdot \underline{u}; \quad y = \underline{c}^T \cdot \underline{x} \tag{9.2}$$

with

$$A = \begin{pmatrix} -\frac{V_0}{T_N} & 0 \\ \frac{1}{T_g \epsilon_n} & -\frac{1}{T_g} \end{pmatrix}; \quad B = \begin{pmatrix} \frac{1}{T_N V_{max}} & \frac{V_0}{T_N} \\ 0 & 0 \end{pmatrix} \tag{9.3}$$

$$\underline{e}_{NL} = \begin{pmatrix} -\frac{1}{T_N V_{max}} \\ 0 \end{pmatrix}; \quad \underline{c} = \begin{pmatrix} 0 \\ 1 \end{pmatrix} \tag{9.4}$$

where T_g ist the time constant of the measurement unit and $T_N = \dfrac{L_N}{V_{max}}$. V_0 is the material velocity in the operating point. The states of the system are the tension ϵ_{12} and the measured (filtered) force F_{12f}. Inputs to the system are the processing velocity V_2 and the unknown tension ϵ_{01}. Thus, we can write:

$$\underline{x} = \begin{pmatrix} \epsilon_{12} \\ F_{12f} \end{pmatrix}; \quad \underline{u} = \begin{pmatrix} V_2 \\ \epsilon_{01} \end{pmatrix} \tag{9.5}$$

The unknown nonlinearity is the radius with respect to the angle position multiplied with $N_W \cdot 2\pi$ and, neglecting the *inherent aproximation error* (chapter 5), it can be described by the following equation:

$$\mathcal{NL} = \underline{\Theta}^T \cdot \underline{\mathcal{A}}(\varphi) \cdot N_W \cdot 2\pi \tag{9.6}$$

where $\underline{\mathcal{A}}(\varphi)$ is the activation vector of the General Regression Neural Network (GRNN, see chapter 4). To get the velocity, the activation vector is multiplied

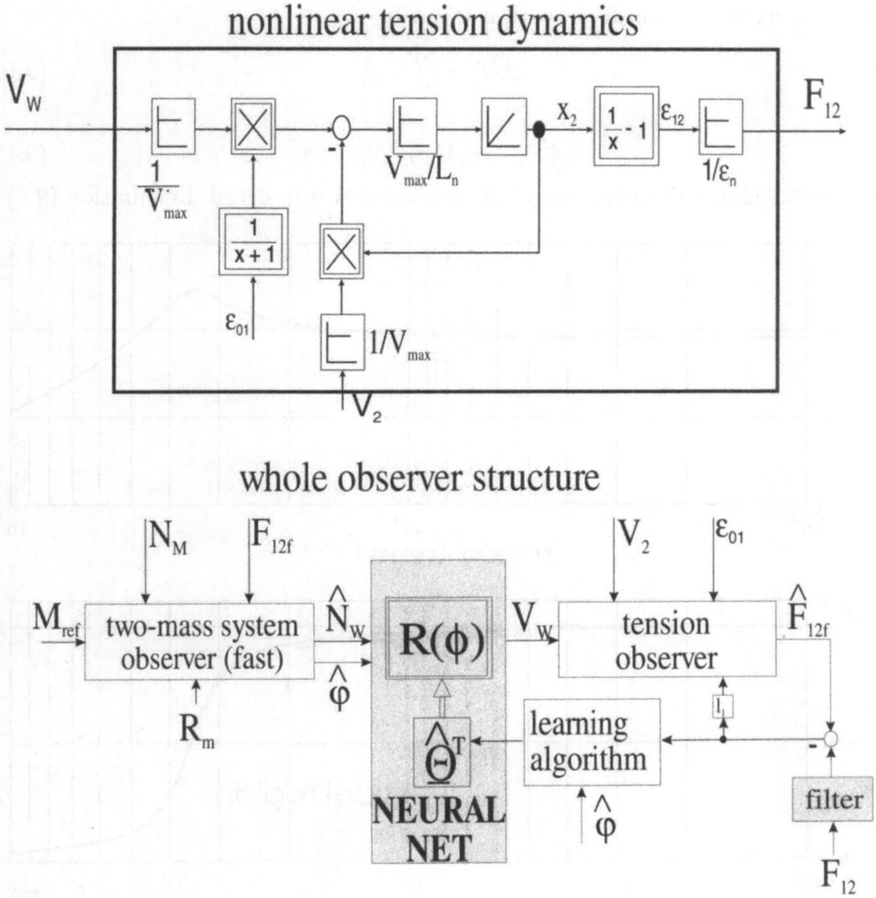

Figure 9.4: Observer concept for the identification of the eccentricity and signal flow chart of tension dynamics

with the winder speed N_W and the factor 2π, which can be seen as a structural a–priori knowledge. Therefore, we get an augmented activation vector:

$$\underline{A}^*(\varphi) = \underline{A}(\varphi) \cdot N_W \cdot 2\pi \tag{9.7}$$

Based on chapter 5 we can write for the observer

$$\dot{\hat{\underline{x}}} = A \cdot \hat{\underline{x}} + \underline{e}_{NL} \cdot \widehat{\mathcal{NL}} + B \cdot \underline{u} + L \cdot \underline{e}; \quad \hat{y} = \underline{c}^T \cdot \hat{\underline{x}} \tag{9.8}$$

with the Luenberger matrix:

$$L = \begin{pmatrix} 0 & -\dfrac{l_1 \epsilon_n}{T_N} \\ 0 & 0 \end{pmatrix} \tag{9.9}$$

The output of the neural network is given by

$$\widehat{\mathcal{NL}} = \underline{\hat{\Theta}}^T \cdot \underline{\hat{\mathcal{A}}}^* \tag{9.10}$$

with

$$\underline{\hat{\mathcal{A}}}^*(\varphi) = \underline{\mathcal{A}}(\varphi) \cdot \hat{N}_W \cdot 2\pi \tag{9.11}$$

Figure 9.6 shows the structure of the force observer in detail. In equation (9.6) it

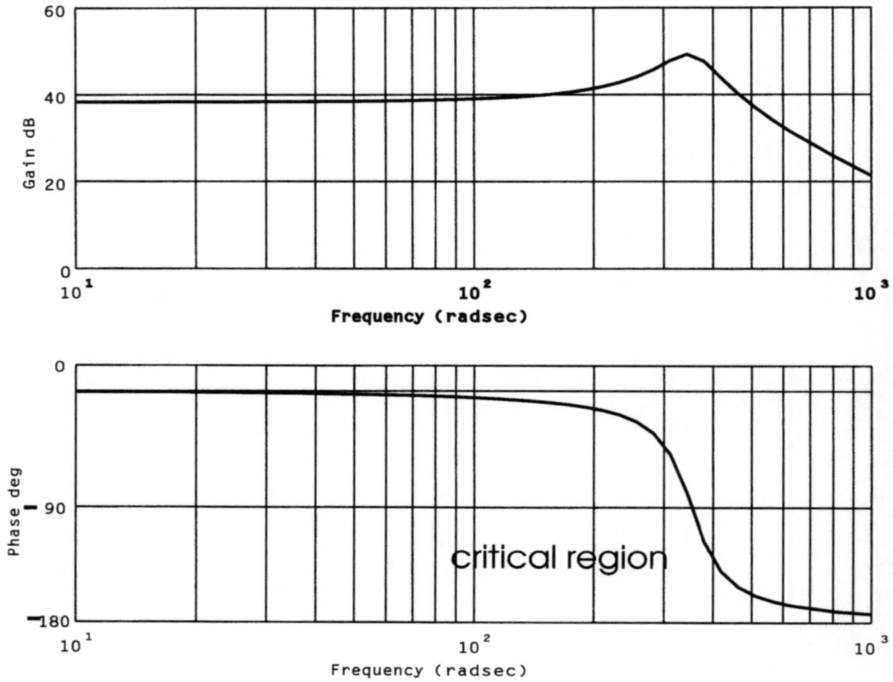

Figure 9.5: Bode-plot of the error transfer function $H(s)$: SPR–condition not fulfilled

is assumed, that the angle position is directly measurable, which is not possible in real plants. In this case the angle position and the winder speed N_W (as inputs of the force observer) are approximated by the fast two–mass system observer. The result of the identified non–circularity is not used for this observer. This means that no closed loop structure exists (no internal coupling) between the two observers, which could lead to additional precautions concerning the learning factor η of the neural net and the Luenberger coefficients (see chapter 5). In this case, information flows only in one direction and results in an easier observer design. Hence, for further explanations it is assumed that

$$N_W \approx \hat{N}_W \tag{9.12}$$

and

$$\varphi_W \approx \hat{\varphi}_W \tag{9.13}$$

Therefore, we can write for the activation vector:

$$\underline{\hat{A}}^*(\varphi) \approx \underline{A}^*(\varphi) \tag{9.14}$$

The identification is done, as mentioned, via the error between real and identified (smoothed) force. Due to the addional smoothing filter caused by measurement, the whole identification system is a system of second order. Thus, the principle of the delayed activation becomes indispensable (see figure 9.5).

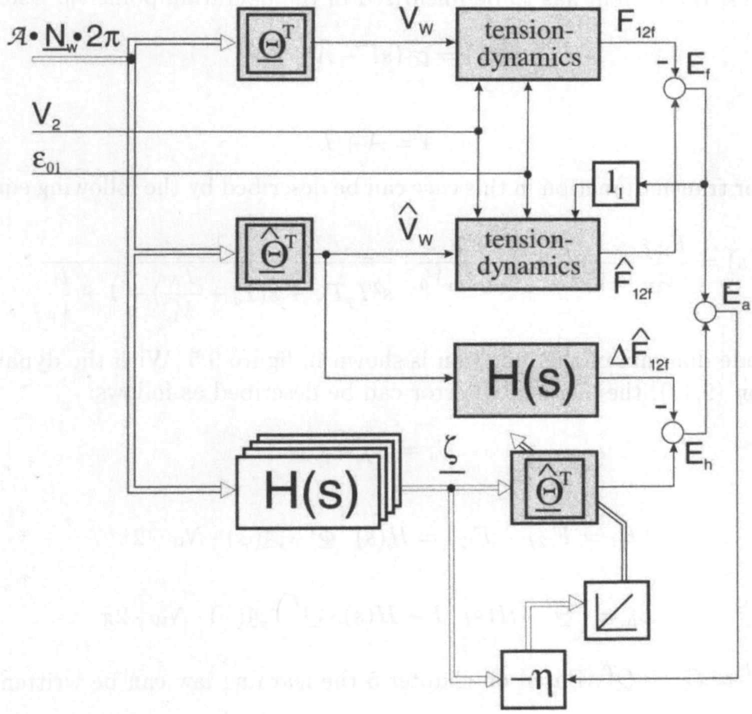

Figure 9.6: Observer structure with delayed activation and additional error (augmented error)

The force observer itself can be divided into two main parts. First the parallel part, which can be seen as a state observer (characterized by equation (9.8)) and a second part, which is responsible for the additional error based on the principle of delayed activation as described in chapter 5. The activation vector $\underline{A}(\varphi)$ is changed (delayed) in such a manner, that stable learning is guaranteed in the whole frequency range (no phase shift between the activation vector and the measurable error). For the adaptation of the neural network parameters $\hat{\underline{\Theta}}^T$, the delayed activation vector and the augmented error are used. $H(s)$ is the error transfer function describing the influence of the nonlinearity on the measurable error signal. It must be noted that the inputs, the velocity of the material V_2 and the stored winder tension ϵ_{01}, must not be considered in the second part of the observer, because these inputs do not affect the error signal. Therefore, the second part of the observer does not represent physically interpretable states. Another point is, that the measurable mean value of the radius can be used as a–priori knowledge of the system. An additional Luenberger coefficient l_1 is used to stabilize the learning behaviour. Because of the nonlinear tension force dynamics, the system has to be linearized in the operating point V_0. Using

$$H(s) = \underline{c}^T(sI - \tilde{A})^{-1}\underline{e}_{NL} \tag{9.15}$$

with

$$\tilde{A} = A + L \tag{9.16}$$

the error transfer function in this case can be described by the following equation:

$$H(s) = \frac{\hat{F}_{12f} - F_{12f}}{\hat{V}_W - V_W} = \frac{-1}{\epsilon_n V_{max} V_0} \cdot \frac{1}{s^2 T_g T_N + s(T_g + \frac{T_N}{V_0}) + 1 + \frac{l_1}{V_0}} \tag{9.17}$$

The Bode diagram of this function is shown in figure 9.5. With the dynamics of equation (9.17), the augmented error can be described as follows:

$$E_a = E_f + E_h \tag{9.18}$$

where

$$E_f = \hat{F}_{12f} - F_{12f} = H(s) \cdot \underline{\Phi}^T \cdot \underline{A}(\varphi) \cdot N_W \cdot 2\pi \tag{9.19}$$

and

$$E_h = \left(\hat{\underline{\Theta}}^T \cdot H(s) \cdot I - H(s) \cdot \hat{\underline{\Theta}}^T\right) \underline{A}(\varphi) \cdot N_W \cdot 2\pi \tag{9.20}$$

with $\underline{\Phi}^T = \hat{\underline{\Theta}}^T - \underline{\Theta}^T$. Based on chapter 5 the learning law can be written in the form

$$\dot{\underline{\Phi}} = \dot{\hat{\underline{\Theta}}} = -\eta \cdot E_a \cdot \underline{\zeta} \tag{9.21}$$

with

$$\underline{\zeta} = H(s) \cdot I \cdot \underline{A}(\varphi) \cdot N_W \cdot 2\pi \tag{9.22}$$

or

$$\underline{\zeta} = H(s) \cdot I \cdot \underline{A}^*(\varphi) \tag{9.23}$$

where I is the unit matrix.

Figure 9.7 shows the simulation results of the proposed method. In this case a simple triangle form of the eccentricity was assumed. After four to five rotations the output of the neural net follows the real radius (left). In case of the rewinder

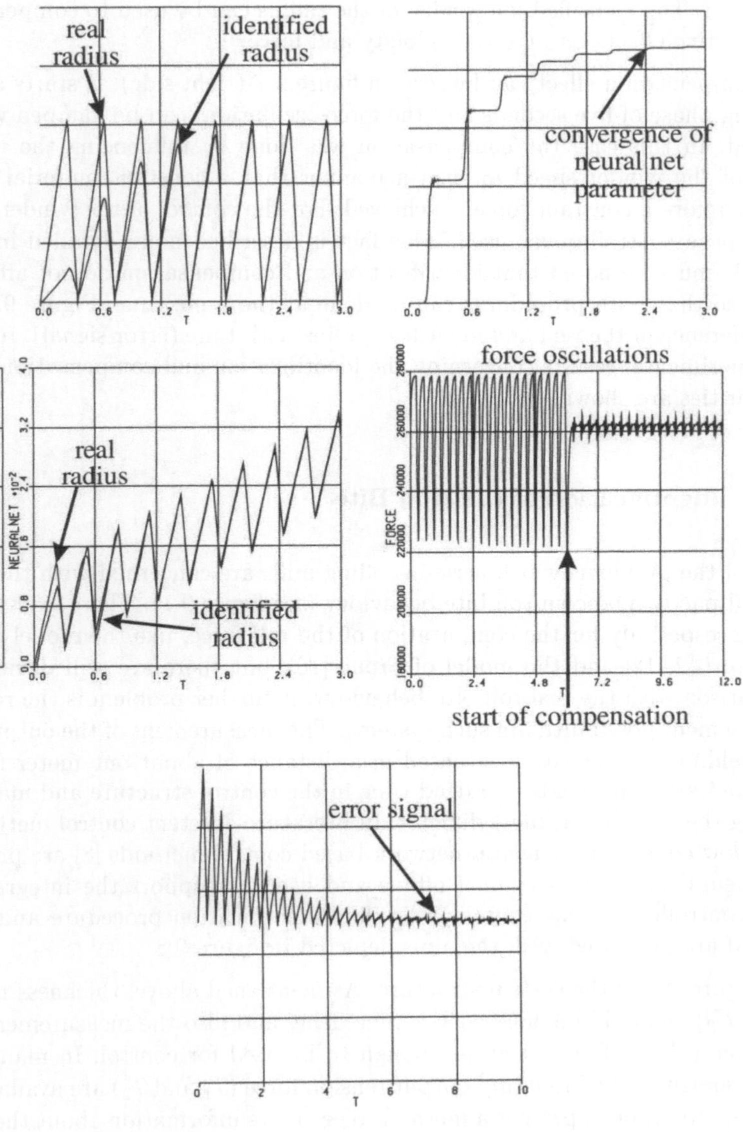

Figure 9.7: Simulation results of eccentricity identification and compensation

(figure below), the time increasing mean value of the radius is also identified, so the observer parameters can also be adapted by this method (i. e. the moment of inertia). Furthermore, it can be seen that the neural network parameters converge to a constant value, if the increasing (or decreasing) mean value is assumed to be known. The identified knowledge of the radius can be used to compensate for the negative effects on material velocity and force.

The compensation effect can be seen in figure 9.7 (right side). It starts after the learning phase of five seconds and the force oscillations can be damped with this method. In this case the compensation was done by influencing the reference value of the winder speed in such a manner that a constant material velocity and therefore a constant force is achieved. For the control of the winder speed a state space controller was used. This fact is described more detailed in section 9.3.3. It must be noted that identification and compensation do not affect each other, so that both procedures can be done at the same time. Figure 9.7 shows the difference of the real and identified radius with time (error signal). In section 9.3 experimental results concerning the identification and compensation of non–circularities are shown.

9.2.3 Identification of the Roll Bite

Most of the problems which arise in rolling mills are concerned with the nonlinear and partly unknown roll bite behaviour (see figure 9.1). There exist different models, especially for the computation of the roll force, like the model of *Bland and Ford* [2, 12] and the model of *Stone* [10], but there are still deviations in comparison with the real roll bite behaviour. A further problem is the restricted measurement possibilitity in such systems. The measurement of the output thickness behind the roll bite is situated in a distance of about one meter from the roll bite itself. This leads to a dead time in the control structure and makes controlling the output thickness difficult. In literature different control methods like *mass flow control* [1] or neural network based control methods [8] are presented. This neural network is trained offline and used to support the integrated roll force controller. In this context an online identification procedure and control method are presented with the aims depicted in figure 9.8.

This figure shows the system structure. As mentioned above, thickness measurement (H_2) can only be done with a time delay and also the measurement of the roll force (P) is often not exact enough to be used for control. In many plants measurement of the input and output tension force (T_1 and T_2) are available, and therefore we want to present a method to get more information about the system based on these signals. The aim is to use the identified thickness $\hat{H}_2$ (directly behind the roll bite) and/or the identified roll force $\hat{P}$ for control purposes. Additionally, the two velocities directly in front of and behind the roll bite ($\hat{V}_1$ and $\hat{V}_2$) are estimated and could be used for example in mass flow control concepts. Furthermore, this example shows, how a system is identified where more than one nonlinearity is unknown.

Figure 9.9 shows the signal flow chart of the system. The roll bite is assumed to be a system of the fifth order described by

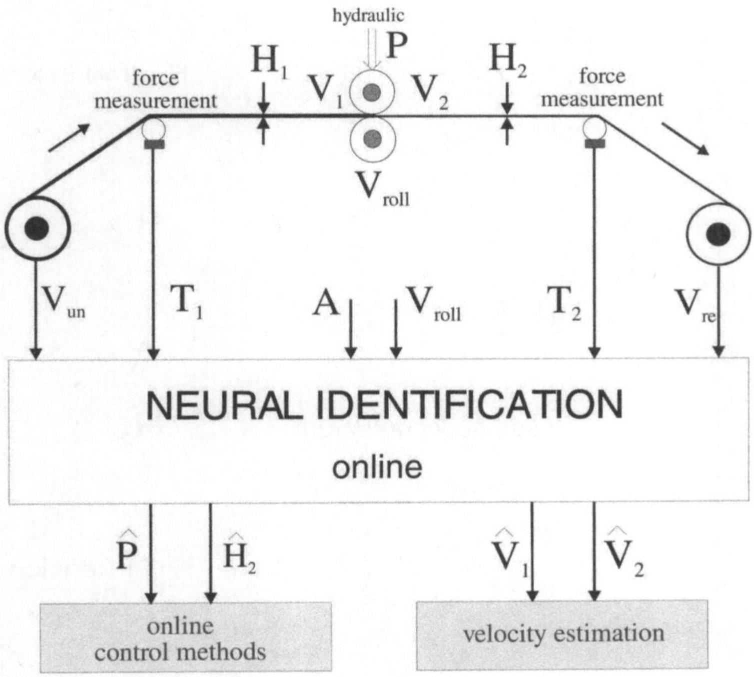

Figure 9.8: The aims of roll bite identification

$$\underline{\dot{x}} = \underline{\mathcal{NL}}(\underline{x}, \underline{u}) \tag{9.24}$$

where the states $\underline{x}$ are the input and output tension force T_1 and T_2, the roll flattening R_f, the mill extension H_{ext} and the hydraulic piston position A. Inputs $\underline{u}$ to the system are the reference value of the hydraulic piston position A_{ref}, the winder velocities V_{un} and V_{re} (in case of a one stand rolling mill), the velocity V_{roll} of the rolls and the input thickness H_1. In the signal flow chart of the whole structure (figure 9.9) the two velocities V_{un} and V_{re} are assumed to be directly measurable at the winder system. In this case, the history of the wound material is not considered and the representation of the tension dynamics can be reduced as it is shown in figure 9.9. In most plants the measurement of the two velocities is not available. An additional observer structure as it was shown in the last section (fast two–mass system observer) can solve this problem (this is not a restriction of the presented method). For reasons of clearness, this fact is neglected in figure 9.9.

It must be noted that the time constant of the roll flattening can be neglected and can be integrated into the roll force block (algebraic loop). In most cases the piston position can be measured directly so that this state can be seen as an input to the system. With these simplifications the whole system can be reduced to a

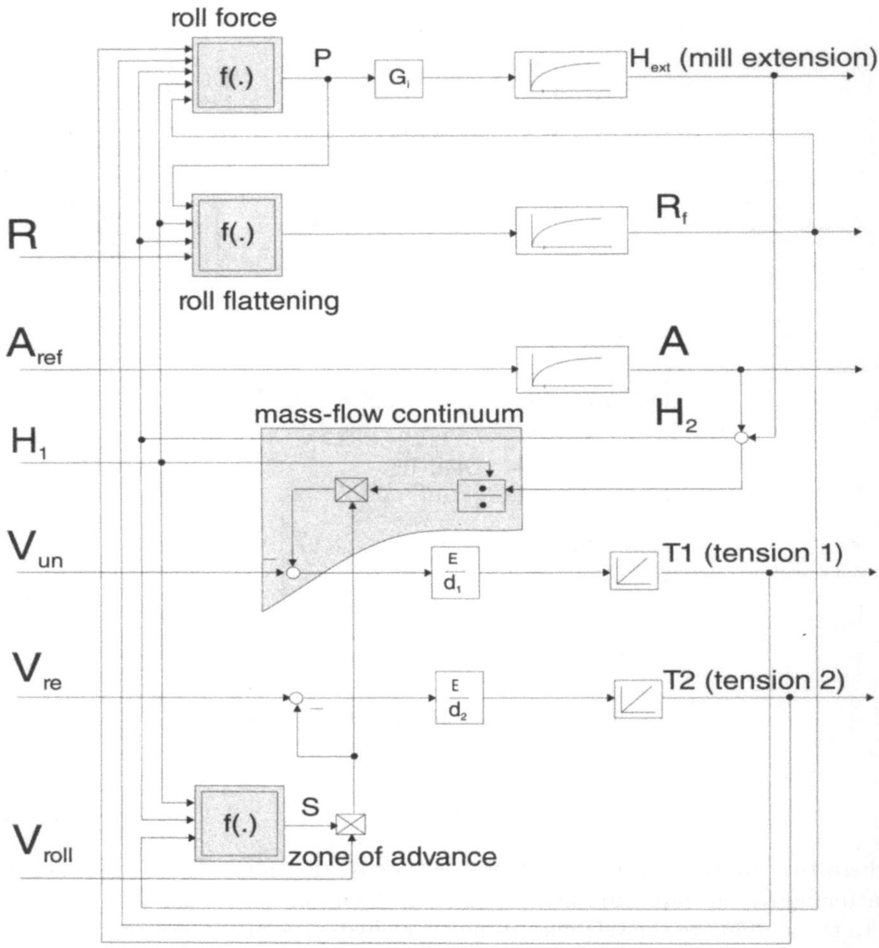

Figure 9.9: Signal flow chart of the roll bite

system of third order. The outputs of this system are the thickness H_2 behind the roll bite and the two tensions T_1 and T_2 which are used for adaptation. The unknown nonlinearities are the roll force P (influencing the mill extension) and the (material–) advance which describes the output velocity with respect to V_{roll} (sliding effect of the zone of advance). The aim now is to identify these two nonlinearities and thus obtain information about the dynamic system behaviour. The system can be described by the following state equations:

$$\dot{H}_{ext} = -H_{ext} + \mathcal{NL}_1 \cdot G \tag{9.25}$$

$$\dot{T}_1 = \frac{E}{d_1} \cdot (\mathcal{NL}_2 \cdot V_{roll} \cdot \frac{A + H_{ext}}{H_1} - V_{un}) \tag{9.26}$$

$$\dot{T}_2 = \frac{E}{d_2} \cdot (V_{re} - \mathcal{NL}_2 \cdot V_{roll}) \tag{9.27}$$

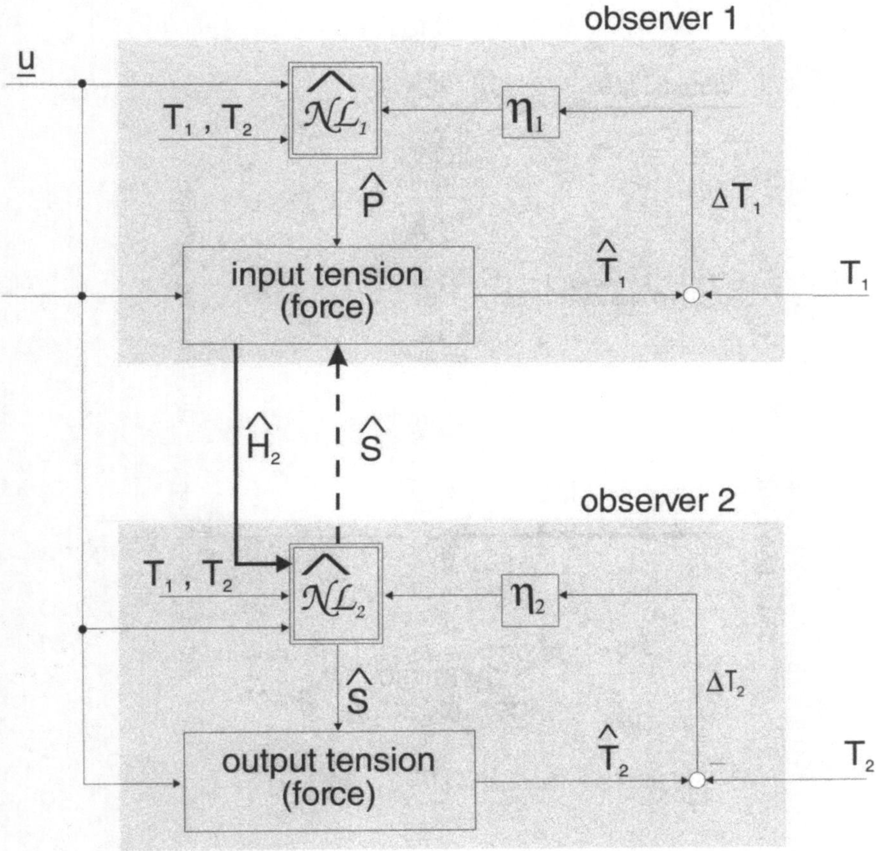

Figure 9.10: Observer structure of roll bite identification

G is the mill stiffness, E is the elastic modulus of the material (steel) and d_1 and d_2 are the material lengths between the winder systems and the roll bite. It is assumed that the two tensions T_1 and T_2 are directly measurable. The first non-linearity $\mathcal{NL}_1$ describes the roll force and $\mathcal{NL}_2$ the material advance in the sliding zone. The second equation is described by the mass flow condition in the roll bite (input and output mass must be the same). As can be seen in the state space representation, the third equation could be decoupled by measurement from the others if the input of $\mathcal{NL}_2$ does not depend on any states of the first equations. With this assumption $\mathcal{NL}_2$ could be identified based on the measurement of T_2. In reality there is a closed loop structure because the zone of advance depends on the output thickness, and the mass flow condition depends on the velocity of the output material. Special assumptions concerning the roll bite behaviour simplify the observer design.

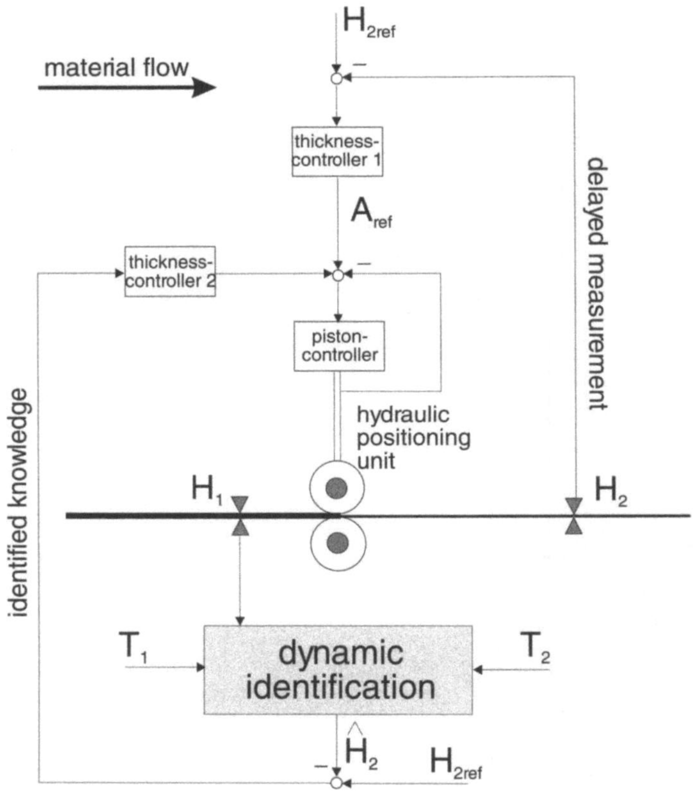

Figure 9.11: Possible concept of neural control of roll bite

Figure 9.10 shows the roll bite observer structure. Two neural nets (GRNN) are used for two different nonlinearities (roll force and material advance). Sufficiently high Luenberger coefficients of the first observer stabilize the learning behaviour and the observation of the output thickness (even if there is a closed loop structure). Additionally, the effect of the material advance to the first observer is very low. This fact is marked in figure 9.10 with dashed lines and leads to *serial decoupled observer structure* because information flows only in one direction. For the identification of $\mathcal{NL}_1$ (roll force) the method of delayed activation has to be used to guarantee stability and convergence for all frequencies. For the second observer the state space representation can be rewritten as follows:

$$\dot{\hat{T}}_2 = \frac{E}{d_2} \cdot (V_{re} - \widehat{\mathcal{NL}}_2 \cdot V_{roll}) + l_2 \cdot (T_2 - \hat{T}_2) \tag{9.28}$$

$$= \underbrace{\frac{E}{d_2} \cdot V_{re}}_{u_1^*} - \underbrace{\frac{E}{d_2} \cdot V_{roll}}_{u_2^*} \cdot \widehat{\mathcal{NL}}_2 + l_2 \cdot (T_2 - \hat{T}_2) \tag{9.29}$$

$$= u_1^* + u_2^* \cdot \widehat{\mathcal{NL}}_2 + l_2 \cdot (T_2 - \hat{T}_2) \tag{9.30}$$

and for the first observer we can write:

$$\dot{\hat{H}}_{ext} = -\hat{H}_{ext} + \widehat{\mathcal{NL}}_1 \cdot G + l_{11} \cdot (T_1 - \hat{T}_1) \tag{9.31}$$

$$\dot{\hat{T}}_1 = \frac{E}{d_1} \cdot (\widehat{\mathcal{NL}}_2 \cdot V_{roll} \cdot \frac{A + \hat{H}_{ext}}{H_1} - V_{un}) + l_{12} \cdot (T_1 - \hat{T}_1) \tag{9.32}$$

$$= \underbrace{\frac{E \cdot \widehat{\mathcal{NL}}_2 \cdot V_{roll} \cdot A}{d_1 \cdot H_1}}_{u_3^*} - \underbrace{\frac{E}{d_1} \cdot V_{un}}_{u_4^*} + \underbrace{\frac{E \cdot \widehat{\mathcal{NL}}_2 \cdot V_{roll}}{d_1 \cdot H_1}}_{u_5^*} \cdot \hat{H}_{ext} + l_{12} \cdot (T_1 - \hat{T}_1) \tag{9.33}$$

$$= u_3^* + u_4^* + u_5^* \cdot \hat{H}_{ext} + l_{12} \cdot (T_1 - \hat{T}_1) \tag{9.34}$$

With this simplified representation we get for the whole observer:

$$\dot{\hat{H}}_{ext} = -\hat{H}_{ext} + \widehat{\mathcal{NL}}_1 \cdot G + l_{11} \cdot (T_1 - \hat{T}_1) \tag{9.35}$$

$$\dot{\hat{T}}_1 = u_3^* + u_4^* + u_5^* \cdot \hat{H}_{ext} + l_{12} \cdot (T_1 - \hat{T}_1) \tag{9.36}$$

$$\dot{\hat{T}}_2 = u_1^* + u_2^* \cdot \widehat{\mathcal{NL}}_2 + l_2 \cdot (T_2 - \hat{T}_2) \tag{9.37}$$

The first error $(T_1 - \hat{T}_1)$ is used for the adaptation of the first neural net $(\widehat{\mathcal{NL}}_1)$ and the second error $(T_2 - \hat{T}_2)$ for the second neural net $(\widehat{\mathcal{NL}}_2)$. l_{11}, l_{12} and l_2 are the Luenberger coefficients. The learning algorithms are used as it was shown in chapter 5 with the additional assumption that one of the observers can be completely decoupled from the other one. This fact is marked in equation (9.32) with the values u_3^* and u_5^* as (measurable) input to the first observer. It must be mentioned that the first observer has a nonlinear connection between the fictitious input u_5^* and the state $\hat{H}_{ext}$, so a linear known part cannot be separated. In this case the influence of u_5^* can be seen as structural a–priori knowledge of the system. Furthermore, the variations of u_5^* in the operating point are very low, so it can be assumed to be constant. Additionally, high Luenberger coefficients stabilize the learning behaviour in the entire operating area. Hence, the whole system can be divided into two systems with isolated nonlinearities in the sense of chapter 5. The observer design for completely coupled systems is shown in chapter 6. In this case additional filters are used for the adaptation of the neural network parameters.

In figure 9.11 a possible control concept can be seen. In this case an additional thickness controller is used to support the superimposed controller based on the thickness measurement (conventional method). The additional controller uses the identified thickness to get better results. To guarantee correct behaviour of the whole system, the neural net(s) of the observer must be initialized in a suitable manner, so that the supporting controller does not influence the superimposed controller, when the neural net is not trained yet. Another possibilty which is not depicted here, is the utilization of both, the identified roll force and the observed thickness, to support conventional thickness control by computing neccessary piston position changes.

In figure 9.12 simulation results of the above described identification and control methods are shown. The first figure (left) shows the real (modelled) roll force and the output of the neural net at the beginning of learning. After some time we can see the correspondence between the two signals (right). The next two figures show an input thickness disturbance (left) and the reaction of the control concept with delayed measurement of the output thickness (conventional control concept). It can be seen that the controller is not able to react fast enough. With the mentioned methods the disturbance in output thickness can be damped (the two figures below). It can also be seen that *identification* and *control* work together and do not disturb each other.

9.3 Experimental Results

9.3.1 Plant Description

For the experimental analysis of the described neural methods an experimental plant consisting of two winder systems and three processing stations is available at the Institute for Electrical Drive Systems at the Technical University of Munich. In this context, eccentricities and friction had to be identified. The used material in this case is paper. The technical data of the plant (see figure 9.13) is:

whole length:	9 m
breadth:	1 m
max. processing speed:	240 $\frac{m}{min}$
max. radius of winders:	0.5 m
mechanical power:	22 kW

The drive systems consist of inverter fed asynchronous motors. The winders are controlled via three cascaded tension, speed and current controllers to get the desired average velocity and the desired force in every subsystem. Because of

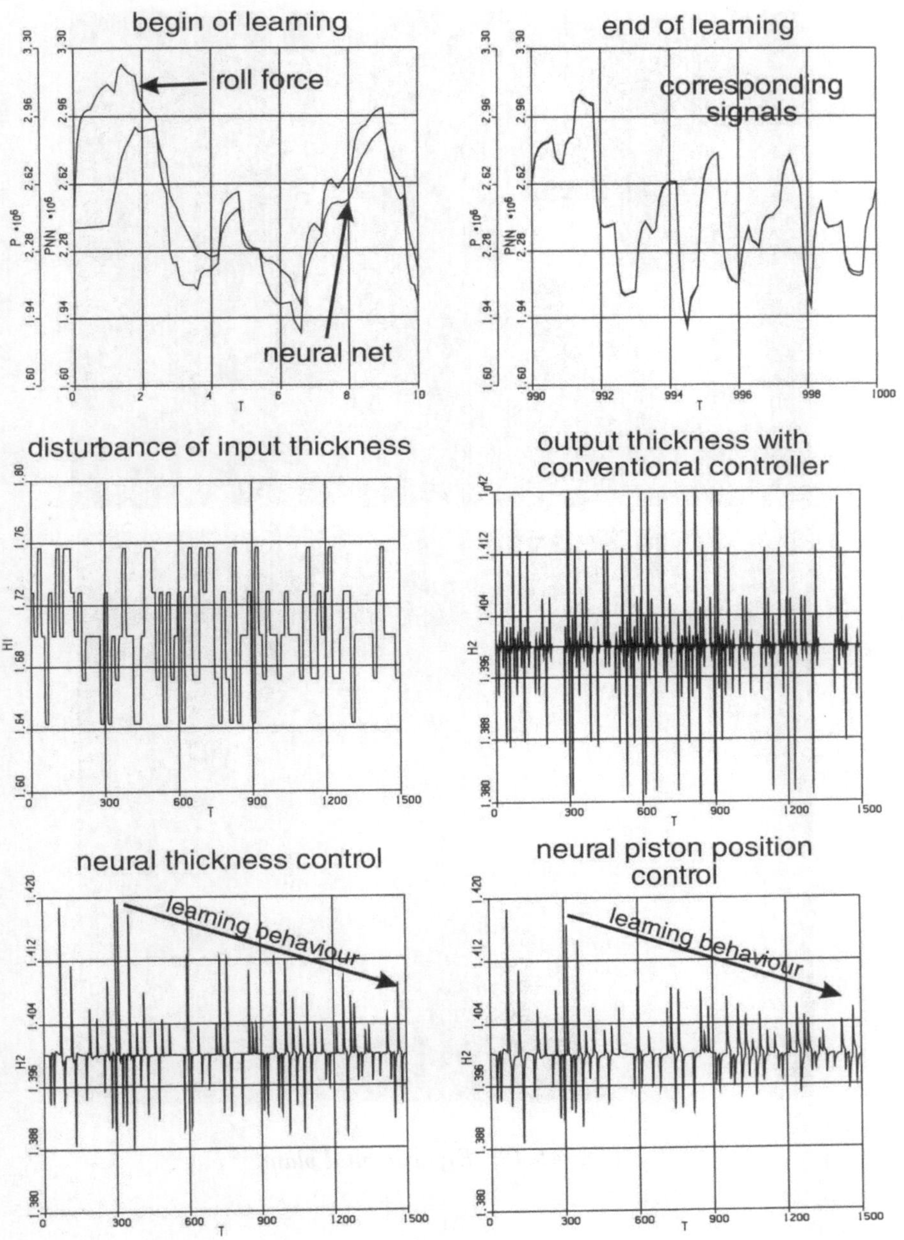

Figure 9.12: Simulation results of roll bite identification

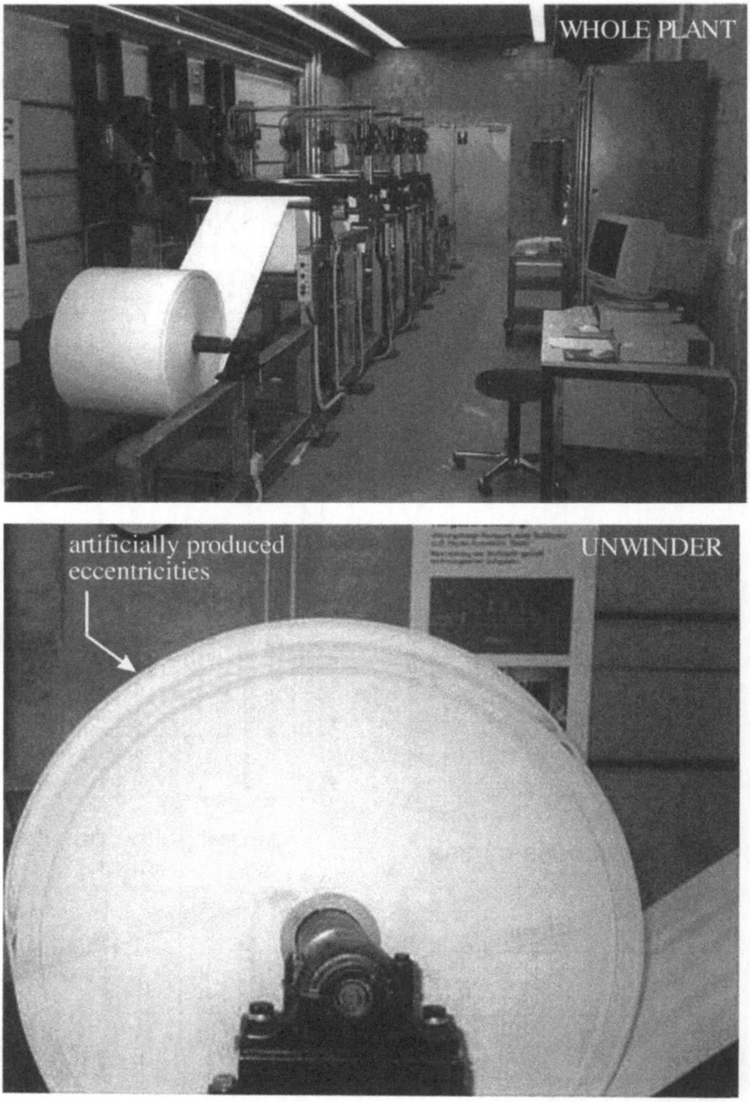

Figure 9.13: Experimental plant

acceleration reasons and friction compensation, there are different input control units implemented in the control loops.

For the identification of eccentricities different, arbitrarily chosen shapes were produced. This can be seen in figure 9.13. Motor speed, reference torque, average velocity and force were recorded with a PC and the whole observer structure with neural nets was implemented in the language C.

9.3.2 Identification Results

As it was mentioned before, the whole winder system can be divided into the drive system, modelled as a two–mass system, and the tension dynamic system. The output of the first subsystem is the non–measurable load speed as an input to the second. Therefore, the quality of the learning result is directly connected with the drive system observer. Due to this, the knowledge of the friction of the drive system would be helpful to achieve better results. The first experiments were done without (coupled) material to get information about the (load–) friction of the winder itself irrespective of the eccentricities. The neural observer, which was used in this case, is not depicted here but has a similar structure as it can be seen in chapter 7. The adaptation was done via the error between the measured and observed motor speed. The observer of the two–mass system including the GRNN demands the method of delayed activation to be used to guarantee stability of the whole system. This method can used for example during putting the plant into operation.

In figure 9.14 the results of the friction identification can be seen. As is was assumed in the last section, the drive system was modelled by a two–mass system of third order. The first two figures show the real and the observed speed based on two different reference value characteristics. In both figures it can be seen that the observed speed is faster than the real one because the a–priori friction of the observer was set to zero. The two figures below show the output of the neural net. The reader may notice a typical dependence of the identified friction with respect to the winder speed; friction increases with higher speed.

Furthermore, an additional dependence on the force is obtained if the system is coupled with the material. This problem can be solved by using the force as a second neural net input. The friction shows a nearly linear dependence on both input values, which is illustrated in figure 9.15. It must be noted, that the identified friction values are dependent on the weight of the winder and of the temperature, thus an online identification is necessary to get the current value of friction (online information).

Figure 9.16 shows the results of the identification of different artificially produced eccentricities. The observer structure used here, was the same as it can be seen in the figures 9.6 and 9.17. A fast drive system observer approximates the non–measurable winder speed as the input to the force observer. The real radius was measured mechanically before the machine was started. On the right side the outputs of the neural nets are shown, and it can be seen that the identified values are corresponding with the real values. Remaining errors come from changes of the real radius while proccessing (soft material used) and from uncertainties like

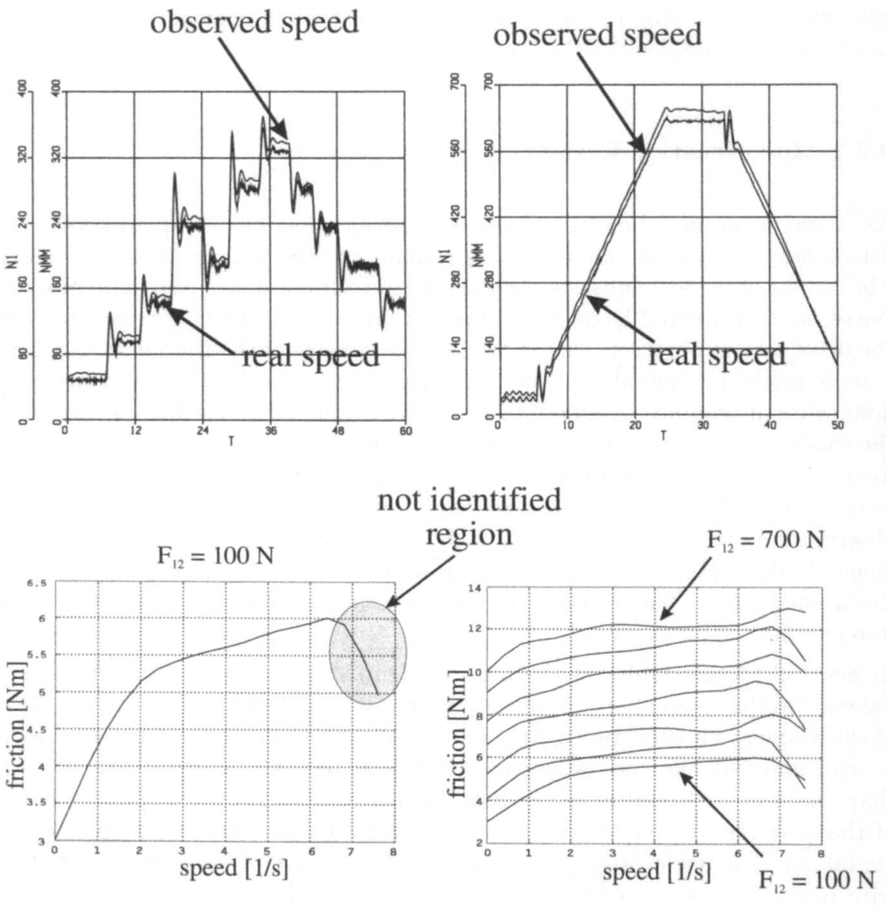

Figure 9.14: Results of friction identification

the elasticity and the unknown tension ϵ_{01} which describes the previous history of the material. The proposed method was also examined at different velocities. Furthermore, the principle of the delayed activation and the augmented error is an indispensable feature to guarantee the stability of the identification procedure. In the next section it will be shown how the identified knowledge can be used for control purposes.

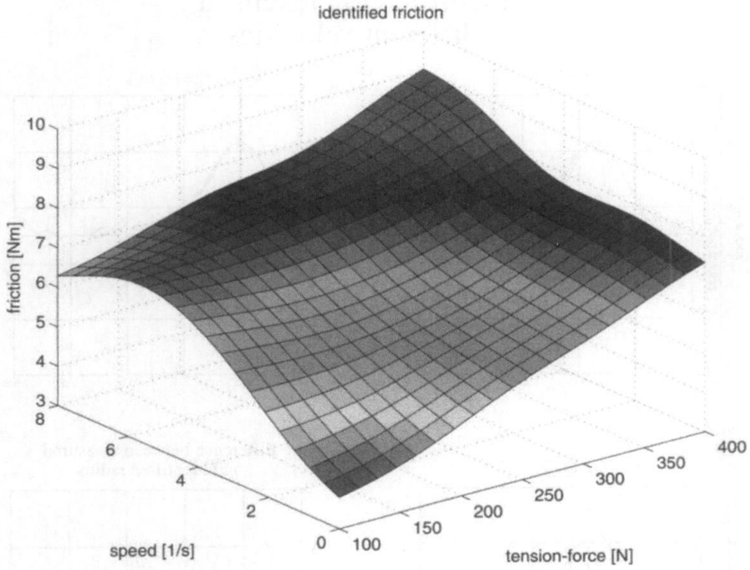

Figure 9.15: Three–dimensional representation of friction with respect to tension force and speed

9.3.3 Compensation Results

As it was mentioned before, the whole indirect control concept is divided into two parts: the identification and the control (utilization of knowledge) itself. The latter one is specific for different applications as it was shown in the previous sections 9.2.2 and 9.2.3.

At this plant a control concept was used as it can be seen in figure 9.17. The identified knowledge, the radius $\Delta\hat{R}$ (deviation of the measurable mean value of the radius R_m), is used to damp the force oscillations caused by the eccentricity. There would be different possibilities to influence the control behaviour. One of them is the compensation via the reference value of the motor speed. As it was expected (see simulation in section 9.2.2), a conventional PI–controller is not sufficient to get the desired behaviour. Figure 9.18 shows the control behaviour of the PI– and the state space controller. The first one follows the reference value very slowly. Furthermore, there is an overshoot of the motor speed. Because of the low stiffness of the two–mass system, the controller behaviour cannot be improved (see chapter 2). As it can be seen in the step response on reference values with higher frequencies and low amplitude (corresponding to a high eccentricity of $15mm$) the controller cannot be used for the compensation of oscillations caused by eccentricities.

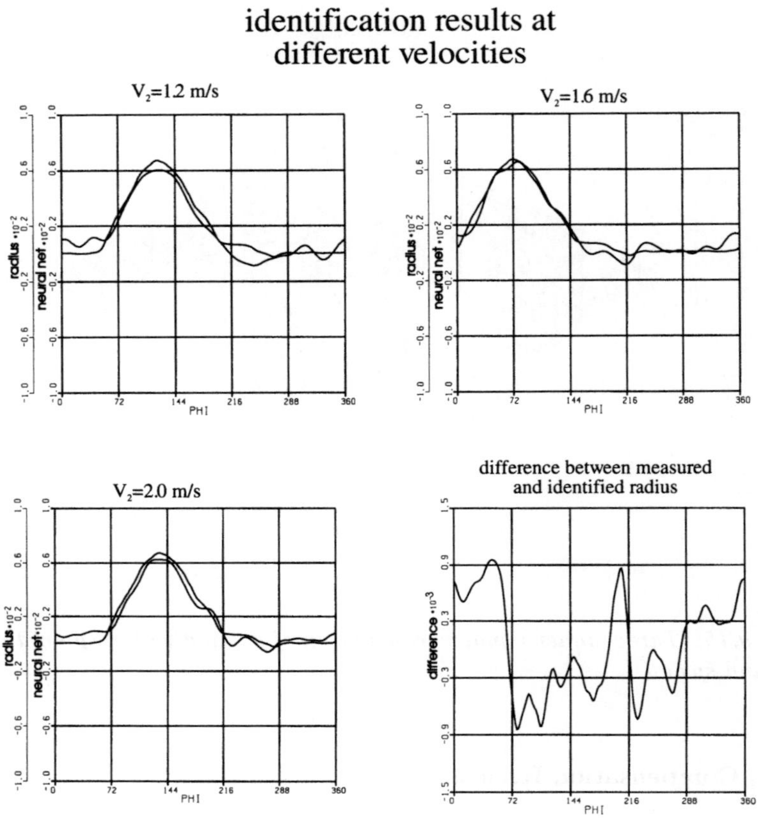

Figure 9.16: Results of eccentricity identification

As a result a state space controller was designed as it was proposed in [9] (*damping optimization criteria*). The advantage is that we obtain a defined rise time T_r for step responses at every operating point (see figure 9.18). Additionally, this controller shows a sufficient disturbance behaviour when the winder is coupled with the plant (disturbance by the variable load torque).

For the compensation, the periodical influence of the nonlinearity (eccentricity) is used. As shown in figure 9.17 the identified output of the neural net $\Delta\hat{R}$ is first delayed by

$$T_{delay} = T_{period} - T_r \tag{9.38}$$

with

$$T_{period} = \frac{2\pi R_m}{V_2} \tag{9.39}$$

where V_2 is the measurable processing velocity. To dampen the oscillations of the force, the output of the force controller has to be added with a modified reference

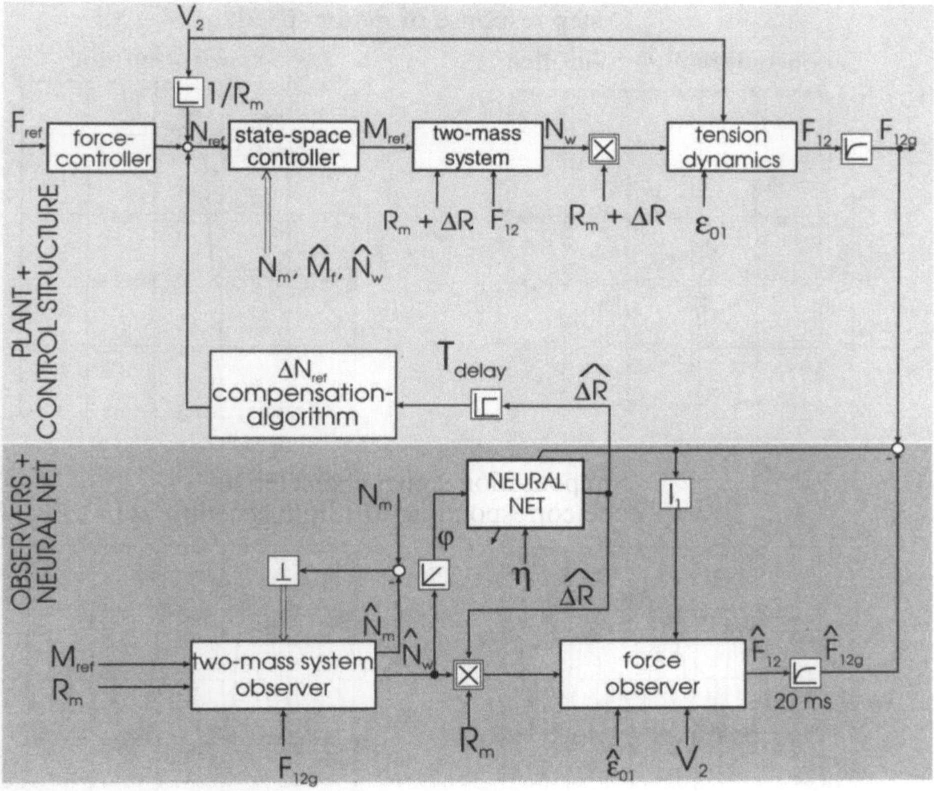

Figure 9.17: Control concept

value ΔN_{ref}. To get a constant value of the winder velocity, the following equation must be considered:

$$(N_W + \Delta N_{ref}) \cdot (R_M + \Delta \hat{R}) = N_W \cdot R_M = V_{ref} \tag{9.40}$$

V_{ref} is the desired (constant) value of the material velocity. For the winder speed we can write:

$$N_W \approx \frac{V_2}{2\pi R_m} \tag{9.41}$$

Then the compensation signal can be expressed by:

$$\Delta N_{ref} = \frac{-\Delta \hat{R} \cdot \dfrac{V_2}{2\pi R_m}}{R_m + \Delta \hat{R}} \tag{9.42}$$

Equation (9.42) shows that both, the modified reference value and the delay time, are dependent on the current operating point concerning the increasing

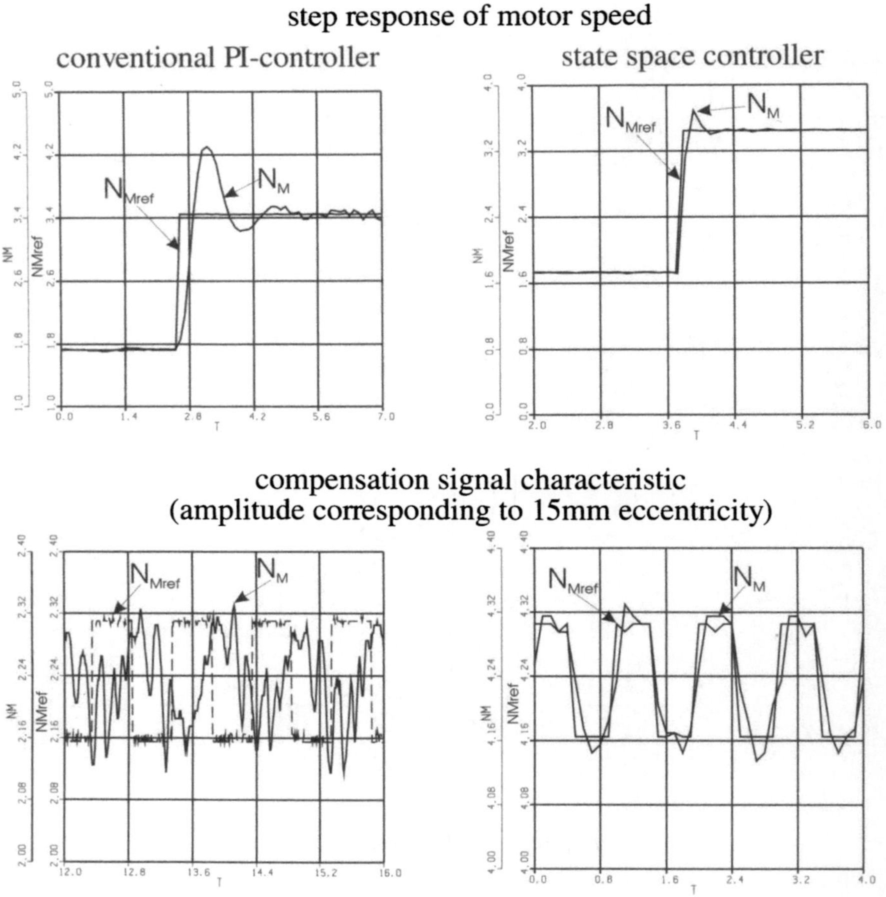

Figure 9.18: Step responses of a conventional and a state space controller

or decreasing mean value of the radius and the processing velocity of the plant. Hence the compensation algorithm must be designed adaptive to fulfill equation (9.40).

Figure 9.19 shows the results of an entire identification and compensation concept of the plant. We can see the error signal tending to zero and the identified radius corresponding with the measured one. The two figures below in 9.19 display the convergence of one neural net parameter and the force before and after compensation. It shows that the oscillations caused by the eccentricity can be damped to low deviations around the reference value of $300N$ in this case. It must be noted, that identification and compensation do not disturb each other, which can be seen in the parameter convergence even if compensation is still

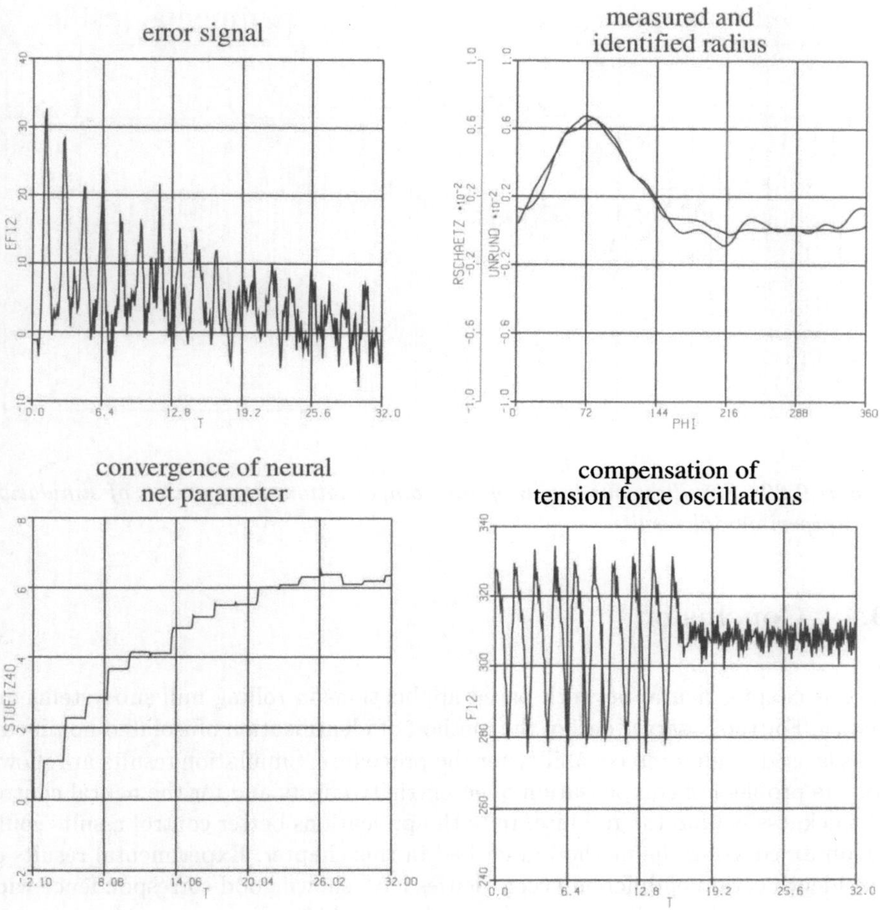

Figure 9.19: Learning results and compensation effect

working (see the time axis). Compensation can already start at the beginning of the learning procedure.

This fact is shown figure 9.20. It can be seen that compensation can be started at $t = 0$ (starting point of identification). The oscillation amplitude decreases with time. Furthermore, a good correspondence between the simulation and the experimental results can be seen which results from the correct choice of the system parameters of the plant's linear part.

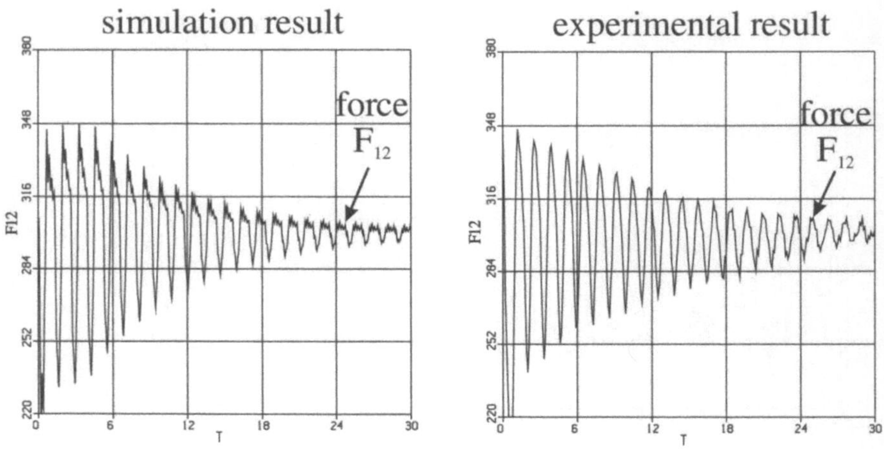

Figure 9.20: Simultaneous learning and compensation: comparison of simulation and experimental results

9.4 Conclusion

In this chapter neural network based applications in rolling mill subsystems are shown. For the observer design the method of identification of isolated nonlinearities is used to guarantee stability for the procedure. Simulation results are shown for the problem of compensation of eccentricity effects and for the neural control of thickness behind the roll bite. In both applications better control results could be obtained using the method described in this chapter. Experimental results of the identification of different eccentricities have shown good correspondence with the theory. Additionally, a controller and a suitable compensation algorithm are designed to use the identified knowledge to dampen force oscillations.

9.5 References

[1] Aeberli, K.:
Japanisches Aluminiumkaltwalzwerk fertigt Dosenbleche höchster Qualität.
Engineering and Automation, Vol. 6, 1994, pp. 20–23.

[2] Bryant, G.F.:
Automation of Tandem Mills.
The Iron and Steel Institute, Carlton House Terrace, London, 1973.

[3] Lenz, U., Schröder, D.:
Local Identification using Artificial Neural Networks.
Proceedings of the Ninth Yale Workshop on Adaptive and Learning Systems, Yale , 1996, pp. 83–88.

[4] Martinez, T., Gramckow, O., Protzel, P.:
Walzwerksteuerung mit neuronalen Netzen.
VDI Berichte No. 1184, 1995, pp. 35–42.

[5] Martinez, T., Gramckow, O., Protzel, P.:
Neuronale Netze zur Steuerung von Walzstraßen.
atp – Automatisierungstechnische Praxis, Heft 38, 1996, pp. 28–42.

[6] Narendra, K., Annaswamy, A.M.:
Stable Adaptive Systems.
Prentice Hall, Englewood Cliffs, New Jersey, 1989.

[7] Schäffner, C., Schröder, D.:
An Application of General Regression Neural Networks to Nonlinear Adaptive Control.
EPE '93, Brighton, UK. Proceedings, Vol. 4, 1993, pp. 219–223.

[8] Schlang, M.:
Neuronale Netze zur Prozessteuerung in der Stahlverarbeitung.
VDI Berichte No. 1282, München und Erlangen, 1996.

[9] Schröder, D.:
Elektrische Antriebe 2.
Springer-Verlag, Berlin, Heidelberg, 1995

[10] Stone, M. D.:
Rolling of thin strip.
Iron & Stell Eng. 30, Part I, 1953, pp. 61–73.

[11] Straub, S., Schröder, D.:
An Example of an Application of Neural Networks in Rolling Mills: Compensation of the Non Circularity of Winders.
Proceedings of IFAC 95, Munich, Germany, 1995, pp. 583–590.

[12] Wusatowski, Z.:
Grundlagen des Walzens.
VEB Deutscher Verlag für Grundstoffindustrie, Leipzig 1963.

10 Input–Output Linearization of Nonlinear Dynamical Systems: an Introduction

Kurt Fischle

The control concepts presented in the previous chapters addressed the control of systems which contain *localized* unknown nonlinearities within an otherwise (mainly) known linear structure. In the this part of the book, we will consider the problem of controlling a more general class of nonlinear plants, using considerably less prior knowledge. This is motivated by the capabilities of biological neural "controllers", which enable humans and animals to control complex nonlinear systems without using any mathematically formulated prior knowledge. However, while the controller itself should use as little mathematical knowledge as possible and acquire the necessary knowledge about the plant by means of a suitable learning law, considerable mathematical background is required to find such a learning law which should guarantee stability and well-defined control performance. In this chapter we will introduce some mathematical tools for the treatment of nonlinear dynamical systems, which will be needed in the following chapter to develop a neural network control concept for a quite general class of nonlinear plants.

10.1 A Useful Canonical Form for Nonlinear Systems

In this chapter we will discuss the task of controlling nonlinear n-th order SISO plants which can be described in state–space form as

$$\dot{\underline{x}} = \underline{f}(\underline{x}) + \underline{g}(\underline{x}) \cdot u \tag{10.1}$$

$$y = h(\underline{x}) \tag{10.2}$$

where u is the input, y is the output, $\underline{x}$ is the state vector, and $\underline{f}(\underline{x})$, $\underline{g}(\underline{x})$ and $h(\underline{x})$ are unknown nonlinear functions.[1] Plants of this structure are said to be *control affine* or *linear in control.* Our goal is to develop a control law

[1]Unlike in the other chapters of this book, the nonlinearities are denoted by small roman letters instead of the symbol $\mathcal{NL}$ here; this will simplify the notation in the following equations.

$$u = f_c^*(y, y_m, \underline{x}) \tag{10.3}$$

which makes the output $y(t)$ of the plant follow a reference signal $y_m(t)$. The reference signal specifies the desired behaviour of the plant and is provided from some external source.

For a known linear system, such a control law can easily be calculated by well-known state space control techniques. For a nonlinear system, however, it is not very obvious whether such a control law even exists. An answer to this question is given by the theory of *input–output linearization.*[2] This theory originated in the area of robotics, e.g. [2]; it was fully developed by Isidori [3] and is still an area of active research today. A good introduction can be found in [7]; here, we will only give a brief introduction to the basic concepts and present some examples for better understanding.

The basic idea of input–output linearization is to find a feedback law which can cancel the nonlinearities of the plant, leading to a linear relationship between the output y and a new artificial input v. If such a feedback law can be found, it is easy to control the linearized plant in such a way that its output shows the desired behaviour. To cancel the nonlinearities of the plant, one has to write the dynamic relationship between u and y in a certain canonical form first. This canonical form, which is somewhat similar to the controllability canonical form for linear systems, will now be derived.

We begin by differentiating $y(t)$ once with respect to time. Applying the familiar chain rule, we obtain

$$\dot{y} \quad = \quad \frac{\partial h}{\partial \underline{x}}\dot{\underline{x}} \quad = \quad \frac{\partial h}{\partial \underline{x}}\underline{f}(\underline{x}) + \frac{\partial h}{\partial \underline{x}}\underline{g}(\underline{x})u \tag{10.4}$$

To provide a convenient notation for the expressions $\frac{\partial h}{\partial \underline{x}}\underline{f}(\underline{x})$ and $\frac{\partial h}{\partial \underline{x}}\underline{g}(\underline{x})$ and other more complex derivatives we will encounter in the following, let us introduce the definition of *Lie derivatives* from differential geometry:

Definition 10.1 *Let $\underline{f} : \mathbb{R}^n \mapsto \mathbb{R}^n$ be a smooth vector-valued function on $\mathbb{R}^n$ and $h : \mathbb{R}^n \mapsto \mathbb{R}$ a smooth scalar function on $\mathbb{R}^n$. Then the Lie derivatives of h with respect to $\underline{f}$ are scalar functions on $\mathbb{R}^n$, which are defined recursively as*

$$L_f h(\underline{x}) \quad \triangleq \quad \frac{\partial h}{\partial \underline{x}}\underline{f}(\underline{x}) \tag{10.5}$$

$$L_f^{i+1} h(\underline{x}) \quad \triangleq \quad L_f(L_f^i h(\underline{x})) = \frac{\partial L_f^i h(\underline{x})}{\partial \underline{x}}\underline{f}(\underline{x}) \tag{10.6}$$

Likewise, if $\underline{g} : \mathbb{R}^n \mapsto \mathbb{R}^n$ is another smooth vector-valued function on $\mathbb{R}^n$, the scalar function $L_g L_f^i h(\underline{x})$ is defined as

[2] Input–output linearization is part of a more general theoretical framework known as *feedback linearization*; see [7] for details.

$$L_g L_f^i h(\underline{x}) \triangleq \frac{\partial L_f^i h(\underline{x})}{\partial \underline{x}} \underline{g}(\underline{x}) \tag{10.7}$$

Note that the Lie derivatives are nothing more than a simple notation for certain expressions which occur when repeatedly differentiating $y(t)$ using the chain rule. Using this notation, we can write (10.4) as

$$\dot{y} = L_f h(\underline{x}) + L_g h(\underline{x})u \tag{10.8}$$

If $L_g h(\underline{x})$ is different from zero, equation (10.4) gives an explicit description of the relationship between u and y. (Note that we assume that there is no direct feedthrough connection between u and y in (10.2); this is the case in most real plants.) If $L_g h(\underline{x})$ is equal to zero for all $\underline{x}$, we can differentiate $y(t)$ again to obtain

$$\ddot{y} = \frac{\partial L_f h(\underline{x})}{\partial \underline{x}} \dot{\underline{x}} = L_f^2 h(\underline{x}) + L_g L_f h(\underline{x})u \tag{10.9}$$

If $L_g L_f h(\underline{x})$ is also equal to zero, this process can be repeated over and over, until for some integer r one obtains

$$y^{(r)} = L_f^r h(\underline{x}) + L_g L_f^{(r-1)} h(\underline{x})u \tag{10.10}$$

with $L_g L_f^{(r-1)} h(\underline{x}) \neq 0$.[3] The integer r is called the *relative degree* of the plant:

Definition 10.2 *The system (10.1) – (10.2) is said to have relative degree r in a region $\Omega \subseteq \mathbb{R}^n$ if*

$$\left.\begin{array}{ll} L_g L_f^i h(\underline{x}) = 0 & \text{for } 0 \leq i < r-1 \\ L_g L_f^i h(\underline{x}) \neq 0 & \text{for } i = r-1 \end{array}\right\} \forall \underline{x} \in \Omega \tag{10.11}$$

Note that for linear systems, this reduces to the familiar definition of the relative degree as the excess of poles over zeros.

The canonical form (10.10) plays a central role in the theory of input–output linearization. We will soon see how it can be used to derive a control law which makes the plant (10.1) – (10.2) behave in a desired way. First, however, let us illustrate the concepts presented so far by some examples:

Example 10.1 Our first example is a two–mass drive system with a nonlinear friction torque acting on the load mass. The system is described by

[3] If one can differentiate $y(t)$ forever without u ever appearing, the system is not controllable; we will exclude this case here and assume the input u has been chosen in a suitable way, i.e. such that it can be used to influence y.

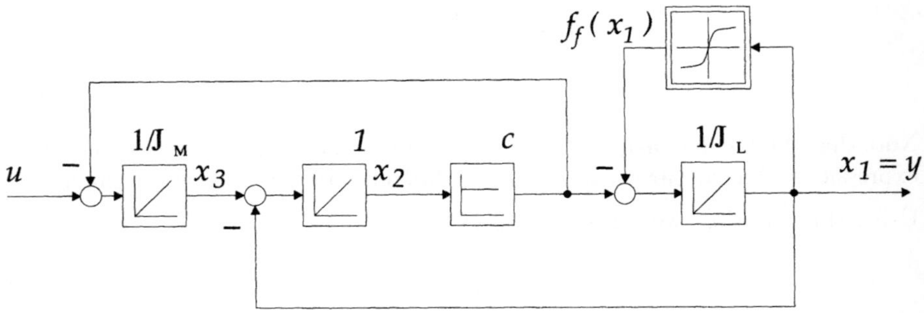

Figure 10.1: Two–mass system with nonlinear friction characteristic

$$\begin{aligned}
\dot{x}_1 &= \frac{1}{J_L}\left[c x_2 - f_f(x_1) \right] && (10.12)\\
\dot{x}_2 &= x_3 - x_1 && (10.13)\\
\dot{x}_3 &= \frac{1}{J_M}(-c x_2 + u) && (10.14)
\end{aligned}$$

$$y = x_1 \qquad (10.15)$$

where $y = x_1$ is the load speed, x_2 is the difference angle between motor and load, x_3 is the motor speed, and u is the air gap torque (which we assume can be controlled by an ideal delay-free torque controller, and thus is directly accessible as the plant input). The signal flow diagram of this plant is shown in figure 10.1. J_M and J_L are the moments of inertia of motor and load, and c is the spring constant of the shaft, which is assumed as a linear spring with zero damping. A nonlinear speed-dependent friction torque $f_f(x_1)$ is acting on the load mass.

Differentiating the load speed $y(t)$, we obtain

$$\begin{aligned}
y &= x_1 = h(x) && (10.16)\\
\dot{y} &= \frac{1}{J_L}[c x_2 - f_f(x_1))] = L_f h(\underline{x}) && (10.17)
\end{aligned}$$

We note that u does not appear in this equation, i.e. $L_g h(\underline{x}) = 0$; thus, we differentiate a second time to obtain

$$\ddot{y} = -\frac{c}{J_L}(x_3 - x_1) - \frac{1}{J_L^2} f_f'(x_1)\left[c x_2 - f_f(x_1)\right] = L_f^2 h(\underline{x}) \qquad (10.18)$$

with

$$f_f'(x_1) \triangleq \frac{\mathrm{d} f_f(x_1)}{\mathrm{d} x_1} \qquad (10.19)$$

Since $L_g L_f h(\underline{x})$ is also equal to zero, we have to differentiate a third time:

$$y^{(3)} = L_f^3 h(\underline{x}) + L_g L_f^2 h(\underline{x}) u \tag{10.20}$$

with

$$\begin{aligned} L_f^3 h(\underline{x}) &= -\frac{1}{J_L^3}[f_f''(x_1) f_f(x_1) + f_f'^2(x_1)] f_f(x_1) - \frac{c}{J_L^2} f_f(x_1) \\ &\quad + \Big[\frac{c}{J_L^3}[2 f_f''(x_1) f_f(x_1) + f_f'^2(x_1)] - \frac{c^2}{J_M J_L}\Big] x_2 \\ &\quad - \frac{c^2}{J_L^3} f_f''(x_1) x_2^2 - \frac{c}{J_L^2} f_f'(x_1)(x_3 - x_1) \end{aligned} \tag{10.21}$$

$$L_g L_f^2 h(\underline{x}) = \frac{c}{J_M J_L} \tag{10.22}$$

$$f_f''(x_1) \triangleq \frac{\mathrm{d}^2 f_f(x_1)}{\mathrm{d}x_1^2} \tag{10.23}$$

Thus, the system is input–output–linearizable with a relative degree $r = 3$. The canonical representation (10.20) of this system is depicted in figure 10.2.

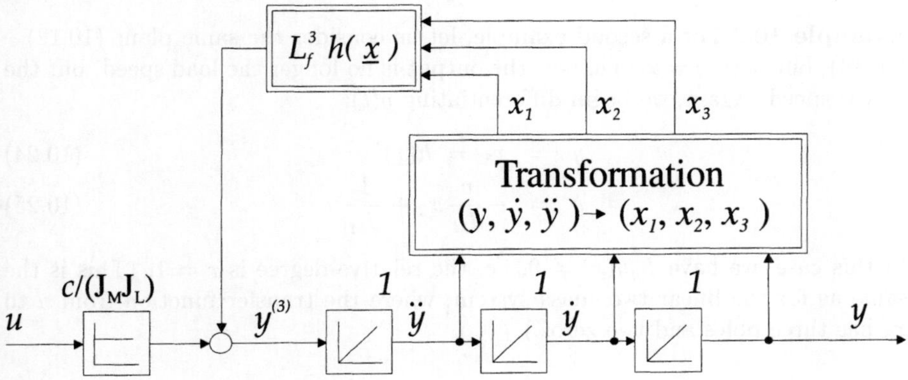

Figure 10.2: Transformed two–mass system with nonlinear friction characteristic

Several points should be noted about this example:

- $L_f^3 h(\underline{x})$ is a rather complex expression, which contains the nonlinearity $f_f(x_1)$ as well as its first and second derivatives with regard to x_1. It is nonlinear in x_1 and x_2, but linear in x_3. $L_f^3 h(\underline{x})$ does only exist if $f_f(x_1)$ can be differentiated twice; this is not the case e.g. for the Coulomb friction characteristic, which is not differentiable at $x_1 = 0$. Note that the friction characteristic would *not* need to be differentiable if the friction was at the motor mass. In that case, the friction could directly be compensated by the input u. However, differentiability of the nonlinearities is required for the transformation in all cases where dynamic elements are located between the input u and the nonlinearity.

- $L_g L_f^2 h(\underline{x})$ is constant, since there are no nonlinearities in the direct feedforward path from u to y.
- The transformed equation (10.20) is valid in the entire state space $[x_1, x_2, x_3]^T \in \mathbb{R}^n$.
- The transformation of the system (10.12) – (10.15) into the form (10.20) is not a true state transformation; the transformed system is described by y, $\dot{y}$ and $\ddot{y}$, but the term $L_f^3 h(\underline{x})$ still depends on the "old" states x_1, x_2, x_3. Note that even if the transformation $[x_1, x_2, x_3] \rightarrow [y, \dot{y}, \ddot{y}]$ (which is given by $[h(\underline{x}), L_f h(\underline{x}), L_f^2 h(\underline{x})]$) is valid globally, the inverse transformation $[y, \dot{y}, \ddot{y}] \rightarrow [x_1, x_2, x_3]$ is not necessarily defined everywhere or unique. However, fortunately, this transformation will not be needed by our control concept.

Example 10.2 For a second example, let us consider the same plant (10.12) – (10.14), but with $y = x_3$; i.e. now the output is no longer the load speed, but the motor speed. Again, we begin differentiating $y(t)$:

$$y = x_3 = h(\underline{x}) \tag{10.24}$$

$$\dot{y} = -\frac{c}{J_M} x_2 + \frac{1}{J_M} u \tag{10.25}$$

In this case, we have $L_g h(\underline{x}) \neq 0$, i.e. the relative degree is $r = 1$. (This is the same as for the linear two–mass system, where the transfer function from u to x_3 has three poles and two zeros.)

Example 10.3 As our last example, we will examine a two–mass system with a nonlinear spring characteristic; i.e. the shaft torque depends on the difference angle x_2 in a nonlinear fashion. The equations of this system are

$$\dot{x}_1 = \frac{1}{J_L} f_s(x_2) \tag{10.26}$$

$$\dot{x}_2 = x_3 - x_1 \tag{10.27}$$

$$\dot{x}_3 = \frac{1}{J_M} (-f_s(x_2) + u) \tag{10.28}$$

$$y = x_1 \tag{10.29}$$

Figure 10.3 depicts the signal flow diagram for this plant.

Differentiating y, one finds that the relative degree of the system is $r = 3$; one has

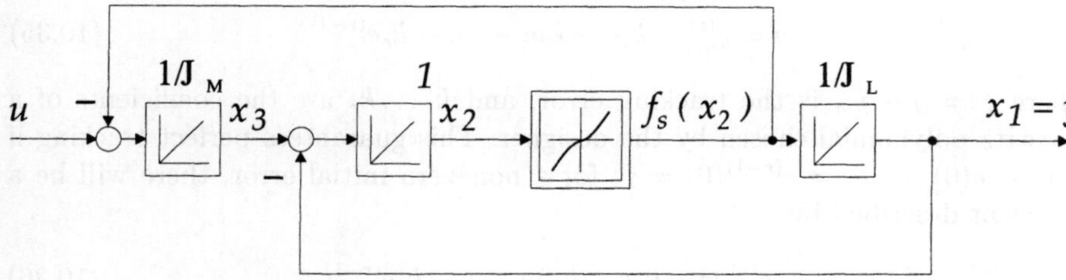

Figure 10.3: Two–mass system with nonlinear spring characteristic

$$y^{(3)} = L_f^3 h(\underline{x}) + L_g L_f^2 h(\underline{x}) u \tag{10.30}$$

with

$$L_f^3 h(\underline{x}) = -(\frac{1}{J_L^2} + \frac{1}{J_L J_M}) f_s'(x_2) f_s(x_2) + \frac{1}{J_L} f_s''(x_2)(x_3 - x_1)^2 \tag{10.31}$$

$$L_g L_f^2 h(\underline{x}) = \frac{1}{J_L J_M} f_s'(x_2) \tag{10.32}$$

In this case, $L_g L_f^{(r-1)} h(\underline{x})$ is not constant, as in the previous examples, but state–dependent. This is due to the fact that the nonlinearity $f_s(x_2)$ lies in the direct feedforward path from u to y in this case. Like in the previous examples, the nonlinearity must be twice differentiable for the transformation to the canonical form (10.30).

10.2 Basic Concept of Input–Output Linearization

In the heading of section 10.1, equation (10.10) has been called "a useful canonical form". So what is it useful for? Equation (10.10) plays a central role in the theory of input–output linearization: if a nonlinear plant can be transformed into this structure, it is obvious that a feedback of the form

$$u = \frac{v - L_f^r h(\underline{x})}{L_g L_f^{(r-1)} h(\underline{x})} \tag{10.33}$$

cancels the nonlinearities of the plant and leads to a linear relationship

$$y^{(r)} = v \tag{10.34}$$

between the plant output y and the new input v. Using this linear relationship, the plant output can easily be made to exhibit a desired behaviour. For example, if one wants y to track a given reference signal y_m, one can choose

$$v = y_m^{(r)} - k_1 e - k_2 \dot{e} - \ldots - k_r e^{(r-1)} \tag{10.35}$$

where $e = y - y_m$ is the tracking error, and $k_1 \ldots k_r$ are the coefficients of a Hurwitz polynomial chosen by the designer. This guarantees perfect tracking if $e(0) = \dot{e}(0) = \ldots = e^{(r-1)}(0) = 0$; for a non–zero initial error, there will be a transient described by

$$e^{(r)} = -k_1 e - k_2 \dot{e} - \ldots - k_r e^{(r-1)} \tag{10.36}$$

We will call the control law

$$u = f_c^*(y_m^{(r)}, e, \dot{e} \ldots e^{(r-1)}, \underline{x}) = \frac{y_m^{(r)} - k_1 e - k_2 \dot{e} - \ldots - k_r e^{(r-1)} - L_f^r h(\underline{x})}{L_g L_f^{(r-1)} h(\underline{x})} \tag{10.37}$$

the *ideal control law* for the system (10.10). The signal flow diagram for this control law is shown in figure 10.4 for the case $r = n = 2$. (For a better understanding of this quite complex structure, it should be kept in mind that the goal is to make $\ddot{y} = \ddot{y}_m$. Theoretically, this control law would also be valid without the feedback terms $k_1 e - k_2 \dot{e}$ if the initial conditions were $e(0) = \dot{e}(0) = 0$.)

Example 10.4 Let us assume that we want to control the two–mass system with nonlinear friction from figure 10.1 (10.12) – (10.15) in such a way that the load speed $y = x_1$ follows a given reference signal y_m. We generate this reference signal from an external reference input $w(t)$ by means of a linear reference model

$$y_m^{(3)} = b_{1m} w - a_{1m} y_m - a_{2m} \dot{y}_m - a_{3m} \ddot{y}_m \tag{10.38}$$

The reference model allows the designer to specify the desired dynamics of the closed–loop system; the selection of the reference model is analogous to the selection of the desired closed–loop poles in linear state controller design. In our example, we use the *method of characteristic ratios*[4] by Naslin [4, 6], a standard rule which allows to design PT_2–like desired dynamics for systems of higher order. Specifying the equivalent time constant of the reference model as $T_m = 0.02\,\text{s}$, the method of characteristic ratios yields the reference model coefficients $a_{1m} = b_{1m} = 8/T_m^3 = 10^6$, $a_{2m} = 8/T_m^2 = 2 \cdot 10^4$ and $a_{3m} = 4/T_m = 200$.

Since we have already transformed this system to the canonical form (10.10) in example 10.1, we can immediately state that the ideal control law (10.37)

$$u = \frac{y_m^{(3)} - k_1 e - k_2 \dot{e} - k_3 \ddot{e} - L_f^3 h(\underline{x})}{L_g L_f^2 h(\underline{x})}$$

with $e = y - y_m$ and $L_f^3 h(\underline{x})$, $L_g L_f^2 h(\underline{x})$ defined according to (10.21), (10.22), will achieve the desired objective. The coefficients $k_1 \ldots k_3$, which describe the dynamics of the initial error, may be chosen as the coefficients of an arbitrary

[4] Also known as the *damping optimum* method.

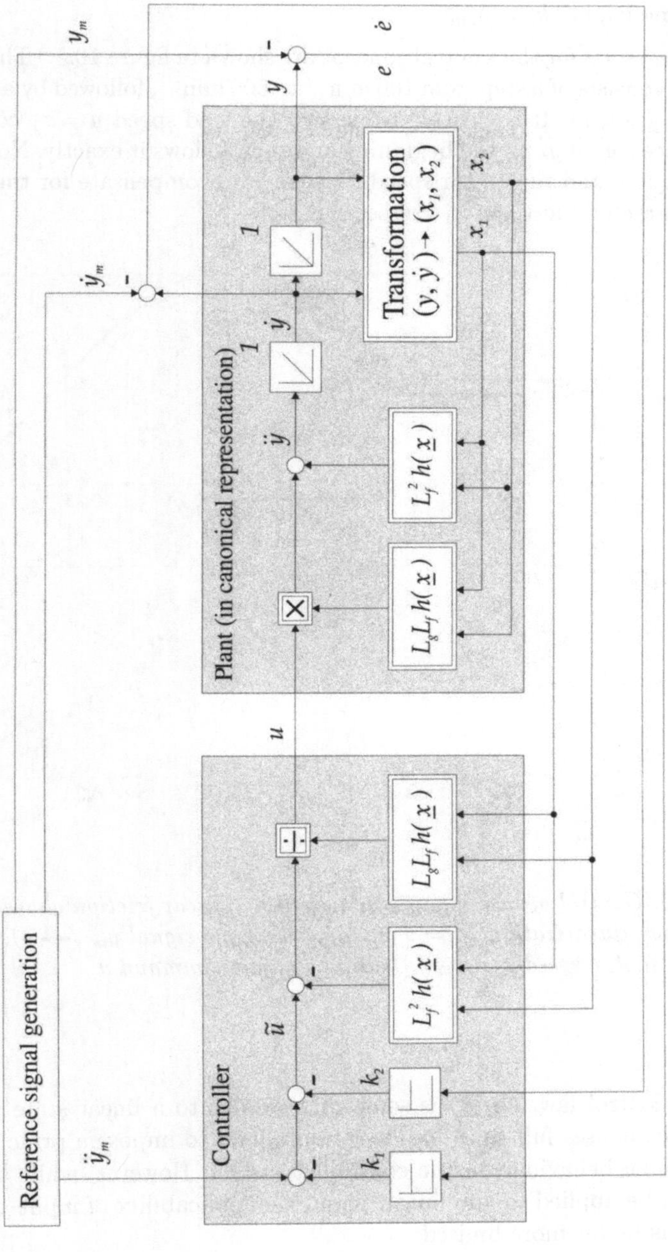

Figure 10.4: Control of a second order plant by input–output linearization

Hurwitz polynomial; for simplicity, we choose the same coefficients as in the reference model, i.e. $k_\nu = a_{\nu m}$.

Simulation results for this control concept are shown in figure 10.5.[5] The reference input $w(t)$ consists of a step from $100\,\text{min}^{-1}$ to $500\,\text{min}^{-1}$, followed by a ramp with a negative gradient. It is clearly visible how the load speed $y = x_1$ converges to the reference signal y_m in the beginning and then follows it exactly. Note how the motor torque u and the motor speed x_3 behave to compensate for the nonlinear friction characteristic when x_1 crosses zero.

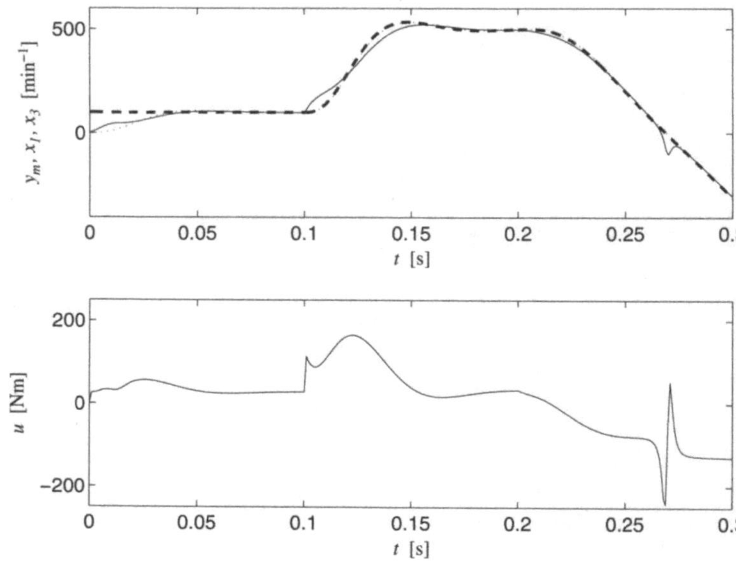

Figure 10.5: Control of two–mass system with nonlinear friction characteristic by input–output–linearization ($y = x_1$); top: reference signal y_m (- - -), load speed x_1 (· · · ·), motor speed x_3 (——); bottom: torque command u

The ideal control law (10.37) is somewhat similar to a linear state space controller, since it uses full state feedback and allows to impose a principally arbitrary dynamic behaviour on the controlled system. However, while state space control can be applied to any linear plant, the applicability of input–output linearization is by far more limited:

[5]The plant parameters used in the simulation were $J_M = J_L = 5 \cdot 10^{-3}\,\text{kgm}^2$, $c = 600\,\text{Nm/rad}$; the friction characteristic was modelled by $f_f(x_1) = 20\,\text{Nm}\,\arctan(x_1/20\,\text{min}^{-1})$, which fulfills the condition of being differentiable two times.

- The first restriction is that it can only be applied to plants which are transformable into the form (10.10). This requires that the Lie derivatives must exist, which in turn implies that $\underline{f}(\underline{x})$, $\underline{g}(\underline{x})$ and $h(\underline{x})$ must be differentiable a sufficient number of times. Thus, plants with hard nonlinearities like Coulomb friction are excluded.
- Furthermore, $L_g L_f^{(r-1)} h(\underline{x})$ must be different to zero everywhere, since it appears in the denominator of (10.37).
- Even if such a transformation exists for a given plant, it may only be valid in a certain region $\Omega \subset \mathbb{R}^n$ of the state space, which implies that the controller only works as long as the state remains inside this region.
- There are even systems which do not have a well-defined relative degree, e.g. if $L_g L_f^i h(\underline{x})$ is zero at one point but non-zero arbitrarily close to that point for some i.

As the attentive reader will probably already have noticed, there is another problem which may arise when using the controller (10.37): The linearized system (10.34) is of order r, while the original system (10.1) – (10.2) was of order n; for the case $r < n$, some states are "lost" in the transformation. (It can easily be shown that, like for linear systems, there is always $r \leq n$.) Thus, if the linearizing feedback law (10.33) is applied, the complete dynamics of the resulting system is described by the linear r-th order relationship (10.34) and an additional subsystem of order $n - r$, which is not observable in y. The dynamics of this subsystem is called the *internal dynamics*, because it has no influence on the external input–output relationship. Input–output linearization can only be applied if the internal dynamics is stable; if it is unstable, the output will show the desired behaviour, but some states of the plant will grow in an unbounded way. Checking the stability of the internal dynamics is generally a difficult task; see [7] for a detailed discussion. In this book we will confine ourselves to the case where the relative degree r of the plant is equal to its order n, and thus there is no internal dynamics.[6]

Example 10.5 As it has been shown in example 10.2, the relative degree of the two–mass system with nonlinear friction characteristic (10.12) – (10.14) is $r = 1$ when the motor speed x_3 is used as the output. Transformation to the canonical form (10.10) yields

$$\dot{y} = -\frac{c}{J_M}\, x_2 + \frac{1}{J_M}\, u$$

Obviously, the linearizing feedback law (10.33)

$$u = J_m v + c\, x_2$$

[6] This assumption is fulfilled e.g. for multi–mass drive systems where u is the torque acting on the first mass (motor), and y is the speed or position of the last mass; see example 10.1.

leads to the linear input–output–dynamics

$$\dot{y} = v$$

which is of first order, while the original system was of third order. The unobservable internal dynamics of the linearized system can be described by the remaining two state equations

$$\dot{x}_1 = \frac{1}{J_L}\left[c x_2 - f_f(x_1) \right] \tag{10.39}$$

$$\dot{x}_2 = y - x_1 \tag{10.40}$$

It can easily be seen that the internal dynamics is marginally stable for $f_f(x_1) \equiv 0$,[7] and stable if $f_f(x_1)$ lies entirely in the first and third quadrant.

Figure 10.6 shows simulation results for the two–mass system with nonlinear friction controlled by the ideal control law

$$u = J_m(\dot{y}_m - k_1 e) + c\,x_2 \tag{10.41}$$

with y_m being the output of a first–order reference model, and $k_1 = a_{1m} = b_{1m} = 50$. (All other simulation parameters are identical to example 10.4.) While the motor speed $y = x_3$ follows the reference signal exactly after a short transient, the load speed x_1 exhibits poorly damped oscillations. This shows that even a stable internal dynamics does not guarantee a satisfactory behaviour of the overall system when input–output linearization is applied; generally, the applicability of input–output linearization has to be checked on a case–by–case basis for plants with $r < n$.

10.3 Simplified Ideal Control Law

The technique of input–output linearization, which has been introduced in the previous sections, is the basis for the class of neural network and adaptive fuzzy control concepts which will be presented in the next chapter. Basically, the objective of these concepts is to approximate the ideal control law (10.37) by a nonlinear adaptive controller of a suitable structure. The first of these concepts, which was developed by Tzirkel–Hancock and Fallside in 1991 [9, 10], uses a controller structure equivalent to figure 10.4, with the unknown functions $L_f^r h(\underline{x})$ and $L_g L_f^{(r-1)} h(\underline{x})$ replaced by neural networks. The same controller structure is also used by the neural network concept by Schäffner and Schröder [5] and an adaptive fuzzy control concept by Spooner and Passino [8]; various other control

[7] For linear systems, stability of the internal dynamics is determined by the location of the zeros of the transfer function. Thus, the marginal stability of the internal dynamics agrees with the fact that the undamped linear two–mass system has two zeros on the imaginary axis.

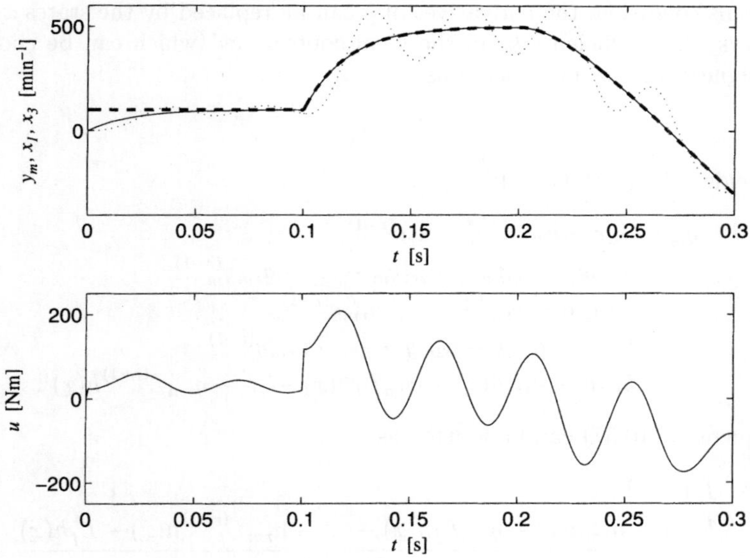

Figure 10.6: Control of two–mass system with nonlinear friction characteristic by input–output–linearization ($y = x_3$); top: reference signal y_m (- - -), load speed x_1 ($\cdots$), motor speed x_3 (——); bottom: torque command u

concepts published in the literature use closely related control laws. However, all these concepts have one major drawback: the control law (10.37) requires $r-1$ derivatives of the tracking error e, and, hence, of the measured plant output y. It is well–known that differentiation of measured signals is very problematic in practice, because it leads to amplification of high–frequency noise. In most practical applications, only a single differentiation is possible; this restricts the applicability of these controller structures to systems of first and second order.

However, this drawback can be avoided: if the reference signal y_m fulfills a certain condition, the ideal control law (and hence the structure of the adaptive controller) can be simplified so that it does not require $e, \dot{e} \ldots e^{(r-1)}$ as inputs any more, but only the states $x_1 \ldots x_n$. The condition is that y_m must be the output of a linear r–th order reference model of numerator degree zero, i.e.

$$y_m^{(r)} = -a_{1m} y_m - a_{2m} \dot{y}_m - \ldots - a_{rm} y_m^{(r-1)} + b_{1m} w \tag{10.42}$$

where $w(t)$ is a bounded external reference input. Note that this is not a hard restriction; (10.42) is a common form for reference models, which can be designed using well–known criteria known from linear state controller theory, e.g. the method of characteristic ratios [4, 6] used in example 10.4.

Under this condition, the derivatives of y can be replaced by the states $x_1 \dots x_n$ as follows: the coefficients k_ν in the ideal control law (which can be chosen by the designer) have to be selected as

$$k_\nu = a_{\nu m} \tag{10.43}$$

This yields (cf. (10.4)–(10.10))

$$\begin{aligned} & y_m^{(r)} - k_1 e - k_2 \dot{e} - \dots - k_r e^{(r-1)} \\ = \; & b_{1m} w - a_{1m} y_m - a_{2m} \dot{y}_m - \dots - a_{rm} y_m^{(r-1)} \\ & - a_{1m} e - a_{2m} \dot{e} - \dots - a_{rm} e^{(r-1)} \\ = \; & b_{1m} w - a_{1m} y - a_{2m} \dot{y} - \dots - a_{rm} y^{(r-1)} \\ = \; & b_{1m} w - a_{1m} h(\underline{x}) - a_{2m} L_f h(\underline{x}) - \dots - a_{rm} L_f^{(r-1)} h(\underline{x}) \end{aligned}$$

Consequently, (10.37) can be written as

$$\begin{aligned} u &= f_c^*(w, \underline{x}) \\ &= \frac{b_{1m} w - a_{1m} h(\underline{x}) - a_{2m} L_f h(\underline{x}) - \dots - a_{rm} L_f^{(r-1)} h(\underline{x}) - L_f^r h(\underline{x})}{L_g L_f^{r-1} h(\underline{x})} \end{aligned} \tag{10.44}$$

Thus, the ideal control law only depends on w and $\underline{x}$ instead of $y_m^{(r)}$, e, $\dot{e} \dots e^{(r-1)}$ and $\underline{x}$. The controller does not need $r-1$ derivatives of the measured quantity y any more, but only a set of state variables $x_1 \dots x_n$ which completely describe the plant state. The structure of this ideal control law for the case $r = n = 2$ is shown in figure 10.7.

Example 10.6 The simplified ideal control law (10.44) for a simple second order plant is computed in section 11.3.1; a graphical representation of the ideal control law is shown in figure 11.5.

Example 10.7 In example 10.4, the two–mass system with nonlinear friction characteristic was controlled by the ideal control law (10.37). This required the feedback of $\ddot{e}$, and hence a double differentiation of the load speed $y = x_1$. In simulation, $\dot{y}$ and $\ddot{y}$ could be computed from the states $x_1 \dots x_3$ via the known equations of the plant; in practice, however, double differentiation of the measured speed would hardly yield a usable result. If a linear reference model of numerator degree zero is used and the coefficients $k_1 \dots k_3$ are selected as $k_\nu = a_{\nu m}$ (as it was already the case in example 10.4), the control law (10.37) can be written as (10.44)

$$u = \frac{b_{1m} w - a_{1m} h(\underline{x}) - a_{2m} L_f h(\underline{x}) - a_{3m} L_f^2 h(\underline{x}) - L_f^3 h(\underline{x})}{L_g L_f^2 h(\underline{x})}$$

with the Lie–derivatives defined according to (10.16) – (10.22). (Note that for $f_f(x_1) \equiv 0$, all nonlinear terms disappear from this control law, and it becomes

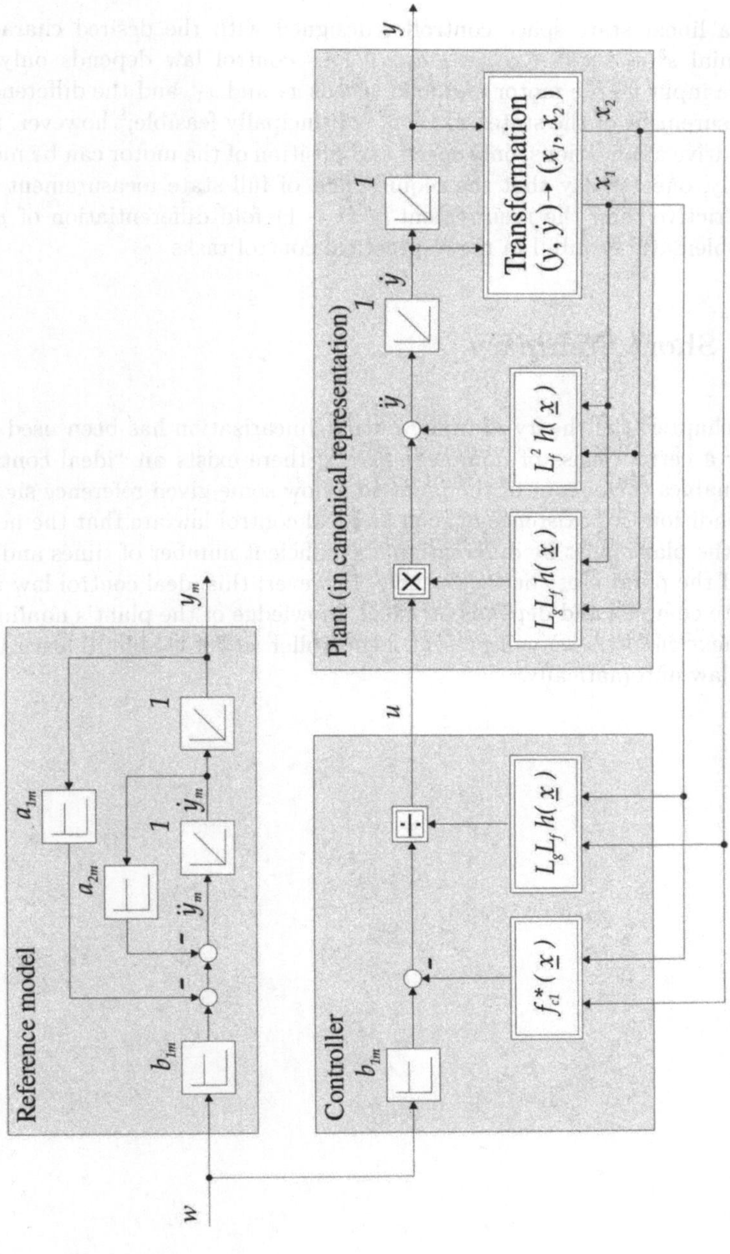

$$f_{c1}^*(\underline{x}) = a_{1m}h(\underline{x}) + a_{2m}L_f h(\underline{x}) + L_f^2 h(\underline{x})$$

Figure 10.7: Control of a second order plant by input–output linearization (simplified control law) simply a linear state space controller designed with the desired characteristic polynomial $s^3 + a_{3m}s^2 + a_{2m}s + a_{1m}$.) This control law depends only on the reference input w, the motor and load speeds x_3 and x_1, and the difference angle x_2. Measurement of the states $x_1 \dots x_3$ is principally feasible;[8] however, in many electric drive applications, only speed and position of the motor can be measured. Generally, one can say that the requirement of full state measurement is much less restrictive than the requirement of $(r-1)$–fold differentiation of y, but is still problematic to fulfill in many practical control tasks.

10.4 Short Summary

In this chapter, the theory of input–output linearization has been used to show that for a certain class of nonlinear plants, there exists an "ideal control law" which enables the output of the plant to follow some given reference signal. The main conditions for existence of such an ideal control law are that the nonlinearities of the plant must be differentiable a sufficient number of times and that all states of the plant must be measurable. However, this ideal control law is generally quite complex and depends on exact knowledge of the plant's nonlinearities. In the next chapter, we will present a controller which is able to learn the ideal control law automatically.

[8]Actually, the control concept presented in the next chapter was tested experimentally using a laboratory setup of a two–mass system, where the states $x_1 \dots x_3$ were measured using incremental encoders at the motor and the load; see [1] for the results.

10.5 References

[1] Fischle, K., Schröder, D.:
Stable Model Reference Neurocontrol for Electric Drive Systems.
Proc. Seventh European Conference on Power Electronics and Applications (EPE), Trondheim, 1997.

[2] Freund, E.:
Decoupling and Pole Assignment in Nonlinear Systems.
Electronics Letter, vol. 9, no. 16, 1973.

[3] Isidori, A.:
Nonlinear Control Systems: An Introduction.
Springer–Verlag, Berlin, Heidelberg, 1989.

[4] Naslin, P.:
Essentials of Optimal Control.
Iliffe, 1968.

[5] Schäffner, C., Schröder, D.:
An Application of General Regression Neural Network to Nonlinear Adaptive Control.
5th European Conference on Power Electronics and Applications EPE '93, Brighton, UK. Proceedings, vol. 4, 1993, pp. 219–223.

[6] Schröder, D.:
Elektrische Antriebe 2: Regelung von Antrieben.
Springer–Verlag, Berlin, Heidelberg, 1995.

[7] Slotine, J., Li, W.:
Applied Nonlinear Control.
Prentice–Hall International, Inc., New Jersey 1991.

[8] Spooner, J.T., Passino, K.M.:
Stable Adaptive Control Using Fuzzy Systems and Neural Networks.
IEEE Transactions on Fuzzy Systems, vol. 4, no. 3, 1996, pp. 339–359.

[9] Tzirkel–Hancock, E., Fallside, F.:
A Direct Control Method for a Class of Nonlinear Systems Using Neural Networks.
Proc. Second IEE International Conference on Artificial Neural Networks, Bournemouth, 1991, pp. 134–138.

[10] Tzirkel–Hancock, E., Fallside, F.:
Stable Control of Nonlinear Systems Using Neural Networks.
Int. J. Robust and Nonlinear Control, vol. 2, no.1, 1992.

10.5 References

[1] Fischle, K., Schröder, D.:
Stable Model Reference Adaptive Control for Electric Drive Systems.
Proc. Seventh European Conference on Power Electronics and Applications (EPE), Trondheim, 1997.

[2] Freund, E.:
Decoupling and Pole Assignment in Nonlinear Systems.
Electronics Letter, vol. 9, no. 16, 1973.

[3] Isidori, A.:
Nonlinear Control Systems: An Introduction.
Springer-Verlag, Berlin, Heidelberg, 1989.

[4] Naslin, P.:
Essentials of Optimal Control.
Iliffe, 1968.

[5] Schäffner, C., Schröder, D.:
An Application of General Regression Neural Network to Nonlinear Adaptive Control.
5th European Conference on Power Electronics and Applications EPE'93, Brighton, UK. Proceedings, vol. 4, 1993, pp. [illegible].

[6] Schröder, D.:
Elektrische Antriebe 2: Regelung von Antrieben.
Springer-Verlag, Berlin, Heidelberg, 1995.

[7] Slotine, J., Li, W.:
Applied Nonlinear Control.
Prentice-Hall International, Inc., New Jersey, 1991.

[8] Spooner, J. T., Passino, K.M.:
Stable Adaptive Control Using Fuzzy Systems and Neural Networks.
IEEE Transactions on Fuzzy Systems, vol. 4, no. 3, 1996, pp. 339–359.

[9] Tzirkel-Hancock, E., Fallside, F.:
A Direct Control Method for a Class of Nonlinear Systems Using Neural Networks.
Proc. Second IEE International Conference on Artificial Neural Networks, Bournemouth, 1991, pp. 134–138.

[10] Tzirkel-Hancock, E., Fallside, F.:
Stable Control of Nonlinear Systems Using Neural Networks.
Int. J. Robust and Nonlinear Control, vol. 2, no. 1, 19[illegible].

11 Stable Model Reference Neurocontrol

Kurt Fischle

11.1 Introduction

In this chapter, we will use the theoretical foundations presented in the previous chapter to introduce the concept of Stable Model Reference Neurocontrol (SMRNC), a neural network control method for a general class of nonlinear plants, which requires relatively little prior knowledge.

Actually, Stable Model Reference Neurocontrol is not a single control concept. Beginning in 1992 with the papers [17, 18] and [12], a whole family of nonlinear adaptive control methods has been published (e.g. [13, 19, 16, 15, 11, 2, 5, 7]) which share the common features

- model reference adaptive control structure,
- use of universal function approximators (e.g. Radial Basis Function Networks, General Regression Neural Networks or adaptive fuzzy controllers) and
- Lyapunov–based stable adaptive laws.

While the concepts using fuzzy controllers as function approximators are generally referred to as "Stable Adaptive Fuzzy Control", a widely accepted name for this group of concepts has not been found yet. "Stable Model Reference Neurocontrol" would be a concise description of its essential features.[1]

The various Stable Model Reference Neurocontrol concepts differ in the controller structure, the adaptive law and the type of the used function approximator (RBF Network, GRNN or fuzzy controller). By combining elements from different publications, a large variety of control concepts can be created. In this book we will confine ourselves to describing one of them — the one with the most simple controller structure — in detail, and giving a brief overview over some of the possible

[1]The term "neurocontrol" is used in its broadest possible sense here, also including adaptive fuzzy control.

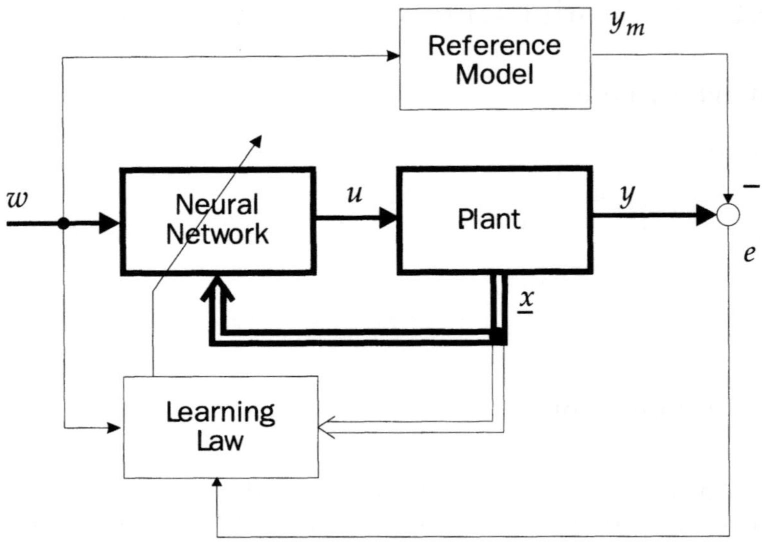

Figure 11.1: Stable Model Reference Neurocontrol: basic system structure

modifications. More detailed information can be found in the source papers and the PhD thesis [8].

The basic structure of a Stable Model Reference Neurocontrol system is shown in figure 11.1. Basically, the method is an extension of linear model reference adaptive control. The control task is to make the output y of an unknown nonlinear plant follow the reference signal y_m, which is generated from an external reference input w by means of a reference model. This model decribes the desired closed–loop dynamics for the inner control loop consisting of the plant and a nonlinear adaptive controller. To achieve the desired dynamics, the parameters of the controller are tuned on–line, based on the error signal e, in an outer adaptation loop.

The main advantage of Stable Model Reference Neurocontrol over most other nonlinear adaptive control concepts is that a mathematical proof of stability can be given for the overall system plant–controller–adaptive law. This proof is independent of the initial conditions; i.e. the adaptation cannot "get stuck" in a local minimum, like it can happen with gradient–based learning laws. A detailed mathematical discussion of the proof would exceed the scope of this book; the interested reader is referred to the source papers. In short, the main result is the following:

It is possible to design the adaptive law in such a way that stability of the control loop and convergence of the output error to zero can be guaranteed mathematically, if

1. **there exists an ideal control law which makes the error e zero (i.e. the plant must physically be able to follow the reference model), and**
2. **the controller is able to approximate this ideal control law with sufficient accuracy.**

At first glance, this is a very remarkable result: it seems that a SMRNC controller can control *any* unknown nonlinear plant in such a way that the closed–loop dynamics matches any arbitrary reference dynamics. However, it has been shown in the previous chapter that requirement 1, while being always fulfilled for linear systems, is only fulfilled under certain conditions for nonlinear systems; this imposes some restrictions on the practical applicability of Stable Model Reference Neurocontrol. We will discuss these issues in detail in the following section.

11.2 Description of the Concept

In the simplest case, the control law for Stable Model Reference Neurocontrol is

$$u = f_c(w, \underline{x}) = \underline{\hat{\Theta}}^T \underline{A}(w, \underline{x}) \tag{11.1}$$

i.e. the controller inputs w and $\underline{x}$ are fed into a universal function approximator (GRNN or adaptive fuzzy controller; see chapter 4) which directly produces the control output u. The goal of SMRNC is to adjust the parameters $\underline{\hat{\Theta}}$ of this function approximator, using the output error $e = y - y_m$, in such a way that it learns to approximate the ideal control law (10.44). To achieve this in a stable manner, the dynamic relationship between the adjustable parameters $\underline{\hat{\Theta}}$ and the output error e has to be taken into account. This requires a rather elaborate learning algorithm; like the learning algorithms presented in the previous chapters, it is based on the theory of stable adaptive systems [10]. More specifically, it uses the *augmented error principle* [9, 10, 14], a well–known mathematical tool from linear adaptive control theory. In this book, the learning algorithm will only be stated as a set of equations; the interested reader is referred to [5] for the mathematical background.

The update law for the parameters $\underline{\hat{\Theta}}$ is

$$\dot{\underline{\hat{\Theta}}} = -\eta \, \epsilon \, \underline{A}_f \tag{11.2}$$

where $\eta > 0$ is the adaptation gain (learning rate), and ϵ is the *augmented error*, defined as

$$\epsilon = e + \hat{k}_p(\underline{\hat{\Theta}}^T \underline{A}_f - u_f) \tag{11.3}$$

The signals $\underline{A}_f(t)$ and $u_f(t)$ are generated from $\underline{A}(t)$ and $\underline{\hat{\Theta}}^T(t)\underline{A}(t)$ $(= u(t))$ by linear filtering according to

$$\underline{A}_f{}^{(n)} = -a_{1m}\underline{A}_f - a_{2m}\dot{\underline{A}}_f - \ldots - a_{nm}\underline{A}_f^{(n-1)} + \underline{A} \tag{11.4}$$

$$u_f{}^{(n)} = -a_{1m}u_f - a_{2m}\dot{u}_f - \ldots - a_{nm}u_f{}^{(n-1)} + \underline{\hat{\Theta}}^T \underline{A} \tag{11.5}$$

$\hat{k}_p$ in (11.3) is an internal estimate of the plant gain, which is computed by the update equation

$$\dot{\hat{k}}_p = -\eta_{\hat{k}_p}\, \epsilon\, (\underline{\hat{\Theta}}^T \underline{A}_f - u_f) \tag{11.6}$$

This learning algorithm has been designed in such a way that

- **stability** of the control loop (i.e boundedness of $\underline{\hat{\Theta}}$, $\hat{k}_p$, u and $y, \dot{y} \ldots y^{(n-1)}$) and
- **convergence** of the output error e to zero

can be proved mathematically, using the method of Lyapunov. As it has already been mentioned, the two basic requirements for this proof are that there must exist a control law which makes y follow y_m and that the controller must be able to approximate this control law with sufficient accuracy. More specifically, the following conditions must be fulfilled:

1. The plant is single input, single output, of the type

$$\begin{aligned} \dot{\underline{x}} &= \underline{f}(\underline{x}) + \underline{g}(\underline{x}) \cdot u & (11.7)\\ y &= h(\underline{x}) & (11.8) \end{aligned}$$

   and has a relative degree r equal to its order n.[2][3]

2. The order n is known.

3. The plant is input–output–linearizable, i.e. its equations can be transformed into the canonical form (10.10)

$$y^{(n)} = L_f^n h(\underline{x}) + L_g L_f^{n-1} h(\underline{x}) \cdot u$$

   where $L_f^n h(\underline{x})$ and $L_g L_f^{n-1} h(\underline{x})$ are the so–called *Lie derivatives* (see chapter 10).

[2] See chapter 10 for the practical implications of this condition. The extension of SMRNC to plants with $r < n$ is discussed in [15].

[3] Unlike in the other chapters of this book, the nonlinearities are denoted by small roman letters instead of the symbol $\mathcal{NL}$ here; this will simplify the notation in the following equations.

4. $L_g L_f^{n-1} h(\underline{x})$ is constant and of known sign.
5. The full state vector $\underline{x} = (x_1 \dots x_n)^T$ can be measured.
6. The reference model is linear and of numerator degree zero, i.e.

$$y_m^{(n)} = b_{1m}w - a_{1m}y_m - a_{2m}\dot{y}_m - \dots - a_{nm}y_m^{(n-1)} \tag{11.9}$$

7. The controller inputs w and $\underline{x}$ are restricted to a certain bounded region, which has to be known for the controller design.
8. The *inherent approximation error* (see chapter 4) is zero, i.e. the controller (11.1) can approximate the ideal control law (10.44) exactly.

What are the practical implications of these conditions? As it has already been pointed out, a main problem of Stable Model Reference Neurocontrol is that some of them are difficult to fulfill in practice:

- Condition 1 is not fulfilled when external disturbances act on the plant during the learning phase.
- The Lie derivatives (condition 3) only exist if the plant's nonlinearities are differentiable (except in the case when direct control access at the nonlinearities is possible); thus, Stable Model Reference Neurocontrol in its present form cannot be applied to most plants with hard nonlinearities.
- Full state measurement (condition 5) is impossible in many cases.

At present, these three problems impose considerable restrictions on the practical applicability of Stable Model Reference Neurocontrol; their solution is still the topic of current research.

Some other conditions may require modifications of the control concept if they are not fulfilled:

- If the control input u is limited, condition 1 is violated. Simulation results show that the presence of control saturation leads to instability; however, this problem can be removed by a relatively straightforward modification of the control concept. Basically, the adaptation must be switched off as long as the saturation is active; a more detailed description will be given later in section 11.4.1. While this modification has shown to work well in simulation, the proof of stability in the presence of control saturation is still a topic of current work.
- Conditions 4 and 6 are specific for the controller structure (11.1). If $L_g L_f^{n-1} h(\underline{x})$ is not constant or the reference model is nonlinear, other controller structures will have to be used; this will be discussed in section 11.4.2.

- Condition 7 is needed because a universal function approximator with a finite number of parameters can only give good approximation in a bounded region of the input space (cf. chapter 4). If condition 7 is not automatically fulfilled by the physical restrictions of the plant, the control law (11.1) has to be extended by an additional supervisory controller [12, 16, 19]. If the plant is known to be stable, condition 7 can be fulfilled by simply stopping control and adaptation if the controller inputs leave the specified region, and to restart them after a certain waiting time.

- Condition 8 (zero inherent approximation error) is normally not fulfilled in practice, since a function approximator with a finite number of rules can only approximate arbitrary nonlinear functions with a limited accuracy. It is known from linear adaptive control that even an arbitrarily small disturbance term in the adaptation loop (such as it is caused e.g. by measurement noise in a linear adaptive control system, or by the inherent approximation error in a SMRNC system) can lead to instability; thus, one cannot rely on the system to remain stable as long as the inherent approximation error is sufficiently small. To guarantee stability in the presence of a nonzero inherent approximation error, the adaptive laws (11.2) and (11.6) have to be modified by introducing a dead zone [12, 10]. Nevertheless, various simulations and experimental results [3, 4, 5, 6, 8] indicate that SMRNC systems can maintain stability in the presence of a nonzero inherent approximation error even without such a modification. This effect seems to be connected to the so-called *persistent excitation* property [10] of certain signals in the system; however, further work is still required to obtain a sound mathematical explanation for these observations.

11.3 Application Example: Nonlinear Second–Order System

11.3.1 Simulation Results

We will now demonstrate the concept of Stable Model Reference Neurocontrol by means of a simulation example. The plant that has been simulated is a simplified model of an electrical drive system with nonlinear friction (see figure 11.2). It is described by

$$\begin{bmatrix} \dot{x}_1 \\ \dot{x}_2 \end{bmatrix} = \begin{bmatrix} -\arctan(80 \cdot x_1) + 5\,x_2 \\ -20\,x_2 \end{bmatrix} + \begin{bmatrix} 0 \\ 20 \end{bmatrix} \bar{u} \qquad (11.10)$$

$$y = x_1 \qquad (11.11)$$

with

$$\bar{u} = \begin{cases} -u_{\max} & \text{for } u < u_{\max} \\ u & \text{for } -u_{\max} \leq u \leq u_{\max} \\ u_{\max} & \text{for } u > u_{\max} \end{cases} \tag{11.12}$$

where x_1 is the speed of the drive, and the relation between the input u and the acceleration torque x_2 is described by a first–order dynamics. The control task is to make the output $y = x_1$ follow a given reference signal.

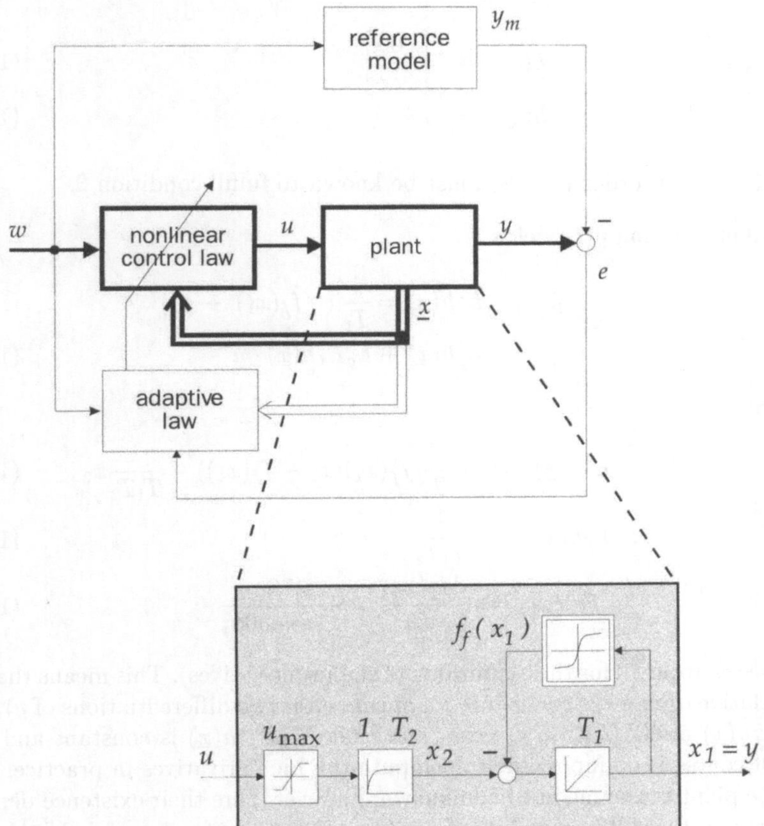

Figure 11.2: Stable Model Reference Neurocontrol for example plant

Let us check whether this plant satisfies the conditions for Stable Model Reference Neurocontrol:

- Condition 1 is not fulfilled because there is control saturation. Thus, the learning algorithm and reference model must be modified in the way described in section 11.4.1. Apart from this, the condition is fulfilled since the plant is single input, single output, time–invariant and linear in u. Note that condition 1 would not be fulfilled if an external disturbance torque was present, i.e. a torque which does not depend on any state of the system. Neglecting the control saturation, the plant equations (11.10) and (11.11) have the form of (11.7) and (11.8) with

$$\underline{f}(\underline{x}) = \begin{bmatrix} \frac{1}{T_1}(-f_f(x_1) + x_2) \\ -\frac{1}{T_2}x_2 \end{bmatrix} \tag{11.13}$$

$$\underline{g}(\underline{x}) = \begin{bmatrix} 0 \\ \frac{1}{T_2} \end{bmatrix} \tag{11.14}$$

$$h(\underline{x}) = x_1 \tag{11.15}$$

- The system order ($n = 2$) must be known to fulfill condition 2.
- Differentiating $y(t)$ yields

$$\dot{y} = L_f h(\underline{x}) = \frac{1}{T_1}(-f_f(x_1) + x_2) \tag{11.16}$$

$$\ddot{y} = L_f^2 h(\underline{x}) + L_g L_f h(\underline{x}) \cdot u \tag{11.17}$$

with

$$L_f^2 h(\underline{x}) = -\frac{1}{T_1^2} f_f'(x_1)(x_2 - f_f(x_1)) - \frac{1}{T_1 T_2} x_2 \tag{11.18}$$

$$L_g L_f h(\underline{x}) = \frac{1}{T_1 T_2} \tag{11.19}$$

$$f_f'(x_1) = \frac{\mathrm{d}f_f(x_1)}{\mathrm{d}x_1} = \frac{16}{1 + 6400x_1^2} \tag{11.20}$$

(see chapter 10 for the definition of the Lie derivatives). This means that the relative degree is $r = 2$ (since u appears after two differentiations of y), that $L_f^n h(\underline{x})$ and $L_g L_f^{n-1} h(\underline{x})$ exist, and that $L_g L_f^{n-1} h(\underline{x})$ is constant and $\neq 0$. Of course it is impossible to compute the Lie derivatives in practice, since the plant is assumed to be unknown; however, here their existence depends only on the differentiability of $f_f(x_1)$. The constant value of $L_g L_f^{n-1} h(\underline{x})$ is due to the fact that there are no nonlinearities in the direct path from u to y. Thus, the fulfillment of conditions 3 and 4 can be inferred from some relatively general knowledge of the plant structure.

- We assume that x_1 and x_2 can be measured, and thus condition 5 is satisfied.

The remaining conditions have to be fulfilled by appropriate design of the controller, which we will discuss now. First, we choose the reference model as

$$\ddot{y}_m = -1600y_m - 80\dot{y}_m + 1600w$$

(i.e. a second–order system with a double pole at $s_{1,2} = -40$); this choice complies with condition 6. Next, we design the universal function approximator. We choose an adaptive fuzzy controller with triangular input membership functions and define two fuzzy sets (N and P) for the input w, ten fuzzy sets (NB, NM, NS, NVS, NZ, PZ, PVS, PS, PM, PB) for x_1 and two fuzzy sets for x_2. The fuzzy sets for x_1 are grouped at closer intervals in the critical area around zero, where the effect of the nonlinear friction is strongest; the outer membership functions are semi–trapezoidal. Generally, the choice of the input membership functions is a trade–off between approximation quality and computational cost. (Another aspect is that the interpretation of a fuzzy controller becomes more difficult with a larger number of rules.) Simulations show that stability is maintained even with a much larger inherent approximation error (i.e. fewer fuzzy sets for x_1), but the performance degrades because friction can no longer be compensated accurately.

The remaining design of the fuzzy controller is done according to section 4.3. The rule base of the controller consists of the 40 rules

$$\begin{array}{l} \text{IF } w = \text{N AND } x_1 = \text{NB AND } x_2 = \text{N THEN } u = \text{U}^1 \\ \text{IF } w = \text{N AND } x_1 = \text{NB AND } x_2 = \text{P THEN } u = \text{U}^2 \\ \vdots \\ \text{IF } w = \text{P AND } x_1 = \text{PB AND } x_2 = \text{P THEN } u = \text{U}^{40} \end{array}$$

The logical AND is implemented as the algebraic product; defuzzification is done by the singleton method (4.10). The positions $u^1 \ldots u^{40}$ of the output singletons $U^1 \ldots U^{40}$ are adjusted online by the adaptive law (11.2). The learning rates are set to $\eta = 5 \cdot 10^6$ and $\eta_{\hat{k}_p} = 10^6$.

To fulfill condition 7, the controller inputs must be restricted to the region where the controller has good approximation capability. This could be achieved by switching off control and learning if the states x_1 or x_2 get into the constant range of the semi–trapezoidal outer membership functions, and restarting them after e.g. one second. However, these conditions never occured in the simulation. To check whether condition 8 is fulfilled, we take a look at the ideal control law (10.44)

$$u = \frac{b_{1m}w - a_{1m}h(\underline{x}) - a_{2m}L_f h(\underline{x}) - L_f^2 h(\underline{x})}{L_g L_f h(\underline{x})}$$

where the Lie derivatives are defined according to (11.15) – (11.19). A graphical representation of this ideal control law is shown in figure 11.5. It is obvious that this control law cannot be approximated exactly by multilinear basis functions, i.e. there is a nonzero inherent approximation error. However, as we will see, the system remains stable and the output error converges to a small neighbourhood of zero even if the unmodified adaptive law (11.2) – (11.6) is used.

Simulation results are shown in the figures 11.3, 11.4 and 11.5. To demonstrate that stability and convergence are guaranteed regardless of the initial conditions, all the controller parameters $u^1 \ldots u^{40}$ are initialized to zero in the beginning. (Of course it would also be possible to construct an initial rule base using qualitative knowledge about the plant.) Adaptation starts after one second. The controller is trained by applying a signal $w(t)$ which changes its value randomly in the interval $(-0.7, 0.7)$ every 0.2 seconds. This signal was chosen because it leads to good exploration of the controller input space. After 31 s and 1801 s a deterministic test signal w is applied to evaluate the learning results. In the initial phase, learning is very fast; the controller needs only about 30 s to make y roughly follow y_m. After that, however, the learning process becomes substantially slower due to the decreasing magnitude of the error signal. Satisfactory control performance is attained after about 1800 s; y follows y_m almost perfectly when the triangular test signal is applied, with only a small error caused by the control saturation when y crosses zero.[4] However, the learning process is not finished after 1800 s; there are still considerable (although very slow) changes in the parameters, accompanied by a further slight decrease of the tracking error. The parameters reach stationary values only after about 10^5 s (figure 11.4); as it can be seen in figure 11.5, these values yield a good approximation of the analytically computed ideal control law (10.44). Figure 11.5 is also a good example of how input–output linearization works: the resulting control law is almost linear except for small values of x_1, where the controller produces extra output to compensate for the fast change of $f_f(x_1)$ as x_1 crosses zero. The dynamics between u and x_2 is taken into account correctly. (The control surface has been cut off at $\pm u_{\max}$ to show the effect of control saturation.)

11.3.2 Experimental Results

To verify the applicability of Stable Model Reference Neurocontrol under real-world conditions, the method was applied to control a laboratory electrical drive system with a nonlinear load characteristic. The experimental setup (figure 11.6) consists of two 4.8 kW industrial servo drives, which are connected by a torsionally stiff coupling. Similar to the previous simulation, the control task was to make the speed of motor I follow a given reference signal, while motor II provided an artificial, arctangent–shaped, nonlinear load characteristic. A software–implemented first–order delay with a time constant of 0.05 s was put between the controller and the actual plant. Both drives were operated in torque control mode, i.e. the built–in field oriented torque controllers of the inverters were used without modification. The SMRNC algorithm, the computation of the load characteristic, and the first–order delay were implemented on a standard 166 MHz Pentium PC. The torque reference signals generated by the PC were converted to $\pm$ 10 V analog signals and fed into the analog inputs of the inverters. As inputs

[4]For the random training signal, the errors caused by the saturation are substantially larger, because the reference model has been made much faster than the achievable dynamics of the plant. This was done to demonstrate that the method works well even if u reaches the saturation frequently.

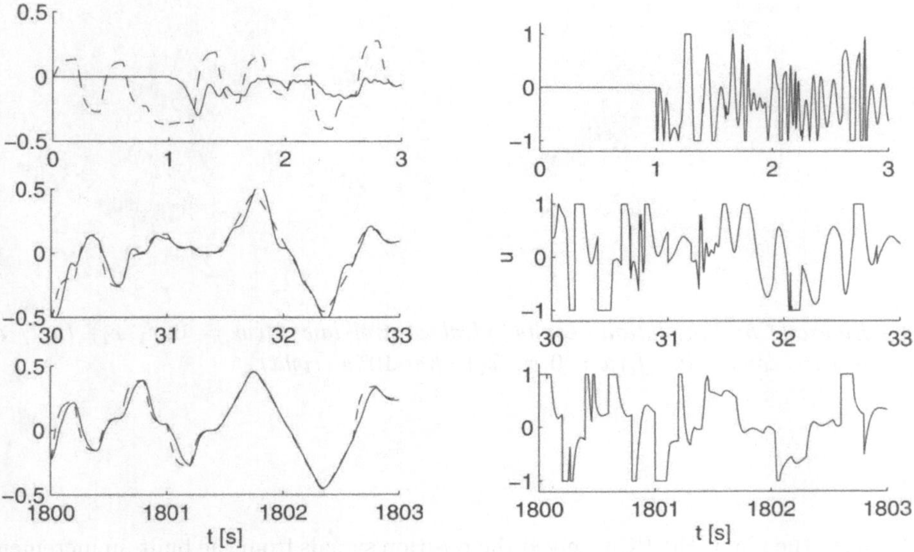

Figure 11.3: Simulation results: plant output y (left, continuous), reference signal y_m (left, dashed) and plant input u (right)

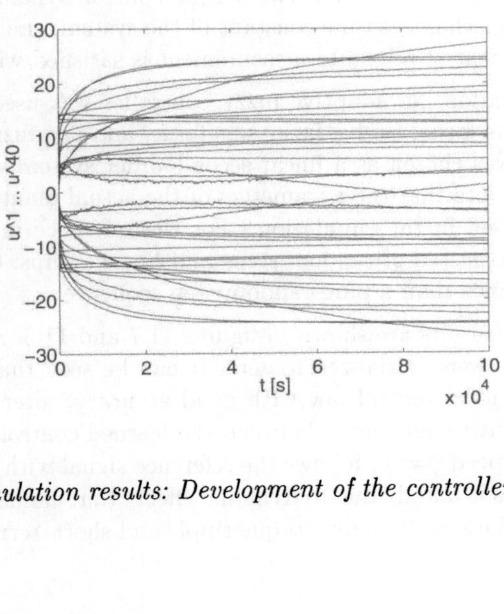

Figure 11.4: Simulation results: Development of the controller parameters u^i

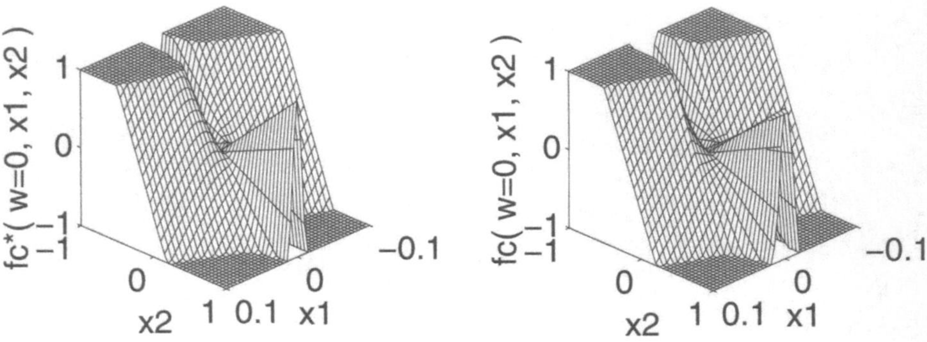

Figure 11.5: Simulation results: ideal control law $f_c^*(w = 0, x_1, x_2)$ (left) and learned control law $f_c(w = 0, x_1, x_2)$ after 10^5 s (right)

from the plant, the PC received the position signals from the built–in incremental encoders; the speed of the drives was obtained by numerical differentiation.

From the view of the controller, the plant is the same as in the simulation example from section 11.3.1: a single rotating mass with a nonlinear load torque plus a first–order delay, which has to be taken into account when compensating the nonlinearity. The control input u is the input of the delay block; the states of the plant are the output x_2 of the delay block (which is the torque reference for drive I) and the speed x_1 of the drive. The torque control dynamics, which is high compared to the mechanical time constant of the system, can be neglected here; thus, the requirement of full state measurement is satisfied with good accuracy.

Like in the simulation, an adaptive fuzzy controller was used; it was designed with two fuzzy sets for w, twelve fuzzy sets for x_1 and two fuzzy sets for x_2. The reference model was chosen as a linear second–order system with a double pole at $s_{1,2m} = -20$. (Note that the parameters of the actual plant are different from the parameters used in the simulation.) For the reference input $w(t)$, a signal was used which consisted of random steps and linear ramps; this yielded better learning performance than a pure random step sequence.

The experimental results are shown in figures 11.7 and 11.8. Again, all the controller parameters were initialized to zero. It can be seen that the controller is able to learn the ideal control law with good accuracy; after 750 s, there is already good qualitative agreement between the learned control law and the ideal control law. The speed $y = x_1$ follows the reference signal with a small remaining tracking error; this is mainly due to dynamic effects that cannot be compensated by input–output linearization, like torque ripple and short–term variations of the bearing friction.

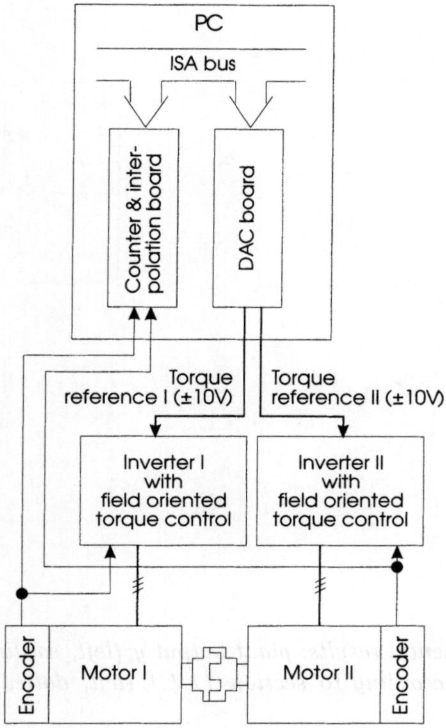

Figure 11.6: Experimental setup

Further experiments were carried out with a two–mass system, i.e. the two drives were coupled by a torsionally flexible shaft of small diameter. The results for this system, which is of third order, are described in [6, 8].

11.4 Modifications

It has been mentioned at the beginning of this chapter that the SMRNC system described in section 11.2 is only *one* possible concept from a large family. In the following section, we will give a brief overview over some other possible SMRNC variants.

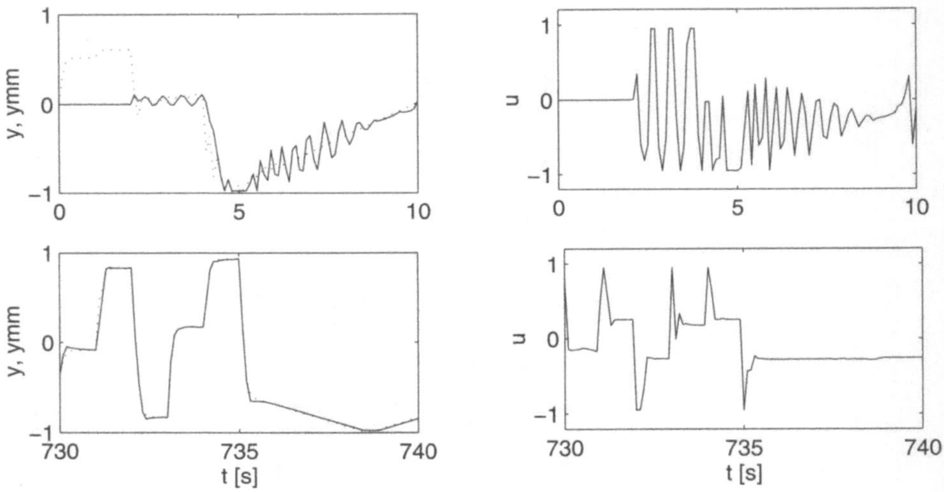

Figure 11.7: Experimental results: plant output y (left, continuous), modified reference signal y_{mm} according to section 11.4.1 (left, dotted) and plant input u (right)

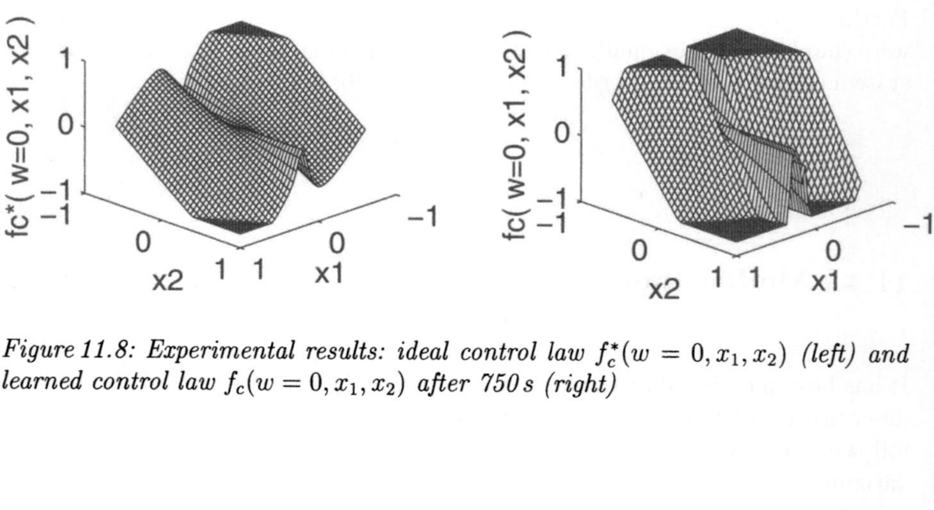

Figure 11.8: Experimental results: ideal control law $f_c^*(w = 0, x_1, x_2)$ (left) and learned control law $f_c(w = 0, x_1, x_2)$ after 750 s (right)

11.4.1 Modification for Plants with Control Saturation

If the control input u is limited, it is no longer possible to make y exactly follow y_m; thus, an error e occurs. This error causes problems to Stable Model Reference Neurocontrol, because it violates an underlying assumption of the adaptive law: the assumption that every error in the output can be traced back to an error in the control law, and thus can be removed by appropriate tuning of the controller parameters. Following this assumption, the adaptive law attempts to counteract the error, which is caused by the saturation, by increasing the controller parameters. However, since the adjustment of the parameters has no effect on this error, the parameters keep increasing in an unbounded fashion. The resulting control behaviour is very unsatisfactory: u is constantly changing between positive and negative saturation.

The solution to this problem is to "switch the adaptation off" if u reaches the saturation, i.e. to set $\dot{\hat{\underline{\Theta}}}$ and $\dot{\hat{k}}_p$ to zero as long as $|u| \geq u_{\max}$. In addition to the modified adaptive law, a modified reference signal y_{mm} must be introduced to achieve stability. The modified reference model has the same dynamics as the original model, but its states $y_{mm}, \dot{y}_{mm}, \ddot{y}_{mm}, \ldots y_{mm}^{(n-1)}$ are reset to $y, \dot{y}, \ddot{y}, \ldots y^{(n-1)}$ every time u leaves the saturation.[5] At the same time, the states of the linear filters (11.4) and (11.5) are reset to zero. Thus, the error e and all its derivatives are forced to zero at the instant u leaves the saturation. The modification of y_m can be seen in figure 11.7. (Note that the *unmodified* reference signal is shown in figure 11.3.) The interested reader is referred to [2, 8] for more details.

11.4.2 Modification for Plants with $L_g L_f^{n-1} h(\underline{x}) \neq$ const.

The controller (11.1) can approximate the ideal control law (10.44) for all plants that are input–output–linearizable, regardless whether $L_g L_f^{n-1} h(\underline{x})$ is constant or not. However, a stable learning algorithm for this controller can only be found under the condition $L_g L_f^{n-1} h(\underline{x})$ = const.; if $L_g L_f^{n-1} h(\underline{x})$ is not constant, the control law

$$u = \frac{\hat{\underline{\Theta}}_1^T \underline{A}_1(w, \underline{x})}{\hat{\underline{\Theta}}_2^T \underline{A}_2(\underline{x})} \tag{11.21}$$

must be used. The corresponding learning law is

$$\dot{\hat{\underline{\Theta}}} = -\gamma \epsilon \underline{A}_f \tag{11.22}$$

$$\epsilon = e + (\hat{\underline{\Theta}}^T \underline{A}_f - (\hat{\underline{\Theta}}^T \underline{A})_f) \tag{11.23}$$

[5] In practice, a discrete-time reference model is used. When u leaves the saturation, its states $y_{mm}(k) \ldots y_{mm}(k-n+1)$ are reset to $y(k) \ldots y(k-n+1)$. Thus, no differentiation of y is required.

with $\underline{\hat{\Theta}}$ and $\underline{\mathcal{A}}$ defined as

$$\underline{\hat{\Theta}} = [\underline{\hat{\Theta}}_1^T, \underline{\hat{\Theta}}_2^T]^T \; ; \quad \underline{\mathcal{A}} = [\underline{\mathcal{A}}_1^T, -\underline{\mathcal{A}}_2^T \cdot u]^T \tag{11.24}$$

Note that the estimate $\hat{k}_p$ of the plant gain is not needed here. To avoid division by zero, the adaptive law must be modified to prevent any element of $\underline{\hat{\Theta}}_2$ from decreasing below a specified boundary.

11.4.3 Method with Differentiation of y

The adaptive law (11.2) – (11.6) is a relatively recent development in Stable Model Reference Neurocontrol and has only been published in 1996 and 1997 [2, 5]. All the methods published up to then used another type of adaptive law which has a simpler structure, but requires the availability of $n-1$ derivatives of the plant output $y(t)$. This requirement is fulfilled e.g. for position control of a single rigid mass ($n = 2$) if position and velocity can be measured; the method can be extended to the control of rigid link manipulators [1]. However, in most practical applications multiple differentiation of y is technically impossible.

The adaptive law with differentiation of y is

$$\dot{\underline{\hat{\Theta}}} = -\eta(p_1 e + p_2 \dot{e} + \ldots + p_n e^{(n-1)})\underline{\mathcal{A}} \tag{11.25}$$

where the coefficients $p_1 \ldots p_n$ must satisfy the so–called *Lyapunov equation*; see e.g. [18, 12, 19, 13] for details.

If $n-1$ derivatives of y are available, it is also possible to use the control law

$$u = \frac{y_m^{(n)} - k_1 e - k_2 \dot{e} - \ldots - k_n e^{(n-1)} - \underline{\hat{\Theta}}_1^T \underline{\mathcal{A}}(\underline{x})}{\underline{\hat{\Theta}}_2^T \underline{\mathcal{A}}(\underline{x})} \tag{11.26}$$

which has the same structure as the first version of the ideal control law (10.37) (cf. figure 10.4). The PD–type feedback gains $k_1 \ldots k_n$ can be chosen as the coefficients of an arbitrary Hurwitz polynomial. Actually, this is the "original" SMRNC control law [17, 18]; the simplified control laws (11.1) and (11.21), which do not require differentiation of y, are later modifications. Compared to (11.21), the control law (11.26) has the advantage that it can approximate the ideal control law not only for linear, but also for nonlinear reference models.

11.4.4 Modifications for Reduction of the Learning Times

The simulation example from the previous section has shown that Stable Model Reference Neurocontrol can require learning times in the order of several thousand seconds for a second–order system.[6] For systems of higher order, the learning times can reach hours and even days. Currently, several modifications of Stable Model Reference Neurocontrol are being developed to accelerate the learning process. Using the two modifications

- use of a–priori knowledge about the plant structure (e.g. that the nonlinear friction depends only on x_1, but not on x_2 for the system in figure 11.2) to divide the control law into several simpler subfunctions, which can be learned separately; and
- reduction of the number of parameters by half by exploiting the symmetry of the control law (applicable only to plants with symmetric dynamics),

reductions of the learning time by over 90% compared to the unmodified control concept have already been achieved [3, 8]. However, further reductions are still required to make Stable Model Reference Neurocontrol applicable to systems of higher order.

11.5 Short Summary

In this chapter, we have presented a class of neural network and neuro–fuzzy control concepts summarized under the name of *Stable Model Reference Neurocontrol*, which are nonlinear extensions of linear model reference adaptive control. In comparison to other nonlinear adaptive control concepts, Stable Model Reference Neurocontrol offers some substantial advantages:

- little prior knowledge about the plant is required
- online learning with guaranteed stability and convergence
- a control strategy that is accessible in linguistic form if a fuzzy controller is used.

Compared to the methods presented in chapters 5 – 9, the advantages of Stable Model Reference Neurocontrol are that the entire control law is learned directly and that much less knowledge about the plant is required. However, these advantages are only gained at the price of some relatively restrictive mathematical assumptions and substantially longer learning times.

Even though significant progress has been made in recent years towards practical application of Stable Model Reference Neurocontrol, some problems still remain

[6] Note that the learning times discussed here are the times required to achieve a specified performance of the output signal. They cannot be compared directly to results from other publications, where learning time is defined e.g. as the time required to merely stabilize a plant.

to be solved. Current research activities deal on the one hand with learning in the presence of external disturbances and on the other hand with the application of Stable Model Reference Neurocontrol to plants with hard (i.e. non–differentiable) nonlinearities and plants where the full state vector cannot be measured. Another objective is a further reduction of the learning times. If these goals can be achieved, Stable Model Reference Neurocontrol has the potential to become a powerful tool for a large variety of nonlinear control tasks.

11.6 References

[1] Beerhold, J.R.:
Stabile adaptive Regelung nichtlinearer Mehrgrößensysteme mit neuronalen RBF–Netzen am Beispiel von Mehrgelenkrobotern.
Automatisierungstechnik, vol. 44, no. 12, 1996, pp. 577–583.

[2] Fischle, K., Schröder, D.:
Stable Adaptive Fuzzy Control of Plants with Control Saturation.
Proc. Fourth European Congress on Intelligent Techniques and Soft Computing (EUFIT'96), Aachen, 1996, pp. 980–984.

[3] Fischle, K., Schröder, D.:
Aktive Dämpfung von Torsionsschwingungen durch adaptive Fuzzy–Regelung.
VDI–Berichte 1285 – Schwingungen in Maschinen, Fahrzeugen und Anlagen. VDI–Verlag, Düsseldorf, 1996, pp. 557–571.

[4] Fischle, K., Schröder, D.:
Stable Adaptive Fuzzy Control of an Electric Drive System.
Proc. Sixth IEEE International Conference on Fuzzy Systems (Fuzz-IEEE'97), Barcelona, 1997, pp. 287–292.

[5] Fischle, K., Schröder, D.:
Stabile adaptive Fuzzy–Regelung ohne Differentiation der Regelgröße.
Automatisierungstechnik, vol. 45, no. 8, 1997, pp. 360–367.

[6] Fischle, K., Schröder, D.:
Stable Model Reference Neurocontrol for Electric Drive Systems.
Proc. Seventh European Conference on Power Electronics and Applications (EPE), Trondheim, 1997.

[7] Fischle, K., Schröder, D.:
An Improved Stable Adaptive Fuzzy Control Method.
(To appear in IEEE Transactions on Fuzzy Systems)

[8] Fischle, K.:
Ein Beitrag zur stabilen adaptiven Regelung nichtlinearer Systeme.
PhD thesis, Technical University of Munich, Department of Electrical Engineering and Information Technology. Herbert Utz Verlag, 1998.

[9] Monopoli, R.V.:
Model Reference Adaptive Control with an Augmented Error Signal.
IEEE Transactions on Automatic Control, vol. 19, 1974, pp. 474–484.

[10] Narendra, K., Annaswamy, A.:
Stable Adaptive Systems.
Prentice–Hall, Englewood Cliffs, New Jersey, 1989.

[11] Ordoñez, R., Zumberge, J., Spooner, J.T., Passino, K.M:
Adaptive Control: Experiments and Comparative Analyses.
IEEE Transactions on Fuzzy Systems, vol. 5, no. 2, 1997, pp. 167–188.

[12] Sanner, R., Slotine, J.:
Gaussian Networks for Direct Adaptive Control.

IEEE Transactions on Neural Networks, vol. 3, November 1992, pp. 837–863.

[13] Schäffner, C., Schröder, D.:
An Application of General Regression Neural Network to Nonlinear Adaptive Control.
5th European Conference on Power Electronics and Applications EPE '93, Brighton, UK. Proceedings, vol. 4, 1993, pp. 219–223.

[14] Slotine, J., Li, W.:
Applied Nonlinear Control.
Prentice–Hall International, Inc., New Jersey 1991.

[15] Spooner, J.T., Passino, K.M:
Stable Adaptive Control Using Fuzzy Systems and Neural Networks.
IEEE Transactions on Fuzzy Systems, vol. 4, no. 3, 1996, pp. 339–359.

[16] Su, C., Stepanenko, Y.:
Adaptive Control of a Class of Nonlinear Systems with Fuzzy Logic.
IEEE Transactions on Fuzzy Systems, vol. 2, 1994.

[17] Tzirkel–Hancock, E., Fallside, F.:
A Direct Control Method for a Class of Nonlinear Systems Using Neural Networks.
Proc. Second IEE International Conference on Artificial Neural Networks, Bournemouth, 1991, pp. 134–138.

[18] Tzirkel–Hancock, E., Fallside, F.:
Stable Control of Nonlinear Systems Using Neural Networks.
Int. J. Robust and Nonlinear Control, vol. 2, no. 1, 1992.

[19] Wang, L.:
Stable Adaptive Fuzzy Control of Nonlinear Systems.
IEEE Transactions on Fuzzy Systems, vol. 1, 1993.

12 Dynamic Neural Network Compositions for Stable Identification of Nonlinear Systems with Known and Unknown Structures

Stephan Straub

12.1 Introduction

The problem of controlling nonlinear dynamic systems has been the topic of many research projects in recent years. Nevertheless, there exist only a few approaches [9, 12, 15] that solve nonlinear control problems for a wide area of applications. Especially, intelligent tools like neural networks or fuzzy approaches can help solve these kinds of problems, but the amount of different learning algorithms, different fields of application and different objects makes it difficult to get a general overview of the features and limits of intelligent compositions.

Control of nonlinear dynamic systems is often directly connected with the **problem of system identification**. Therefore, this chapter is concerned first with the **neural control and identification** in general. The aims and restrictions of the method proposed in chapter 5 are repeated and connected with definitions of static and dynamic neural networks. Furthermore, the difference between static and dynamic neural networks, as well as between **nonlinear observers** and **dynamic identificators** is shown. Based on this, a method is presented, which enables identification of **nonlinear dynamic systems with unknown structure** using a dynamic identificator. In this context, the mathematical background and the proposed method will be shown. Three simulation examples are included to illustrate the theory.

12.2 Classification of Identification Methods

12.2.1 Motivation

Because of their nonlinear approximation ability, today neural nets are used more and more as controllers, support for a controller or as an identification tool [19, 12, 4, 14]. Different control methods and applications have been presented in chapters 4–11. For most applications, not the question of a suitable kind of a network arises, but the utilization of a robust and stable learning algorithm. In the following sections, the General Regression Neural Network (GRNN) is used as a universal approximator for nonlinear (static) characteristics. This type of network shows advantages concerning convergence and stability as described in chapters 4 and 5. Another possibility would be a fuzzy approach which is described in detail in chapter 11 and will not be a part of these considerations.

In adaptive and neural control theory, two general approaches are distinguished: **direct control** and **indirect control**. Both methods are depicted in figure 12.1 and explained in chapter 2. One disadvantage of the direct method is the fact that all states must be measured to be available for the neural net input (see chapter 11). In many technical applications, this assumption is not given. On the other hand, no information about nonlinear characteristics, which occur in the system, can be extracted.

In the case of indirect control, the processes of identification and control can be separated. The utilization of the knowledge for control depends on the application and cannot be generalized. In chapters 5 and 6, two neural identification procedures are shown, for which stability and parameter convergence can be derived. In chapter 5, a method for the identification of isolated nonlinearities is shown. The entire identification system can be called a **nonlinear observer** where additional system states are observed and used (if necessary) for control. In chapter 6, a method is described to design an observer (based on chapter 5) where more than one (distributed) nonlinearity can be extracted. Both methods require the existence of **structural a–priori knowledge**. Possible applications and validations can be seen in chapters 7, 8 and 9. There are different methods to influence the controller based on the identified knowledge of the system. Possible control (compensation) methods are shown in chapters 7 and 9. The disadvantage in this case is that no general control concept can be derived to use the identified knowledge.

To avoid the restrictions of the described control and identification methods, another approach has been developed called **dynamic identificator** (see figure 12.1) and described in section 12.3. The method combines **elements of direct and indirect control**. For this approach **dynamic neural structures** are used. Due to this fact, possible net compositions are explained at the beginning of the next section. To get a better overview, the methods described in the last

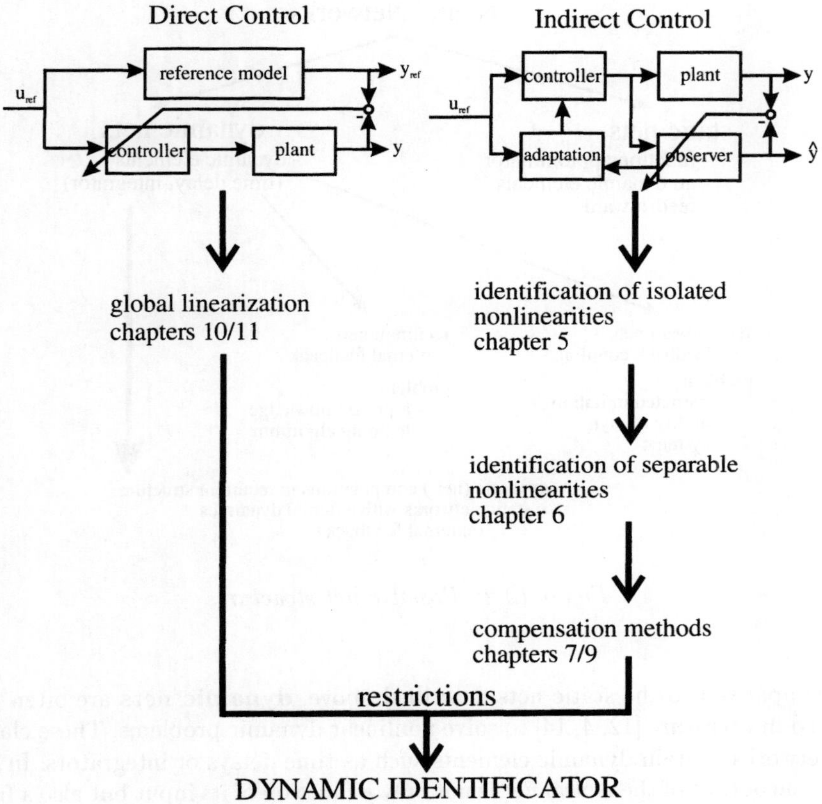

Figure 12.1: Indirect and direct control methods

chapters are classified by definitions concerning observation, identification and recurrent structures.

12.2.2 Different Net Structures

For the identification of dynamic systems, the distinctive marks depicted in figure 12.2 should be mentioned. Generally, neural nets can be divided into static and dynamic nets. Since **static** nets are used for the mapping of static functions, they could be called function approximators. No time delays or integrators are implemented into the network. There is no feedback from the output to the input of the network, hence the network is a feedforward net. (Feedbacks would lead to algebraic loops.)

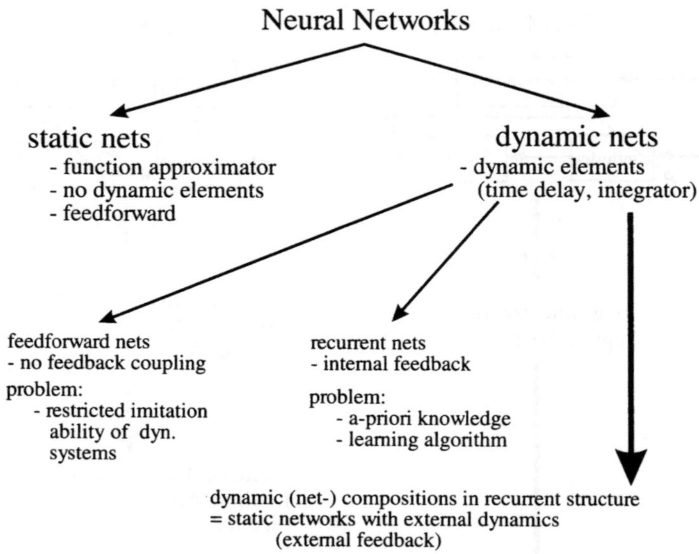

Figure 12.2: Possible net structures

In comparison to the static nets described above, **dynamic nets** are often presented in literature [12, 4, 14] to solve nonlinear dynamic problems. These classes of networks contain dynamic elements such as time delays or integrators. In this case the output of the neural net is not only a function of its input but also a function of the system's history. This means that the net includes internal states. In the dynamic case, feedforward and recurrent nets can be distinguished. Dynamic feedforward nets are classified by (intentionally designed) time delays from the input to the output with no internal or external feedback. Information flows only in one direction. In contrast to dynamic feedforward nets, the so–called recurrent nets are provided with additional internal feedbacks to solve a wider range of dynamic problems[1]. Because of this, the activation of each neuron can depend on the history of all other neurons. The utilization of this kind of coupling leads to structures of high complexity and of low degree of transparency. Another disadvantage is the great difficulty to implement a–priori knowledge and to derive a general learning algorithm for the adaptation of the neural net parameters.

Because of this, **dynamic nets (net compositions) with external feedback** are often used. These types of networks consist of a static network with additional dynamic elements (time delays or integrators). In this case only the output of the net is connected with its input. The number of dynamic elements depends on the identification problem and the requirements of the network. Examples for such

[1]In this context the notion *recurrent networks* only marks neural nets with an internal feedback.

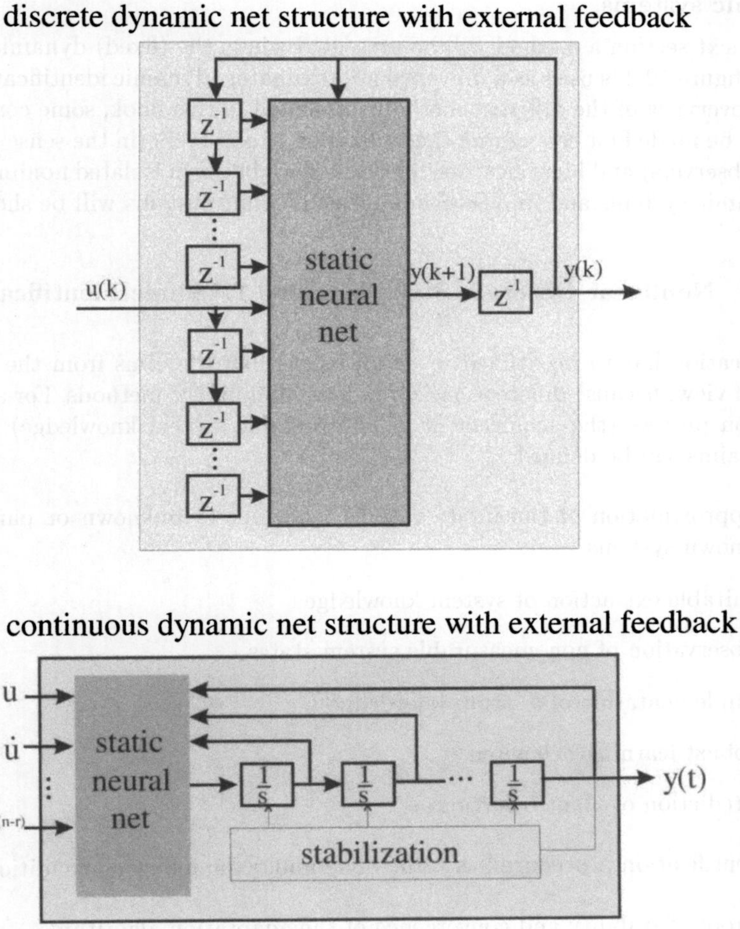

Figure 12.3: Discrete and continuous dynamic neural networks with external coupling

kinds of networks can be found in [12, 14]. Figure 12.3 shows two dynamic nets with external feedback. These networks can be designated as **net–compositions in recurrent structure**. In this case, dynamic elements are not part of the neural network definition itself. As shown later, the network structures depicted in figure 12.3 can be used for the identification of a **wide class of nonlinear dynamic systems**.

In the next section a method will be presented where the (fixed) dynamic structure of figure 12.3 is used as a universal approximator (dynamic identificator). To get an overview of the different methods presented in this book, some comments have to be made first concerning the utilization of observers (in the sense of state space observers) and identificators. The difference between isolated nonlinearities in dynamic systems and universal nonlinear dynamic systems will be shown.

12.2.3 Nonlinear Observer Structures and Dynamic Identificators

Identification has to be defined in detail concerning the aims from the control point of view, because different aims can demand different methods. For an identification process (the acquirement of interpretable system knowledge) the following aims can be defined:

- approximation of the input–output behaviour of unknown or partly unknown systems
- suitable extraction of system knowledge
- observation of non–measurable system states
- implementation of a–priori knowledge
- robust learning behaviour
- prediction of plant behaviour

The identification procedure has to be done under the following conditions:

- proof of stability and convergence of the adaptation algorithm
- high accuracy
- fast adaptation for time–variant systems

Based on the necessity of each point described above, a suitable method has to be chosen to obtain the desired information about the system. If states have to be extracted for control purposes, the system has to be designed as an observer. From the control point of view the proof of stability of the learning algorithm is necessary to get a predictable behaviour in every operating point. For applications like

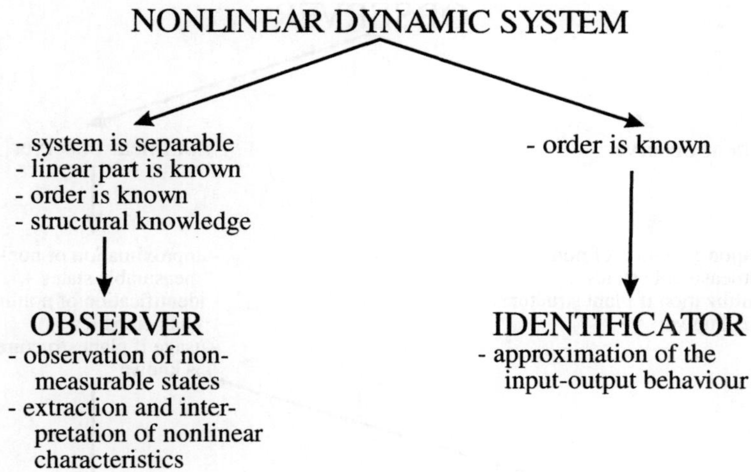

Figure 12.4: Observer and identificator

image or pattern recognition, this assumption is less important (classification). It must be noted that the choice of a suitable method does not only depend on the aims but also on the preconditions like a–priori system knowledge. This fact will be described by the following explanations.

Besides the extraction of nonlinear characteristics, in most cases the observation of internal, i.e. non–measurable states, is requested (see chapters 7 and 9). In this case the identificator becomes an observer[2]. With this assumption, the observer is an extension of the identificator because additional information, which can be used for control, is available in the form of interpretable system states. The exact state observation is based on the approximation of nonlinearities in the system; hence, two aims are obtained in parallel: on the one hand the observed states can be used for an augmented state controller, on the other hand compensation of the negative effects of the nonlinearities can be conducted based on the identified knowledge of the system. The influence of the nonlinearities can be damped for example via a suitable change of the control reference value (see chapter 9). For the construction of an observer, some conditions have to be fulfilled as depicted in figure 12.4. One of them is structural knowledge of the system, which means that the number and the exact positions of the nonlinearities must be known to design an observer (see chapters 5, 6, 7, 8 and 9). In many applications a state observation is neglected; in this case only the input–output behaviour is approximated [4]. In chapter 5 a method was shown to obtain a nonlinear observer with a suitable combination of dynamic elements, one or more neu-

[2]It must be noted that an identificator in general can be seen as a tool for the approximation of an input–output behaviour of an unknown plant.

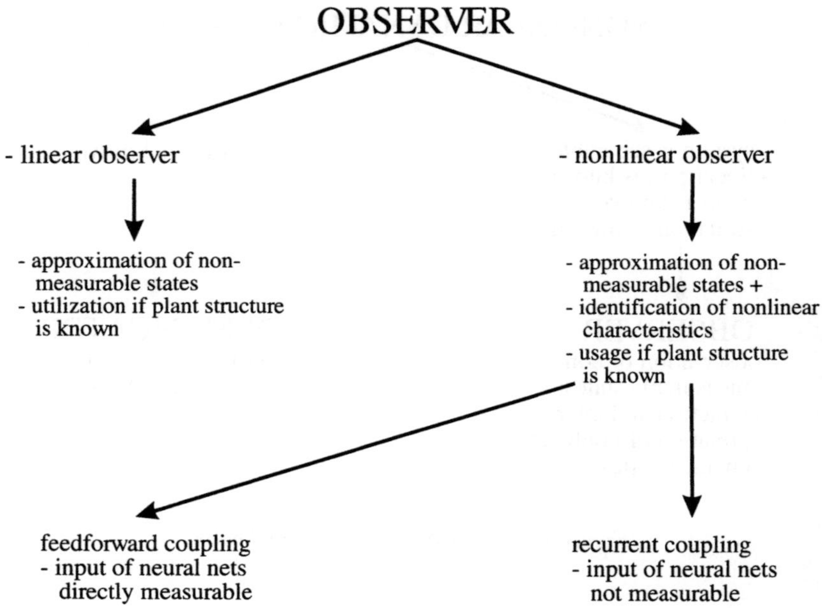

Figure 12.5: Observer structures

ral networks and additional a–priori knowledge to meet the demands described above. This method of **identification of nonlinear dynamic systems with isolated and/or separable nonlinearities** is based on a local consideration of the nonlinear terms in the system. Definitions can be found in chapters 5 and 6. In this case **static neural networks** are used for the approximation of the nonlinear characteristics of the plant (see figure 12.5).

First, the proposed methods of chapters 5, 6, 7, 8 and 9 are arranged into the definitions as they are explained in section 12.2.2. It is the aim of the nonlinear observer to approximate the following system:

$$\dot{\underline{x}} = \underline{\mathcal{NL}}(\underline{x}, \underline{u}) = A\underline{x} + B\underline{u} + \underline{\mathcal{NL}}^*(\underline{x}_1, \underline{u}_1) \tag{12.1}$$

$$y = \underline{c}^T \underline{x} + \underline{d}^T \underline{u} \tag{12.2}$$

Besides the observation of $\underline{x}$, this includes the approximation of the unknown vector $\underline{\mathcal{NL}}^*(\underline{x}_1, \underline{u}_1)$ with the matrices A and B assumed to be known (a–priori knowledge). The output y is assumed to be a scalar. It must be mentioned that the components of $\underline{\mathcal{NL}}^*(\underline{x}_1, \underline{u}_1)$ must be seen as **isolated and separable nonlinearities** in the sense of the definitions in chapters 5 and 6, which means that the point of activation of each nonlinearity must be known. This assumption is called **structural knowledge**. The equations of the observer can be written as follows:

$$\dot{\hat{\underline{x}}} = \widehat{\underline{NN}}(\underline{x}, \underline{u}) + \underline{l}(y - \hat{y}) = A\hat{\underline{x}} + B\underline{u} + \widehat{\underline{NN}}^{*}(\underline{x}_1, \underline{u}_1) + \underline{l}(y - \hat{y}) \tag{12.3}$$

or

$$\dot{\hat{\underline{x}}} = \widehat{\underline{NN}}(\hat{\underline{x}}, \underline{u}) + \underline{l}(y - \hat{y}) = A\hat{\underline{x}} + B\underline{u} + \widehat{\underline{NN}}^{*}(\hat{\underline{x}}_1, \underline{u}_1) + \underline{l}(y - \hat{y}) \tag{12.4}$$

The two equations show differences in the input of the neural nets $\widehat{\underline{NN}}^{*}$. In a qualitative sense the first observer (12.3) can be divided into following parts:

observer =
dynamic elements + static nets + structural (a–priori) knowledge

The observer can be seen as an observer in feedforward structure, because the network inputs are directly measurable. The entire observer structure can be defined as

dynamic feedforward (net–)composition with
plant–dependent and fixed structure

There is no feedback of the internal observer states into the neural nets. The static neural nets have a fixed position within the entire structure. The second observer (12.4) can be divided as follows:

observer =
dynamic elements + static nets + structural (a–priori) knowledge + external feedbacks

In this case a closed loop structure arises, because the inputs of the neural net must also be observed (assuming they are non–measurable). In the sense of the definitions made above the whole structure now can be described as a

dynamic (net–)composition with recurrent coupling and
plant–dependent and fixed structure

Figure 12.6 shows the difference between the two approaches. In the first case the input of the neural net is not dependent on the learning behaviour of the whole observer structure. In the second case the output of the neural net is directly connected with its inputs. There are similar approaches in the discrete case; further comments can be found in literature [12].

In the case of identifying systems with isolated and/or separable nonlinearities (see chapters 5 and 6), the states on which the nonlinearity acts is known (examples: unknown friction in electrical drives, influence of noncircularity in winder systems ect.). Furthermore, for each nonlinearity one neural net is implemented in the observer for the approximation (see chapter 6). A stable learning law can

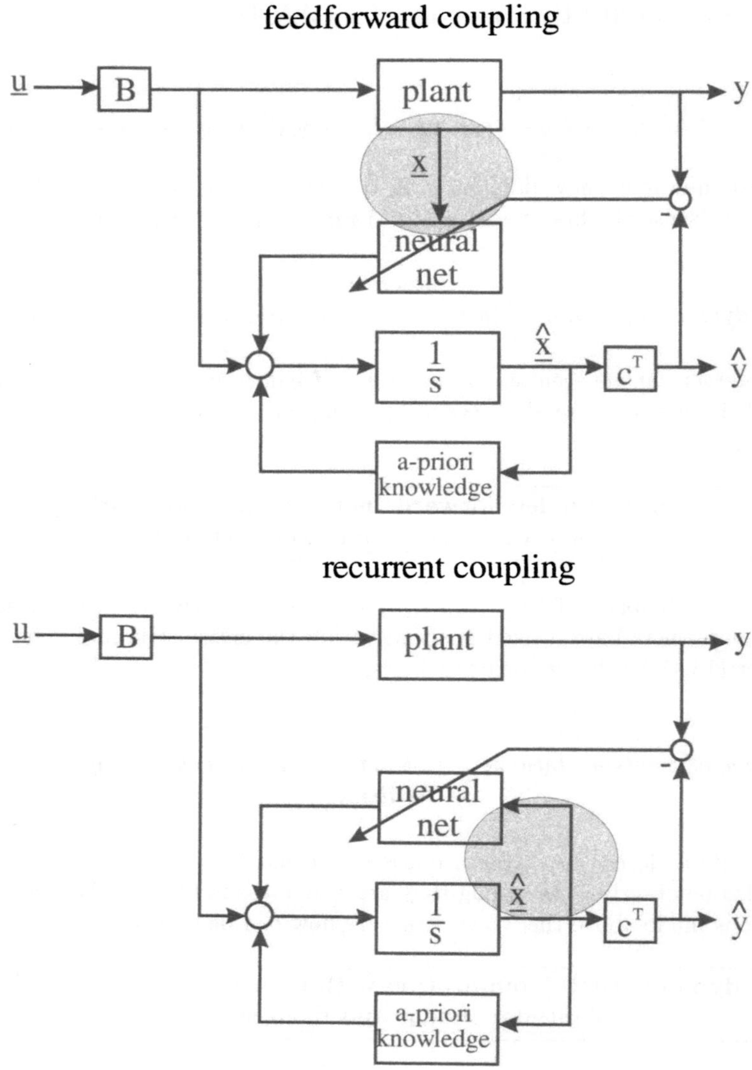

Figure 12.6: Feedforward and recurrent observer structures

be derived for both approaches (equations (12.3) and (12.4)) and have been discussed in chapter 5. Practical applications and validations can be seen in chapters 7, 8 and 9.

Although these methods solve a wide class of nonlinear problems concerning identification and observation, there are applications on where these approaches cannot be used. Examples for these classes of problems are **systems with unknown structure**. This includes systems with an unknown number of nonlinearities and/or systems with reduced state measurement. For problems like this, the method of **identification of systems with unknown structure**, as it will be explained in detail in the next section, can be useful. In contrast to the above described nonlinear observer, the whole (nonlinear) part describing the system dynamics is combined into **one (!) static neural net**. The dynamic elements represent internal system states which, as it can be seen in the next section, can no longer be seen as physically interpretable states. The advantage of state observation is lost in this case. Nevertheless, the transformed states can be useful for control in combination with a direct control method like the one presented in chapter 10. The connection of a static neural net and dynamic elements with external coupling can be referred to as a

dynamic neural (net–)composition with external feedback and plant–independent structure

The whole structure is the same as it can be seen in figure 12.3. The dynamic structure approximates the input–output behaviour of a (structurally) unknown plant [12]. The differences between the nonlinear observer and the **dynamic identificator** are depicted in figure 12.4. The details are shown in figure 12.7. It must be mentioned in this context that the application fields for both methods are different and depend on the conditions given by measurement and structural knowledge. In the next section the mathematical background of this method will be described.

12.3 Identification of Systems with Unknown Structure Using a Dynamic Identificator

As mentioned above, there are cases where the method of the identification of isolated and/or separable nonlinearities cannot be applied. Examples are:

- the number of nonlinearities in the system is unknown
- the position in the system, on which a unknown nonlinearity acts, is unknown
- the system is not separable in the sense of chapter 6
- the linear part is unknown in the sense of chapter 5

In this section a method is presented to handle such systems where some of the above described problems arise. The aim is to identify the input–output

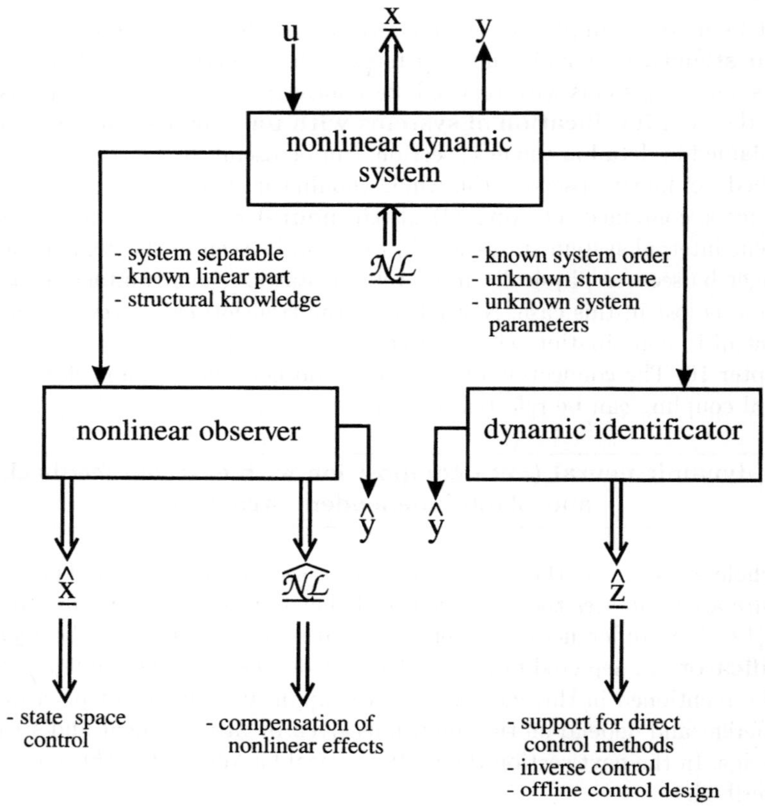

Figure 12.7: Differences between observer and identificators

behaviour in order to use it for control or control design. Unlike the method of global linearization described in chapter 10, **no whole state measurement** is assumed. In the following section the theoretical background of the so–called **dynamic identificator** is presented. As it will be shown, the term *observer* in the sense of section 12.2.3 can no longer be used, but nevertheless, transformed states can be used for control under certain conditions.

12.3.1 Motivation and Theoretical Approach

The aim of this method is to get an input–output approximation of the system

$$\dot{\underline{x}} = \underline{\mathcal{NL}}(\underline{x}, u) \tag{12.5}$$

$$y = h(\underline{x}) \tag{12.6}$$

using a dynamic neural (net–)structure with $\underline{x}, \underline{f} \in \mathbb{R}^n$. With a state transformation, the system (12.5) can be transformed into

$$y^{(n)} = \mathcal{NL}_x(\underline{x}, u) \tag{12.7}$$

The method is based on the method of feedback linearization. The mathematical background of such a transformation can be seen in [12] and in chapter 10. The basic condition for this transformation is that the system is input–output linearizable. With this assumption the system can be linearized by a suitable control law (chapter 10). The system then can be described by the equation

$$y^{(n)} = v \tag{12.8}$$

where y is the output and v is the new input of the system. This equation represents an integrator chain for which it is easy to design a controller. This kind of linearization has to be applied directly at the system's input. This (modified) input compensates for all nonlinear terms affecting the system's behaviour. That means that a representation of equation (12.7) will be possible if the conditions of feedback linearization are given. The whole system then can be described by **one** nonlinear term and linear transfer elements. This is the advantage of the proposed transformation.

One necessary point is the so–called relative degree r of the system (12.5). This value describes the number of necessary differentiation steps of the measurable output y until the input u arises explicitly in equation (12.7). This means that the input–output behaviour of a system of n-th order can be represented by a system of the order r neglecting the internal dynamics. The disadvantage of this is that the control based on a reduced representation cannot stabilize unstable internal dynamics. A suitable choice of the output y (in the case of full state measurement) can possibly solve this problem [17]. Another possibility would be to differentiate the output y n times, which leads to derivatives of u in the input–output representation of the system. Further comments can be seen in [17] and in chapters 10 and 11.

In this context it is assumed that $r = n$. Based on this assumption, we can differentiate the output y of the system (12.5):

$$\dot{y} = \mathcal{L}_{\underline{NL}} h(\underline{x}) = f_1(\underline{x}) \tag{12.9}$$

where $\mathcal{L}_{\underline{NL}} h(\underline{x})$ is the so–called Lie derivative of the function h along the vector field $\underline{\mathcal{NL}}$ (see also [17] and the chapters 10 and 11). Further differentiation leads to:

$$\begin{array}{rcccl} \ddot{y} & = & \mathcal{L}^2_{\underline{NL}} h(\underline{x}) & = & f_2(\underline{x}) \\ \vdots & \vdots & \vdots & \vdots & \vdots \\ y^{(n-1)} & = & \mathcal{L}^{n-1}_{\underline{NL}} h(\underline{x}) & = & f_{n-1}(\underline{x}) \\ y^{(n)} & = & \mathcal{L}^{n}_{\underline{NL}} h(\underline{x}) & = & f_n(\underline{x}, u) \end{array} \tag{12.10}$$

The last equation describes a system of n–th order with the input u and the output y, where $\mathcal{L}^n_{\underline{NL}} h(\underline{x}, u) = \mathcal{NL}_x(\underline{x}, u)$ is the only nonlinearity describing the whole system behaviour. The existence of the Lie derivatives in the equations (12.10) is assumed in this case. Thus, we can write for the transformed system

$$\dot{\underline{z}} = A\underline{z} + \underline{e}_{NL}\mathcal{NL}_x(\underline{x}, u) \tag{12.11}$$

$$y = z_n \tag{12.12}$$

with $\underline{e}_{NL} = \begin{bmatrix} 1 & 0 & 0 & \dots & 0 \end{bmatrix}^T$ and

$$A = \begin{bmatrix} 0 & 0 & 0 & \dots & 0 \\ 1 & 0 & 0 & \dots & 0 \\ 0 & 1 & 0 & \dots & 0 \\ 0 & 0 & 1 & \dots & 0 \\ \vdots & \vdots & \vdots & \vdots & \vdots \\ \dots & \dots & \dots & 1 & 0 \end{bmatrix} \tag{12.13}$$

The vector $\underline{z}$ describes the **transformed state vector**. This leads to the following coherence:

$$\begin{pmatrix} y \\ \dot{y} \\ \ddot{y} \\ \vdots \\ y^{(n-1)} \end{pmatrix} = \begin{pmatrix} z_n \\ z_{n-1} \\ z_{n-2} \\ \vdots \\ z_1 \end{pmatrix} = \underline{\Phi}(\underline{x}) \tag{12.14}$$

Here the matrix A is chosen as proposed in equation (12.13) (integrator chain). As it can be seen later, the choice of an asymptotically stable matrix A can lead to advantages concerning stabilization of the learning and interpretation procedure.
Equation (12.14) describes the dynamic behaviour of the states z_i with respect to $\underline{x}$, which is not a closed form. Additionally, this representation is not suitable for identification based on reduced state measurement. Using

$$\underline{x} = \underline{\Psi}(\underline{z}) = \underline{\Phi}^{-1}(\underline{z}) \tag{12.15}$$

equation (12.11) converts to

$$\dot{\underline{z}} = A\underline{z} + \underline{e}_{NL}\mathcal{NL}_z(\underline{z}, u) \tag{12.16}$$

The utilization of A in (12.13) is not a necessary condition [17]. The choice of an asymptotically stable matrix can lead to better results in learning and interpretation. This fact is marked in figure 12.3 by additional stabilization factors.

Concerning the above described transformation from equation (12.11) to (12.15) the following definition must be noted [17]:

Def.: *A function (vector) $\underline{\Phi}$: $\mathbb{R}^n \rightarrow \mathbb{R}^n$, defined in a region Ω, is called a* <u>*diffeomorphism*</u> *if it is smooth, and if its inverse $\underline{\Phi}^{-1}$ exists and is smooth.*

If the region is the whole space $\mathbb{R}^n$ then the function $\underline{\Phi}(\underline{x})$ is called a global diffeomorphism. If it only exists in a neighbourhood of a given operating point $\underline{x} = \underline{x}_0$, $\underline{\Phi}(\underline{x})$ is called a local diffeomorphism. This leads to the following lemma [17]:

Lemma: *Let $\underline{\Phi}(\underline{x})$ be a smooth function defined in a region Ω in $\mathbb{R}^n$. If the Jacobian matrix $\nabla\underline{\Phi}(\underline{x})$ is non–singular at a point $\underline{x} = \underline{x}_0$ of Ω, then $\underline{\Phi}(\underline{x})$ defines a local diffeomorphism in a subregion of Ω.*

This Lemma is used for the transformation of a nonlinear system into another with respect to a new (transformed) set of states. If this condition is true the system (12.5) can be transformed into system (12.16). This implies the existence of $\underline{x} = \underline{\Psi}(\underline{z})$. It is characteristic of this representation that the whole system dynamics can be described by a single nonlinearity plus an integrator chain (for example), with the state z_n assumed to be measurable. The thoughts presented here should help the reader to get a better understanding of the identification procedure, because the employed dynamic net with external coupling depicted in figure 12.3 has the same structure as the transformed plant and can therefore be called a **transformed parallel observer**.

The assumption for the declarations above was $r = n$, which cannot be assumed in general. As mentioned before, the input–output behaviour of systems with $r < n$ can be described by a system of lower order than n. From the control point of view, neglecting internal states is not advisable because the stability of these states is not guaranteed (see also chapter 10). Further differentiation of the output y leads to derivatives of the input u which must be taken into account for the design of the identificator.

The utilization of dynamic neural net structures for the identification of this type of nonlinear dynamic systems has been the topic of different research projects in literature [18, 1, 12, 4]. Important points are the choice of the network itself and the deviation of suitable learning algorithms for the adaptation of the network parameters. Methods like *Backpropagation, Dynamic Backpropagation* and *Extended Kalman Filter* have been presented [19, 12, 4]. In the following section, an approach for the stable adaptation of the net parameters is presented. This approach in general is based on the methods of chapter 5.

12.3.2 Design of a Dynamic Identificator

Based on the methods presented in chapter 5, the identification of systems with unknown structure is done in a similar manner. The aim is to identify the following system:

$$y^{(n)} = \mathcal{NL}(\underline{z}, \underline{u}) \tag{12.17}$$

or

$$\dot{\underline{z}} = A\underline{z} + \underline{\mathcal{NL}}(\underline{z}, \underline{u}) \tag{12.18}$$

$$y = z_n \tag{12.19}$$

with $\underline{\mathcal{NL}} = \begin{bmatrix} \mathcal{NL} & 0 & 0 & \dots & 0 \end{bmatrix}^T$. This includes the approximation of the input–output behaviour and the observation of the state vector $\underline{z}$. Nevertheless, the term *observer* cannot be used in the sense of chapters 5 and 6 (identification of isolated and/or separable nonlinearities) because equation (12.17) is based on a transformed state space representation. In analogy to chapter 5, a parallel observer can be designed, which has the following form:

$$\dot{\underline{\hat{z}}} = A\underline{\hat{z}} + \underline{\widehat{\mathcal{NL}}}(\underline{z}, \underline{u}) + \underline{l}(y - \hat{y}) \tag{12.20}$$

with $\underline{\widehat{\mathcal{NL}}} = \begin{bmatrix} \widehat{\mathcal{NL}} & 0 & 0 & \dots & 0 \end{bmatrix}^T$. The vector $\underline{l}$ includes the Luenberger coefficients which are necessary for the stabilization of the learning behaviour. Using the system matrix A as described in equation (12.13) (integrator chain), a suitable choice of the Luenberger coefficients is indispensable to guarantee asymptotic stability of the error transfer function (see chapter 5). First of all, it is assumed that the input variables of the nonlinearity $\underline{\mathcal{NL}}$ are directly measurable. Neglecting the inherent approximation error, the nonlinearity can be replaced by the following term:

$$\mathcal{NL}(\underline{z}, \underline{u}) = \underline{\Theta}^T \underline{A}(\underline{z}, \underline{u}) \tag{12.21}$$

$$\widehat{\mathcal{NL}}(\underline{z}, \underline{u}) = \underline{\hat{\Theta}}^T \underline{A}(\underline{z}, \underline{u}) \tag{12.22}$$

The vector $\underline{A}$ is the activation of the GRNN. Also in this case the advantage of a multiplicative connection of activation and neural net parameters can be seen. With (12.21), (12.22) and (12.20) the error differential equation can be written as:

$$\dot{\underline{z}} - \dot{\underline{\hat{z}}} = \dot{\underline{e}} = A\underline{e} + \underline{e}_{NL}(\underline{\Theta} - \underline{\hat{\Theta}})\underline{A} - \underline{l}(y - \hat{y}) \tag{12.23}$$

Using $y = \underline{c}^T \underline{z}$ and $\hat{y} = \underline{c}^T \underline{\hat{z}}$ equation (12.23) can be transformed into:

$$\dot{\underline{e}} = (A - \underline{l}\underline{c}^T)\underline{e} + \underline{e}_{NL}(\underline{\Theta} - \underline{\hat{\Theta}})\underline{A} \tag{12.24}$$

with $\underline{c}^T = \begin{bmatrix} 0 & 0 & \dots & 0 & 1 \end{bmatrix}$ and $\underline{e}_{NL}^T = \begin{bmatrix} 1 & 0 & \dots & 0 & 0 \end{bmatrix}$.
The aim of this approach is to minimize the error $\underline{e}$ between plant and identificator. A comparison of the dynamic net structure of figure 12.3 with the parallel structure of equation (12.23) shows the correspondence. The structure therefore can be seen as a **universal dynamic approximator** for systems with a local or global diffeomorphism, because, apart from the system's order, **the structure does not have to be adapted to different classes of unknown plants.** The deviation of the learning algorithm can be done in the same manner as it was done in chapter 5. The error transfer function can be written as follows:

$$H(s) = \underline{c}^T(sE - (A - \underline{l}\underline{c}^T))^{-1}\underline{e}_{NL} \tag{12.25}$$

Therefore the learning law can be expressed by:

$$\dot{\hat{\underline{\Theta}}} = -\eta\epsilon \underbrace{H(s)\underline{A}(\underline{z}, \underline{u})}_{delayed\ activation} \tag{12.26}$$

with the **augmented error** used for adaptation:

$$\epsilon = e_y + \hat{\underline{\Theta}}^T H(s)\underline{A}(\underline{z}, u) - H(s)\hat{\underline{\Theta}}^T \underline{A}(\underline{z}, \underline{u}) \tag{12.27}$$

where $e_y = y - \hat{y}$. η is the learning factor of the neural net. For the identification procedure, asymptotic stability of $H(s)$ is assumed. For the choice of A, the form of (12.13) is not necessary, which is described in detail in [17]. An asymptotic stable matrix A prevents the identificator from drifting away after decoupling from the plant. Additionally, better learning behaviour can be obtained. The utilization of the delayed activation and the augmented error is necessary, if the error transfer function does not meet the SPR–condition.

The above derived learning law (12.27) was based on the full state measurement of the transformed vector $\underline{z}$ as the neural net input. This assumption is not always fulfilled because the transformation vector $\underline{\Psi}$ is usually not known. Therefore the identification has to be done using observed states as neural net inputs. In this case the error equation has the following form:

$$\dot{\hat{\underline{z}}} = A\hat{\underline{z}} + \underline{e}_{NL}\widehat{\mathcal{NL}}(\hat{\underline{z}}, \underline{u}) + \underline{l}(y - \hat{y}) \tag{12.28}$$

This equation shows a closed loop representation of the identificator (recurrent structure). There is a connection between the plant and the identificator only via the error of the real and identified output. This means that the observation of the transformed states $\underline{z}$ influence the learning behaviour. As shown in chapter 5, this problem can be solved by using a sufficiently high learning factor:

$$0 < \eta_{min} < \eta < \eta_{max} \tag{12.29}$$

An additional stabilization of the learning behaviour can be obtained by a suitable choice of the Luenberger coefficients. This can be shown with the following equations. The error equation of the system has the following form using the observed states as neural net inputs:

$$\begin{aligned} \dot{\underline{e}} &= (A - \underline{l}\underline{c}^T)\underline{e} + \underline{\Theta}^T\underline{A} - \hat{\underline{\Theta}}^T\hat{\underline{A}} && (12.30)\\ &= (A - \underline{l}\underline{c}^T)\underline{e} + (\underline{\Theta}^T(\hat{\underline{A}} + \Delta\underline{A}) - \hat{\underline{\Theta}}^T\hat{\underline{A}}) && (12.31)\\ &= (A - \underline{l}\underline{c}^T)\underline{e} + (\underline{\Theta}^T - \hat{\underline{\Theta}})\hat{\underline{A}} + \underline{\Theta}^T\Delta\underline{A} && (12.32) \end{aligned}$$

where $\Delta\underline{A}$ is the difference between the activation with real and observed states. Sufficiently high values of the Luenberger coefficients can keep the influence of this term near zero which can be seen as a decoupling of state observation and identification itself [2] ($\Delta\underline{A} = f(\underline{e})|_{x_0}$!).

In figure 12.8 the whole structure of the dynamic identificator can be seen. With this constellation, a **global dynamic identification** can be realized. In this

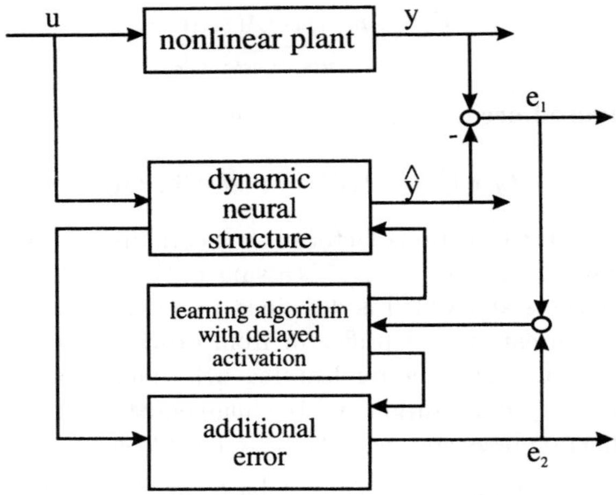

Figure 12.8: Whole identification structure

case only system input, system output and the system's order are known. If the latter one is not exactly known, one has to try to get convergent behaviour with additional dynamic elements implemented in the identification structure. Further problems concerning the relative degree of the system can be solved with the input derivatives. The identified knowledge now can be used for control purposes and is described in the next section.

12.3.3 Possible Control Concepts

In the last section, a method for the identification of nonlinear dynamic systems with unknown structure based on the utilization of a dynamic identificator, and its mathematical background were described. There are different possibilities how this knowledge can be used for control purposes:

- offline control design
- online interpretation of the neural net
- inverse control
- utilization of the transformed states (see figure 12.9) for reference control

For uncertain systems (systems with different unknown parameters and nonlinear influences) an offline control design can be useful. The control design is done on the basis of completely identified system behaviour. Various control strategies

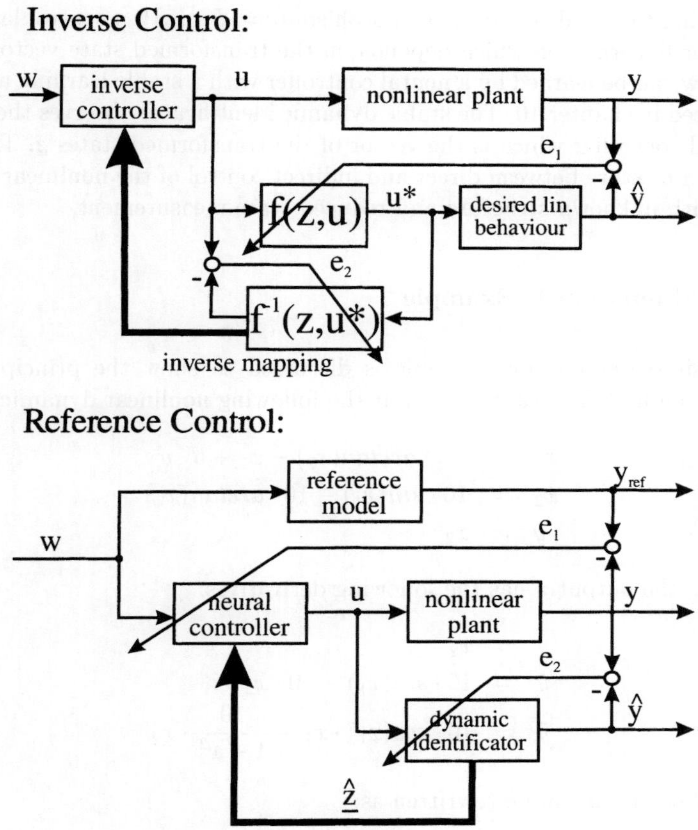

Figure 12.9: Possible control structures based on dynamic identification

can be examined without using the real plant, which is a noncritical way of control design.

The second item is specific for every application and difficult to realize. In the case of isolated or separable nonlinearities, the identification result can be used directly for control (compensation) because it is interpretable. In the case of the dynamic identificator, interpretation is difficult because the neural net describes not only one nonlinear effect but the whole nonlinear dynamic behaviour. A possibilty could be to design an additional identification procedure to get the inverse mapping of the identified term (inverse control). This could be used to compensate for the whole (identified) nonlinearity (Direct Inverse Control).

In analogy to the direct control method presented in chapters 10 and 11, another possibility to use the identified knowledge is shown in figure 12.9. As mentioned before, the transformed states can no longer be interpreted physically. Nevertheless, meeting the condition of diffeomorphism ($\underline{z} = \underline{\Phi}^{-1}(\underline{x})$), a control law can be derived for linearization which depends on the transformed state vector $\underline{z}$. This control law can be learned by a neural controller with a stable learning algorithm as presented in chapter 10. The stable dynamic identificator observes the input of the neural controller which is the vector of the transformed states $\underline{\hat{z}}$. The whole system is a mixture between direct and indirect control of the nonlinear dynamic system with unknown structure and reduced state measurement.

12.3.4 Simulation 1: Example

First, a simple simulation example is discussed to show the principle of the proposed method. We want to look at the following nonlinear dynamic system:

$$\begin{aligned} \dot{x}_1 &= -5 \cdot arctan(x_2) - x_1 + 5 \cdot u && (12.33)\\ \dot{x}_2 &= 10 \cdot sin(x_1) - 10 \cdot arctan(x_2) && (12.34)\\ y &= x_2 && (12.35) \end{aligned}$$

Therefore, the output y has the following derivatives:

$$\begin{aligned} y &= x_2 && (12.36)\\ \dot{y} &= 10 \cdot sin(x_1) - 10 \cdot x_2 && (12.37)\\ \ddot{y} &= 10 \cdot cos(x_1) \cdot \dot{x}_1 - \frac{10}{1 + x_2^2} \cdot \dot{x}_2 && (12.38) \end{aligned}$$

The third equation can be rewritten as:

$$\ddot{y} = 10 \cdot cos(x_1)(-5 \cdot arctan(x_2) - x_1 + 5 \cdot u) - \frac{10}{1 + x_2^2} \cdot (10 \cdot sin(x_1) - 10 \cdot arctan(x_2)) \qquad (12.39)$$

Therefore, we can write:

$$\ddot{y} = \mathcal{NL}_x(x_1, x_2, u) \qquad (12.40)$$

Equation (12.40) still shows a dependence on the real states of the plant. For an identification this means that the real state vector must be available to get $\mathcal{NL}_x(x_1, x_2, u)$. The following equation is used for transformation:

$$\begin{aligned} y &= z_2 \rightarrow && (12.41)\\ y &= x_2 && (12.42)\\ \dot{y} &= z_1 \rightarrow && (12.43)\\ \dot{y} &= 10 \cdot sin(x_1) - 10 \cdot arctan(x_2) && (12.44) \end{aligned}$$

or

$$\underline{z} = \underline{\Phi}(\underline{x}) \qquad (12.45)$$

The inverse of this equation can be written as follows:

$$x_2 = z_2 \tag{12.46}$$

$$x_1 = arcsin(\frac{z_1 + 10 \cdot arctan(z_2)}{10}) \tag{12.47}$$

or

$$\underline{x} = \underline{\Psi}(\underline{z}) \tag{12.48}$$

The equation (12.48) can be put into (12.39) and in (12.40) and we get:

$$\ddot{y} = \mathcal{NL}_z(z_1, z_2, u) \tag{12.49}$$

which is an independent representation of the same input–output behaviour. The aim is to identify this item and therefore we do not know whether the function $\underline{\Phi}(\underline{x})$ is a diffeomorphism which means that the function $\underline{\Psi}$ exists in an area around the operating point. This fact is a restriction of the proposed method. In this example the activation u was chosen to meet the condition locally. This means especially:

$$\frac{-\pi}{2} \leq x_1 \leq \frac{\pi}{2} \tag{12.50}$$

For the dynamic net structure we can write:

$$\ddot{\hat{y}} = \widehat{\mathcal{NL}}_z(\hat{z}_1, \hat{z}_2, u) = \underline{\hat{\Theta}}^T \underline{A}(\hat{z}_1, \hat{z}_2, u) \tag{12.51}$$

The applied learning law was proposed in section 12.3.2. Figure 12.10 shows the simulation results for this example. The first two figures show the real and the identified output at the beginning and at the end of the learning phase for sinusoidal activation. It can be seen that the output of the identificator follows the real output after a learning phase. Furthermore, the influence of the Luenberger coefficients tends to zero, which shows that the output of the (static) neural net represents the dynamic behaviour of the plant (it does not just follow the output). The remaining error is due to the *inherent approximation error* because of the finite number of neural net parameters. This error can be reduced by increasing the number of neurons. The error between the real and the identified plant tends, as expected, to zero. The figure below shows the learning result of the static part of the dynamic structure where we can see the nonlinear dependence of the neural net with respect to z_1. This fact shows the nonlinear approximation ability of the neural net in general. The ambiguous representation is based on the additional dependency upon u and z_2.

12.3.5 Simulation 2: Two–Mass System

The second simulation example shows the effectiveness of the proposed method. The aim is to approximate the input–output behaviour of a nonlinear dynamic system without any a-priori knowledge. We will look at a two–mass system with the (measurable) motor torque as the system's input (see figure 12.11). The

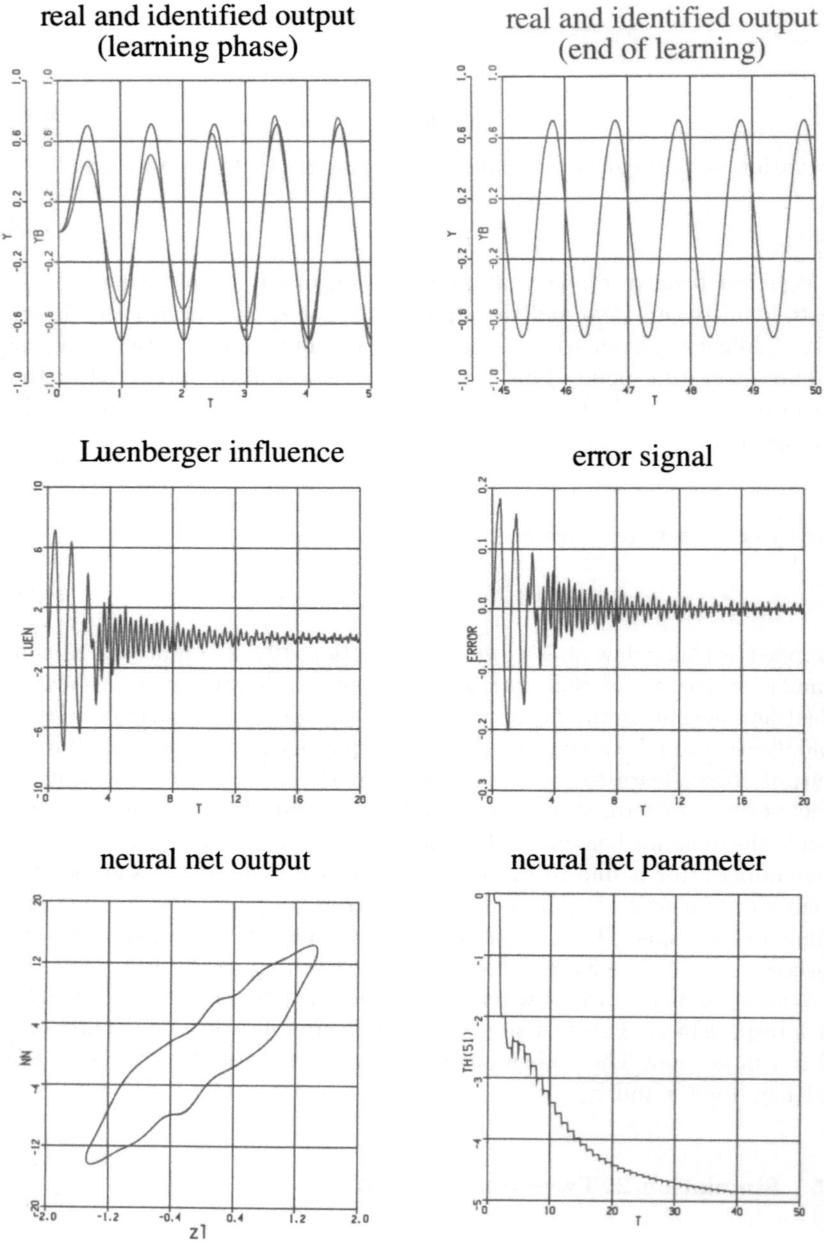

Figure 12.10: Simulation results of the first example

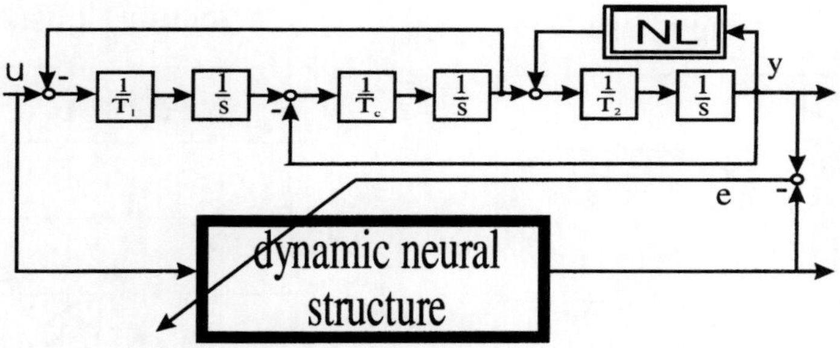

Figure 12.11: Two–mass system and dynamic neural net structure

output y is assumed to be the load speed of the system (arbitrarily chosen). We get the following state space representation of the whole system (the damping factor was neglected):

$$\begin{bmatrix} \dot{x}_1 \\ \dot{x}_2 \\ \dot{x}_3 \end{bmatrix} = \begin{bmatrix} 0 & -\frac{1}{T_1} & 0 \\ \frac{1}{T_C} & 0 & -\frac{1}{T_C} \\ 0 & \frac{1}{T_2} & 0 \end{bmatrix} \begin{bmatrix} x_1 \\ x_2 \\ x_3 \end{bmatrix} + \begin{bmatrix} \frac{1}{T_1} \\ 0 \\ 0 \end{bmatrix} u + \begin{bmatrix} 0 \\ 0 \\ 1 \end{bmatrix} \mathcal{NL}(x_3) \tag{12.52}$$

$$y = x_3 \tag{12.53}$$

with

$$\mathcal{NL}(x_3) = -arctan(2 \cdot x_3) \tag{12.54}$$

The input–output behaviour has to be approximated using the dynamic net structure depicted in figure 12.3. No a–priori knowledge was implemented into the structure, which means that all values of the (static) neural net were set to zero before learning started ($\underline{\hat{\Theta}} = \underline{0}$). For the dynamic identificator we can write:

$$\hat{y}^{(3)} = \widehat{\mathcal{NL}}_z(\hat{z}_1, \hat{z}_2, \hat{z}_3, u) = \underline{\hat{\Theta}}^T \underline{\mathcal{A}}(\hat{z}_1, \hat{z}_2, \hat{z}_3, u) \tag{12.55}$$

We can see, as shown before, a closed system with internal (dynamic) feedbacks. The number of inputs of the net is sufficient because

$$r = n \tag{12.56}$$

Therefore, no derivative of the input u is necessary. If this information is not available the neural net input can be increased, which leads to a definite identification. The learning law was chosen as shown in the last section. Simulation results are shown in figure 12.12. We see results for two different activations of the two–mass system: a sinusoidal activation and a signal generated by a random–check generator. We can see the augmented learning time in the second case. Furthermore, it must be distinguished between the end of the learning phase with plant coupling (bottom left) and without plant coupling (bottom right).

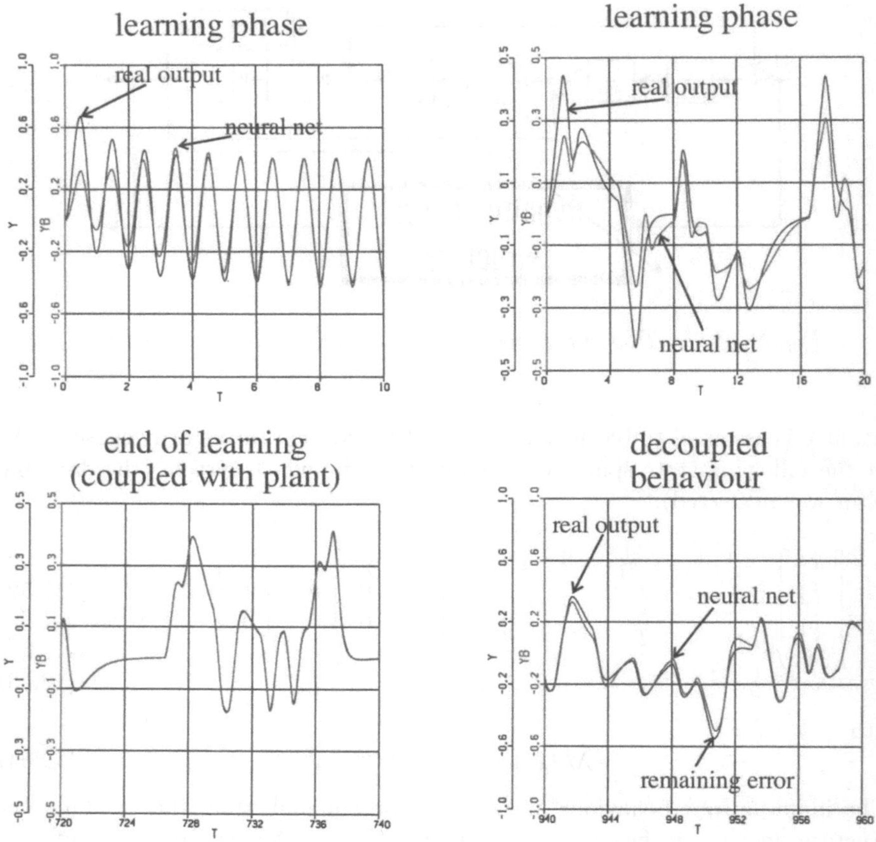

Figure 12.12: Results of the identification of a unknown two–mass system

We can see a remaining error between plant and identificator in the second case, which originates firstly from the low number of network parameters and secondly from a rather short adaptation time. Nevertheless, a good correspondence of the dynamic behaviour can be seen. It must be noted that the remaining error is kept low by the Luenberger coefficients if the identificator is still coupled with the plant.

12.3.6 Simulation 3: Inverse Control

As it was mentioned before one possible control concept of the proposed method is the inverse control approach. In this example we consider an arbitrarily chosen system of second order:

$$\begin{aligned} \dot{x}_1 &= a_1 \mathcal{NL}_1(x_1) + a_2 x_2 + a_3 \mathcal{NL}_2(u) \\ \dot{x}_2 &= a_4 x_1 \end{aligned} \tag{12.57}$$

$$y = x_2 \tag{12.58}$$

y is the measurable output. It is assumed that the parameters a_i and the nonlinearities $\mathcal{NL}_i$ are not (exactly) known, which makes conventional control design difficult. The control aim is to compensate for all nonlinear effects in order to achieve a desired linear plant behaviour. For this reason we will design a transformed observer structure with the neural net (GRNN) at the observer input:

$$\begin{aligned} \dot{\hat{z}}_1 &= a_{1W} \hat{z}_1 + a_{2W} \hat{z}_2 + \widehat{\mathcal{NL}}_z(\underline{\hat{z}}, u) \\ \dot{\hat{z}}_2 &= a_{3W} \hat{z}_1 + a_{4W} \hat{z}_2 \end{aligned} \tag{12.59}$$

$$\hat{y} = \hat{z}_2 \tag{12.60}$$

The linear part of this system represents the desired dynamic, controlled behaviour of the plant characterised by the parameters a_{iW} (opposite of the integrator chain). Hence, the whole concept is divided into two parts: First the identification of the nonlinearity $\mathcal{NL}_z$ and second the derivation of the inverse mapping of $\widehat{\mathcal{NL}}_z$. This knowledge can be used for control to linearize the system as depicted in figure 12.9 (inverse control). So we can write for the input u

$$u = \widehat{\mathcal{NL}}_z^{-1}(\underline{\hat{z}}, w) \tag{12.61}$$

where w is the reference value of the control loop and $\widehat{\mathcal{NL}}_z^{-1}(\underline{\hat{z}}, w)$ the inverse mapping of the identified nonlinearity $\widehat{\mathcal{NL}}_z(\underline{\hat{z}}, u)$. If we assume an optimal learning behaviour we get for the controlled system:

$$\begin{aligned} \dot{\hat{z}}_1 &= a_{1W} \hat{z}_1 + a_{2W} \hat{z}_2 + w \\ \dot{\hat{z}}_2 &= a_{3W} \hat{z}_1 + a_{4W} \hat{z}_2 \end{aligned} \tag{12.62}$$

These equations represent the desired linear system with the input w.

Figure 12.13 shows the simulation results of the proposed control method. The aim in this case is to damp the oscillations of the plant via a suitable control law and to get a desired (controlled) plant behaviour in every operating point. Both curves can be seen in figure 12.13a. Figure 12.13b shows the identification error $y - \hat{y}$ which tends to zero; the dynamic net structure described by equation (12.59) has identified the unknown dynamic behaviour of the plant. This means that the neural net has learned the deviation between the real plant behaviour and the desired linear behaviour.

After the compensation of the nonlinear (static) effects at the plant input the negative oscillations can be damped and the desired step response is achieved (figure 12.13c, start of compensation $t = 300s$). Figure 12.13d shows that the second neural net (see also figure 12.9) learns the inverse mapping of the nonlinear static characteristic at the input of the plant. This mapping is used as a controller and linearizes the nonlinear effects of the plant; the behaviour of the unknown plant follows the desired linear system characterized by the parameters a_{iW}.

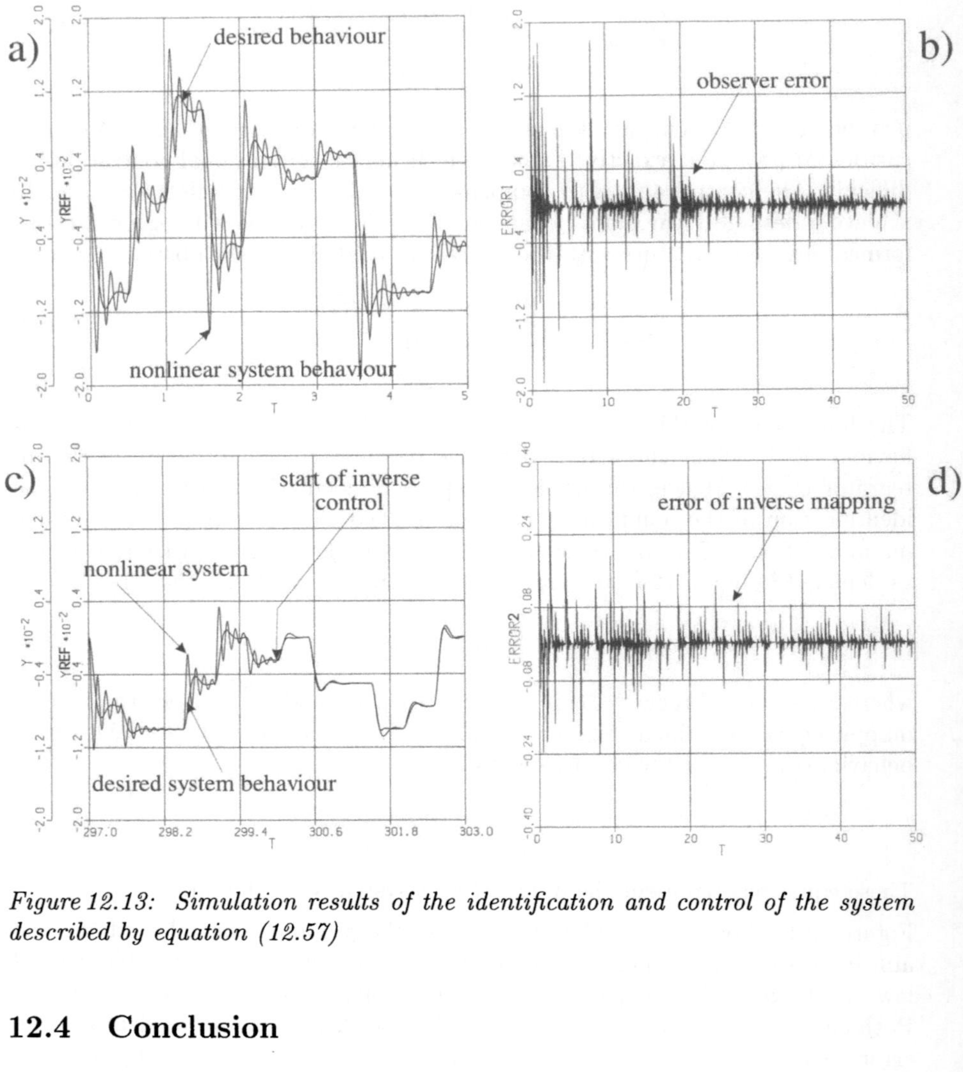

Figure 12.13: Simulation results of the identification and control of the system described by equation (12.57)

12.4 Conclusion

In this chapter, a method for the identification of nonlinear dynamic systems with unknown structure was presented. In connection with the proposed control concept we obtain a mixture between direct and indirect control. To get a delimitation to other methods like the nonlinear observer proposed in the last chapters, characteristical points of observers and dynamic identificators had to be shown. This includes the difference between static and dynamic net structures to give the reader an overview of the possiblities of identification methods for different applications.

12.5 References

[1] Campolucci, P., Uncini, A., Piazza, F:
Causal back Propagation Through Time For Locally Recurrent Neural Networks.
ISCAS '96, Atlanta, USA. Proceedings, Vol. 3, 1996, pp. 531–534.

[2] Engell, S.:
Entwurf nichtlinearer Regelungen.
R. Oldenbourg Verlag, München, 1995.

[3] Fabri, S., Kadirkamanathan, V.:
Dynamic Structure Neural Networks for Stable Adaptive Control of Nonlinear Systems.
IEEE Transactions on Neural Networks, Vol. 7, No. 5, September 1990.

[4] Feldkamp, A., Davis, L. I.:
Dynamic Neural Network Methods to On–Vehicle Idle Speed Control.
ICNN '96, Washington D.C., USA, 1996.

[5] Fischle, K.:
Ein Beitrag zur stabilen adaptiven Regelung nichtlinearer Systeme.
Dissertation, TU München, 1997.

[6] Frenz, Th.:
Stabile neuronale online Identifikation und Kompensation statischer Nichtlinearitäten am Beispiel von Werkzeugmaschinenvorschubantrieben.
Dissertation, TU München, 1998.

[7] Hangl, F., Lenz, U., Schröder, D.:
Theorie des systematischen Entwurfs lernfähiger Beobachter für eine Klasse nichtlinearer Strecken.
Workshop GMA–Ausschuss 1.4. Theoretische Verfahren der Regelungstechnik, Interlaken, Schweiz, 28. September – 1. Oktober 1997.

[8] Kiendl, H.:
Fuzzy Control im Rahmen der Computational Intelligence.
Fachtagung Computational Intelligence, VDI Berichte 1381, 1998, pp. 25–32.

[9] Lenz, U., Schröder, D.:
Local Identification using Artificial Neural Networks.
Proceedings of the Ninth Yale Workshop on Adaptive and Learning Systems, Yale, 1996, pp. 83–88.

[10] Meyer–Bäse, A.:
Dynamische neuronale Netze und neuronale Radialbasisnetze mit Anwendung in der Sprachverarbeitung.
VDI–Fortschrittberichte, Reihe 10, No. 405, Düsseldorf, 1996.

[11] Misava, E. A., Hedrick, J. K.:
Nonlinear Observers – A State–of–the–Art Survey.
Journal of Dynamic Systems, Measurement and Control, Vol. 111, 1989, pp. 344–352.

[12] Narendra, K., Parthasarathy, K.:
Identification and Control of Dynamical Systems Using Neural Networks.
IEEE Transactions on Neural Networks, Vol. 1, No. 1, March 1990.

[13] Narendra, K. S.:
Neural Networks for Control: Theory and Practice.
IEEE Proceedings, Vol. 84, No. 10, 1996, pp. 1385–1405.

[14] Obradovic, L. D.:
On–Line Training of recurrent Neural Networks with Continuous Toplogy Adaptation.
IEEE Transactions on Neural Networks, Vol. 7, No. 1, January 1996.

[15] Schäffner C.:
Analyse und Synthese neuronaler Regelungsverfahren.
Herbert Utz Verlag Wissenschaft, München, 1996.

[16] Schäffner, C., Schröder, D.:
An Application of General Regression Neural Networks to Nonlinear Adaptive Control.
EPE '93, Brighton, UK. Proceedings, Vol. 4, 1993, pp. 219–223.

[17] Slotine, J. J. E., Li, W.:
Applied Nonlinear Control.
Prentice Hall International Editions, New Jersey, 1991.

[18] Straub, S., Schröder, D.:
Identification of Nonlinear Dynamic Systems with Recurrent Neural Networks and Kalman Filter Methods.
ISCAS '96, Atlanta, USA. Proceedings, Vol. 3, 1996, pp. 341–345.

[19] Straub, S., Schröder, D.:
An Example of an Application of Neural Networks in Rolling Mills: Compensation of the Non–Circularity of Winders.
IFAC 95, Munich, Germany. Proceedings, 1995, pp. 583–590.

[20] Straub, S., Schröder, D.:
Neuronalbasierter Beobachterentwurf für eine breite Klasse nichtlinearer Systeme.
Fachtagung Computational Intelligence, VDI Berichte 1381, 1998, pp. 361–380.

[21] Tzirkel–Hancock, E., Fallside, F.:
A Direct Control Method For a Class of Nonlinear Systems Using Neural Networks.
Technical Report CUED/F–INFENG/TR.65, Cambridge University Engineering Department, Cambridge, England, 1991.

13 Further Strategies for Nonlinear Control with Neural Networks

Martin Rau, Anne Angermann

13.1 Introduction

The previous chapters of this book mainly focused on the identification of nonlinearities in electromechanical systems by intelligent observers based on neural networks. These observers have the advantage that they provide, apart from the knowledge about the nonlinearity, knowledge about the non–measurable states. Only in two cases was the identified knowledge used for the compensation of the nonlinearity (chapter 7 and section 9.2.2), whereas no use was made of the complete state observation.

In this chapter, we would like to concentrate on the *control* of nonlinear electromechanical systems. Of course, there is a great variety of approaches to this type of problem; therefore, we will present only three possibilities in this chapter:

1. Starting from the above mentioned situation, the first control concept utilizes the knowledge of the system's states provided by the observer. By this a nonlinear state space controller can be designed which takes the nonlinearity of the system into consideration.

2. The second approach is similar to the concept of *Stable Model Reference Neurocontrol* presented in chapter 11. In contrast to chapter 11 we now assume that the plant has a well–defined and –known structure with an isolated nonlinearity. Then we are able to design an observer, and there is no longer the restriction of full state measurement. The adaptation of the neurocontroller can be achieved with the observed states.

   Since in this approach we know the structure, parameters, nonlinearity and states of the plant (due to observation), we are able to design an optimal controller (e.g. time– or energy–optimal). Using such a control concept, we are also able to take into consideration possible limitations of the system's input value.

3. The concept of time–optimal control is the basis of the last approach presented in this chapter. This control concept is also able to consider limitations of the input value. However, the only similarity with the previous approaches is that the neural network is integrated in the controller.

13.2 Compensation and State–Space Control Strategies for a Class of Nonlinear Systems

In current industrial applications, electrical drives are usually controlled by cascaded P– or PI–controllers. In more advanced systems, a composition of a state–space observer and state–space controller is implemented. All common design rules for linear controllers do not take into account nonlinear effects, which are existent in many real world electromechanical drives, e.g. stick–slip friction, backlash or nonlinear spring characteristics.

In chapters 5 and 6, it was shown that it is possible to identify systems with isolated nonlinearities. These systems can be represented by the following equations in the case of SISO (Single Input Single Output):

$$\begin{aligned} \dot{\underline{x}} &= A \cdot \underline{x} + \underline{b} \cdot u + \underline{e}_{NL} \cdot \mathcal{NL}(\underline{x}, u) \\ y &= \underline{c}^T \cdot \underline{x} + d \cdot u \end{aligned} \quad (13.1)$$

where A is the system matrix, $\underline{b}$ the coupling vector for the input u, $\underline{e}_{NL}$ the coupling vector for the nonlinearity $\mathcal{NL}(\underline{x}, u)$, and $\underline{c}$ and d are necessary to describe the output y of the system.

The *systematic intelligent observer design* in chapter 5 provides knowledge about the exact characteristic of the nonlinearity $\mathcal{NL}(\underline{x}, u)$ and, in addition, about estimates of the states $\underline{x}$. But how can one take advantage of this knowledge when designing controllers for such a system? There are probably many answers to this question, but one possible solution will be discussed in the following sections of this book. We will not develop a completely new theory for controller design, but will present an approach to improve a linear controller's performance in the presence of isolated nonlinearities. The strategy is based on the theory of disturbance rejection in linear systems. The principle of its functionality was already shown in chapter 7 in the case of stick–slip friction ([3]). Compared to linear systems, the isolated nonlinearity cannot be considered as an additional independent input of the system, since it depends on internal states (equation (13.1)). As a consequence, the compensation algorithm depends on the nonlinearity and also on the internal states. This results in a nonlinear state–space control system. Finally, we will derive unified design rules for the control of this class of nonlinear systems.

13.2.1 Systems with Isolated Nonlinearities and Nonlinear Observer

Earlier in this book, we have designed observers for dynamic systems with isolated nonlinearities. The nonlinearity in general can depend on all states and the input of the system. In real plants, it is usually a smaller subset of the whole state $\underline{x}$. To apply the *systematic intelligent observer design* as explained in chapter 5, all states the nonlinearity depends on need to be measurable in order to guarantee stability of the learing law. Here, the input vector of the nonlinearity is also assumed to be measurable and is called $\underline{x}_M$. $\underline{x}_M$ is a subset of the whole state vector $\underline{x}$ and often consists only of one component.

In general, the systems under consideration in this section are described by the following equations:

$$\begin{aligned} \dot{\underline{x}} &= A \cdot \underline{x} + \underline{b} \cdot u + \underline{e}_{NL} \cdot \mathcal{NL}(\underline{x}_M) \\ y &= \underline{c}^T \cdot \underline{x} \end{aligned} \tag{13.2}$$

The input u does not directly affect the output y ($d = 0$). The nonlinearity $\mathcal{NL}(\underline{x}_M)$ is any scalar function of $\underline{x}_M$ which does not necessarily have to be known in advance, except its input. The matrix A and the vectors $\underline{b}$ and $\underline{c}$ have to be known to design a nonlinear observer based on the Luenberger observer and a *General Regression Neural Network* (chapter 5). This intelligent observer provides information about

- all states $\underline{x}$ of the plant
- the nonlinear characteristic $\mathcal{NL}(\underline{x}_M)$ in the range of interest of its input $\underline{x}_M$

Both, state vector $\underline{x}$ and nonlinearity $\mathcal{NL}(\underline{x}_M)$ will be used for controller design. Even if all state variables can be measured, the observer is still necessary to determine $\mathcal{NL}(\underline{x}_M)$.

13.2.2 Exact Compensation of Isolated Nonlinearities

Many possibilities for designing linear controllers for nonlinear systems described in equation (13.2) are known in literature. The method of *linearization* near a *fixed point of operation* is well–known. If several points of operation have to be considered, the procedure is repeated for every point of operation. Between the nominal points of operation, a linear combination of two or more controllers is used. This method is called *gain scheduling* and is successfully applied to aircraft control ([12]). However, there is no general proof of stability available so far.

The idea of exact compensation of the *isolated nonlinearity* is to apply a nonlinear feedback to the system, such that the output of the compensated plant is equal to the output of a reference model containing no nonlinearity. The reference model is the goal of the compensation algorithm. We want the compensated nonlinear system to show the same input–ouput behaviour as the reference model. Note that this behaviour can only be valid for the output, not for all internal states of the nonlinear system. We want to point out, that the reference model does not have to be simulated in real–time while the controller is active. It is only the goal of the compensation algorithm, and the nonlinear feedback is designed that way. Here, we will choose the reference model to be the linear part of the nonlinear system. Therefore, it can be written in state–space notation:

$$\begin{aligned} \dot{\underline{x}}_R &= A \cdot \underline{x}_R + \underline{b} \cdot u \\ y_R &= \underline{c}^T \cdot \underline{x}_R \end{aligned} \tag{13.3}$$

A, $\underline{b}$, and $\underline{c}$ are the same matrices as in equation (13.2). Designing state–space controllers for this linear system is well–known (e.g. pole placement or Riccati–optimization).

The compensation algorithm has to compensate the isolated nonlinearity of the nonlinear system in equation (13.2). The compensation condition can be expressed easily, since the objective is that the output y_R of the reference model is equal to the output y of the compensated system

$$y(t) = y_R(t) \qquad \forall \quad t > 0 \tag{13.4}$$

13.2.2.1 Transfer Function Description of the Nonlinear System

The state–space description of the nonlinear system in equation (13.2), depicted in figure 13.1, and the asumption, that the input $\underline{x}_M$ of the nonlinearity is measurable, enables us to consider $\mathcal{NL}(\underline{x}_M)$ to be a second input of the remaining linear system. $\mathcal{NL}(\underline{x}_M)$ is either known or already identified with the neural observer in chapter 5. Introducing the Laplacian notation and using the superposition

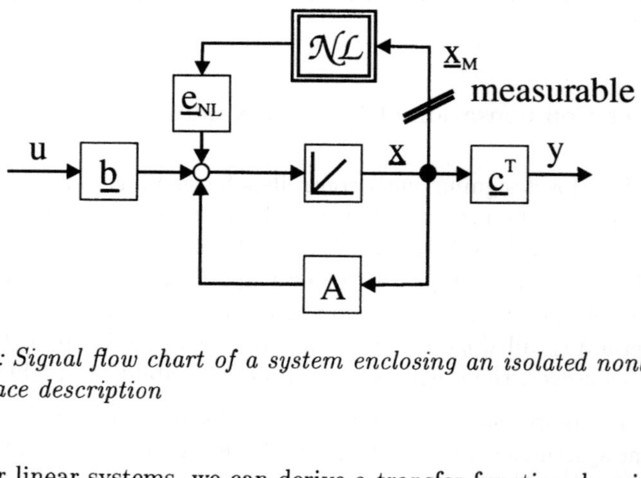

Figure 13.1: Signal flow chart of a system enclosing an isolated nonlinearity $\mathcal{NL}$ in state–space description

principle for linear systems, we can derive a transfer function description of the system with isolated nonlinearity

$$\begin{aligned} y(s) &= \underbrace{\left[\underline{c}^T \cdot (sI - A)^{-1} \cdot \underline{b}\right]}_{G_1(s)} \cdot u(s) + \\ &+ \underbrace{\left[\underline{c}^T \cdot (sI - A)^{-1} \cdot \underline{e}_{NL}\right]}_{G_2(s)} \cdot \mathcal{NL}(\underline{x}_M)(s) \\ &= G_1(s) \cdot u(s) + G_2(s) \cdot \mathcal{NL}(\underline{x}_M)(s) \end{aligned} \tag{13.5}$$

$G_1(s)$ and $G_2(s)$ are defined as

$$G_1(s) = \underline{c}^T \cdot (sI - A)^{-1} \cdot \underline{b} = \frac{\alpha_{n-1}s^{n-1} + \alpha_{n-2}s^{n-2} + \cdots + \alpha_0}{s^n + \gamma_{n-1}s^{n-1} + \cdots + \gamma_0} \tag{13.6}$$

$$G_2(s) = \underline{c}^T \cdot (sI - A)^{-1} \cdot \underline{e}_{NL} = \frac{\beta_{n-1}s^{n-1} + \beta_{n-2}s^{n-2} + \cdots + \beta_0}{s^n + \gamma_{n-1}s^{n-1} + \cdots + \gamma_0} \tag{13.7}$$

α_i, β_i and γ_i are the resulting coefficients of the numerator and denominator of the transfer functions $G_1(s)$ and $G_2(s)$.

The transfer function of the linear reference model can be obtained from equation (13.3)

$$\frac{y_R(s)}{u(s)} = \underline{c}^T \cdot (sI - A)^{-1} \cdot \underline{b} = G_1(s) = \frac{\alpha_{n-1}s^{n-1} + \alpha_{n-2}s^{n-2} + \cdots + \alpha_0}{s^n + \gamma_{n-1}s^{n-1} + \cdots + \gamma_0} \tag{13.8}$$

It is equal to $G_1(s)$. The objective is to compensate $G_2(s) \cdot \mathcal{NL}(\underline{x}_M)$ by an adequate compensation algorithm. Note that the numerator degree is $n-1$, since $d = 0$ was assumed in equation (13.1).

A signal flow chart description of the nonlinear system is depicted in figure 13.2. The output y is composed from the linear portion $G_1(s) \cdot u(s)$ and the nonlinear portion $G_2(s) \cdot \mathcal{NL}(\underline{x}_M)(s)$. Due to the nonlinearity, a simplification of equation (13.5) is not generally possible.

13.2.2.2 Compensation Algorithm

The compensation algorithm can be derived from the transfer function description

$$y(s) = G_1(s) \cdot u(s) + G_2(s) \cdot \mathcal{NL}(\underline{x}_M)(s) \tag{13.9}$$

and the compensation condition

$$y(s) \stackrel{!}{=} y_R(s) = G_1(s) \cdot u(s) \tag{13.10}$$

The new input of both, the nonlinear system and the linear reference model, will be v. Thus, we can calculate the compensating feedback law and therefore can calculate u from the condition in equation (13.10). That means, we want

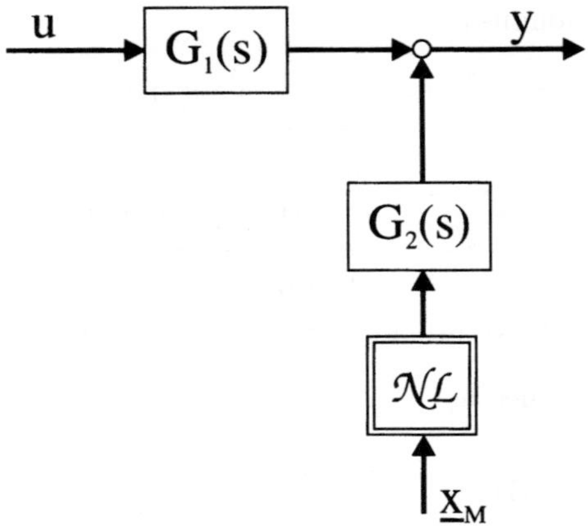

Figure 13.2: Signal flow chart of a system enclosing an isolated nonlinearity $\mathcal{NL}$ in transfer function description

to calculate u such that the compensated nonlinear system behaves like $y(s) = G_1(s) \cdot v(s)$:

$$\begin{aligned} y_R(s) = G_1(s) \cdot v(s) &\overset{!}{=} y(s) = G_1(s) \cdot u(s) + G_2(s) \cdot \mathcal{NL}(\underline{x}_M)(s) \\ \Longrightarrow u(s) &= v(s) - \frac{G_2(s)}{G_1(s)} \cdot \mathcal{NL}(\underline{x}_M)(s) \end{aligned} \tag{13.11}$$

Introducing the compensation filter $K(s)$, we can express the linearizing feedback law in the compact form

$$\begin{aligned} K(s) &= \frac{G_2(s)}{G_1(s)} = \frac{\beta_{n-1}s^{n-1} + \beta_{n-2}s^{n-2} + \cdots + \beta_0}{\alpha_{n-1}s^{n-1} + \alpha_{n-2}s^{n-2} + \cdots + \alpha_0} \\ \Longrightarrow u(s) &= v(s) - K(s) \cdot \mathcal{NL}(\underline{x}_M)(s) \end{aligned} \tag{13.12}$$

The signal flow chart of the compensation algorithm is illustrated in figure 13.3. The whole system with input v and output y is equivalent to the linear reference model. Again, this is only valid for the input–output behaviour, not for internal states. The transfer function of the compensated system is now equivalent to the linear reference model

$$y(s) = F(s) \cdot v(s) \tag{13.13}$$

13.2.2.3 Realization of the Compensation Filter $K(s)$

The numerator $N(\alpha_i, s)$ of the transfer function $G_1(s)$ appears in the denominator of the compensation filter $K(s)$.

$$N(\alpha_i, s) = \alpha_{n-1}s^{n-1} + \alpha_{n-2}s^{n-2} + \cdots + \alpha_0 \tag{13.14}$$

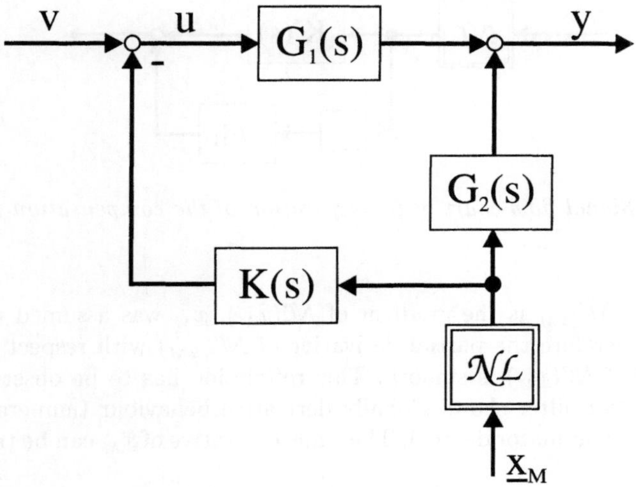

Figure 13.3: Signal flow chart of the compensation algorithm for a system with isolated nonlinearity $\mathcal{NL}$

To guarantee the realizability of $K(s)$, the real part of all zeros of $N(\alpha_i, s)$ has to be negative. Equivalently, the transfer function $G_1(s)$ and hence the system in equation (13.2), has to be *minimum phase.* This implies stability of all internal dynamics.

Furthermore, it is not obvious that the denominator degree of $K(s)$ is greater or equal to the numerator degree. Depending on the system, it is possible that $\alpha_{n-1} = 0$ where $\beta_{n-1} \neq 0$ (see the simulation example later in this chapter). This means $K(s)$ shows globally derivative behaviour and needs to be separated into a transfer function $K'(s)$ with the same degree of numerator and denominator and the derivative of $\mathcal{NL}(\underline{x}_M(t))$. This separation can be calculated by polynomial division.

$$K(s) = K'(s) + k_d \cdot s \tag{13.15}$$

The signal flow chart representation of equation (13.15) is depicted in figure 13.4, where the Laplacian–operator s was substituted by the derivative with respect to time. The derivative part of equation (13.15) does not have to be calculated explicitly by differentiating the output of $\mathcal{NL}(\underline{x}_M(t))$. It can be composed of all states $\underline{x}$ of the nonlinear plant in equation (13.2). These states are either measurable or observable with the *nonlinear observer* presented in chapter 5. The derivative of $\mathcal{NL}(\underline{x}_M(t))$ is

$$\dot{\mathcal{NL}}(\underline{x}_M(t)) = \frac{d\mathcal{NL}(\underline{x}_M(t))}{dt} = \frac{\partial\mathcal{NL}(\underline{x}_M(t))}{\partial\underline{x}_M} \cdot \frac{\partial\underline{x}_M}{\partial t} \tag{13.16}$$

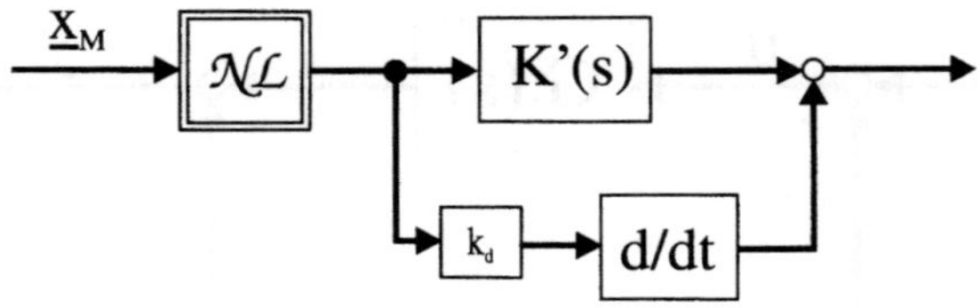

Figure 13.4: Signal flow chart of the separation of the compensation filter $K(s)$

where $\partial \mathcal{NL}(\underline{x}_M)/\partial \underline{x}_M$ is the gradient of $\mathcal{NL}(\underline{x}_M)$. $\underline{x}_M$ was assumed to be measurable and therefore the partial derivative of $\mathcal{NL}(\underline{x}_M)$ with respect to $\underline{x}_M$ can be calculated if $\mathcal{NL}(\underline{x}_M)$ is smooth. This restriction has to be observed only if the compensation filter shows globally derivative behaviour (numerator degree greater than denominator degree). The time–derivative of $\underline{x}_M$ can be transformed into

$$\dot{\underline{x}}_M = \frac{\partial \underline{x}_M}{\partial t} = C_M \cdot \dot{\underline{x}} = C_M \cdot [A \cdot \underline{x} + \underline{b} \cdot u + \underline{e}_{NL} \cdot \mathcal{NL}(\underline{x}_M)] \tag{13.17}$$

where C_M is a matrix that describes the subset $\underline{x}_M$ of the whole state vector, that is input of $\mathcal{NL}(\underline{x}_M)$

$$\underline{x}_M = C_M \cdot \underline{x} \tag{13.18}$$

The whole derivative contribution of the compensation filter is

$$\dot{\mathcal{NL}}(\underline{x}_M) = \frac{\partial \mathcal{NL}(\underline{x}_M(t))}{\partial \underline{x}_M} \cdot [C_M \cdot [A \cdot \underline{x} + \underline{b} \cdot u + \underline{e}_{NL} \cdot \mathcal{NL}(\underline{x}_M)]] \tag{13.19}$$

When the state $\underline{x}$ of the system is accessible for measurement, this transformation can be made directly. Otherwise, the estimate of the whole state vector from the nonlinear observer in chapter 5 can be used. This transformation shows a realization of the compensation filter without explicit differentiation, even if $K(s)$ shows globally derivative behaviour. The compensation algorithm has become a *nonlinear state feedback law*. The derivate of $\mathcal{NL}(\underline{x}_M)$ with respect to $\underline{x}_M$ can either be calculated numerically or by approximating the nonlinearity by an analytic function, which can be differentiated.

If the difference between numerator and denominator degree of $K(s)$ is greater than one, this method can be extended without explicit differentiation until the input u appears explicitly in the k–th derivative of the output $\underline{x}_M$ in equation (13.17). This limit is similar to the relative degree of a nonlinear system.

Restrictions for Exact Compensation As we have seen in this section, exact compensation of isolated nonlinearities is not always possible. The necessary requirements are

- $\mathcal{NL}(\underline{x}_M)$ has to be known (e.g. via nonlinear observer)

- The nonlinear system in equation (13.2) has to be *minimum phase*; all zeros of $G_1(s)$ have to be in the negative complex half plane
- If $\mathcal{NL}(\underline{x}_M)$ is not smooth
  - Numerator and denominator degree of $K(s)$ have to be equal
- If $\mathcal{NL}(\underline{x}_M)$ is smooth and $K(s)$ shows globally derivative behaviour
  - The state vector $\underline{x}$ has to be known, if explicit differentiation is impossible (e.g. measurement or nonlinear observer)
  - The number of existent derivatives of $\mathcal{NL}(\underline{x}_M)$ with respect to $\underline{x}_M$ has to be greater or equal to the difference between the numerator and denominator degree of the compensation filter $K(s)$

Many of these requirements are naturally fulfilled in real systems. The main restriction is the smoothness of $\mathcal{NL}(\underline{x}_M)$, which is not fulfilled in presence of stick–slip friction in mechanical systems. In this case, a smooth approximation of the nonlinearity can solve the problem.

13.2.3 State–Space Control of the Compensated System

The input–output behaviour of the compensated system can be expressed now by a simple linear transfer function

$$y(s) = G_1(s) \cdot v(s) \tag{13.20}$$

This is equal to the linear reference model $y_R(s)$. Any linear controller structure, requiring only the output y, can be applied to the compensated system, e.g. PI– or PID–controllers. When designing state–space controllers, some considerations concerning the internal states are necessary.

Due to the compensation filter and the isolated nonlinearity, the internal states of the linear reference model and the plant are not identical. The measured or observed states of the plant are not useful for linear state–space control. To implement the well–known theory of pole placement or Riccati–optimization, the states of a linear system, producing the same input–output relation, are necessary.

This can be achieved by utilizing a parallel linear Luenberger observer, containing only the linear part of the system (without an isolated nonlinearity). These estimated states $\hat{\underline{x}}_R$ correspond to a linear system with the same input–output behaviour.

This controller design is similar to the controller design for a linear system. It consists of two steps

- Design of a linear Luenberger observer for the system with input v and output y
- Controller design with pole placement or Riccati–optimization utilizing the observed states $\underline{\hat{x}}_R$

The observer's differential equation is linear. Its dynamics is directly influenced by the Luenberger coefficients $\underline{l}$.

$$\begin{aligned} \dot{\underline{\hat{x}}}_R &= A \cdot \underline{\hat{x}}_R + \underline{b} \cdot v + \underline{l} \cdot (\hat{y}_R - y) \\ \hat{y}_R &= \underline{c}^T \cdot \underline{\hat{x}}_R \end{aligned} \tag{13.21}$$

Like in linear control theory, the state feedback law is calculated as

$$v = w - \underline{r}^T \cdot \underline{\hat{x}}_R \tag{13.22}$$

All characteristics of linear systems are valid. Figure 13.5 shows the structure of the controlled system. In figure 13.5 the compensation filter is realized without explicit differentiation and the states $\underline{x}$ are measured. In figure 13.6 the compensation and control structure are depicted for observed states $\underline{\hat{x}}$ with a neural observer (chapter 5 and [6]).

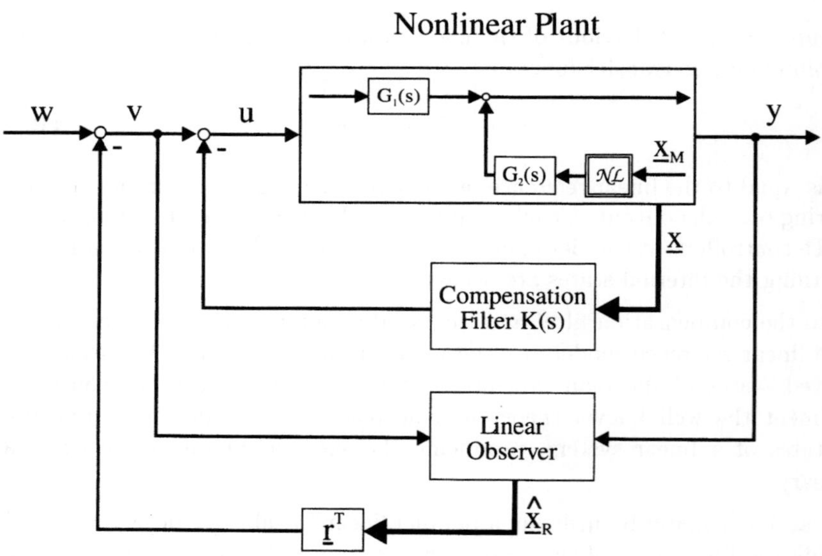

Figure 13.5: State–space control structure for the compensated system with an isolated nonlinearity; explicit differentiation is not necessary

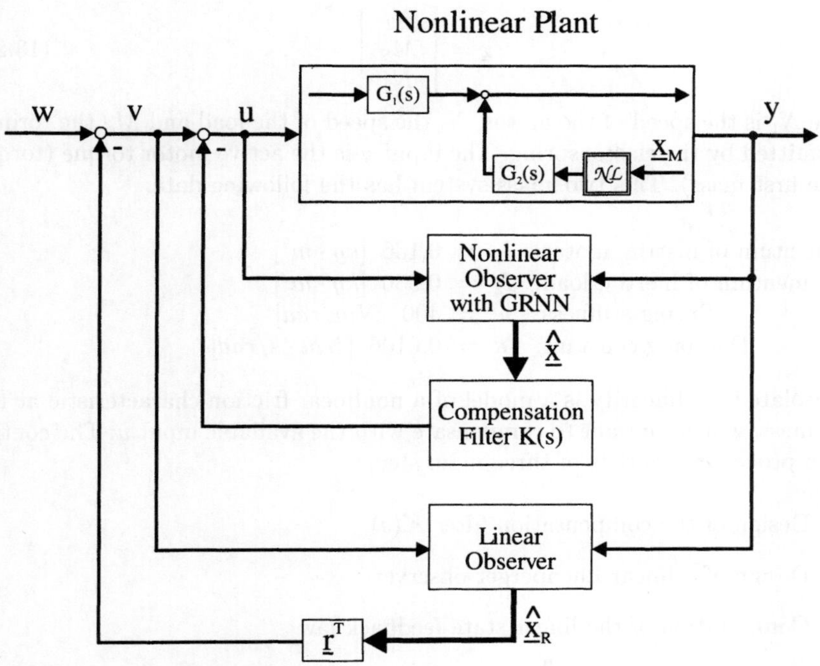

Figure 13.6: State–space control structure for the compensated system with isolated nonlinearity including a nonlinear observer; explicit differentiation is not necessary

13.2.3.1 Simulation Example

So far, we presented a unified theory for the compensation of isolated nonlinearities in dynamic systems. In this section, we will show that the proposed theory can be applied to real systems. We will also show all necessary steps for a simulation of the structure in figure 13.5.

We will use a two–mass system coupled by an elastic spring as the system under consideration. One mass is the motor; the other one is the load. The control objective is the speed of the load. A state–space representation is given by

$$\begin{aligned}\dot{\underline{x}} &= \underbrace{\begin{bmatrix} -\frac{d}{J_1} & -\frac{c}{J_1} & \frac{d}{J_1} \\ 1 & 0 & -1 \\ \frac{d}{J_2} & \frac{c}{J_2} & -\frac{d}{J_2} \end{bmatrix}}_{A} \cdot \underline{x} + \underbrace{\begin{bmatrix} \frac{1}{J_1} \\ 0 \\ 0 \end{bmatrix}}_{\underline{b}} \cdot u + \underbrace{\begin{bmatrix} 0 \\ 0 \\ \frac{1}{J_2} \end{bmatrix}}_{\underline{e}_{NL}} \cdot \underbrace{3\arctan(10x_3)}_{\mathcal{NL}(x_3)} \\ y &= \underbrace{\begin{bmatrix} 0 & 0 & 1 \end{bmatrix}}_{\underline{c}^T} \cdot \underline{x} \end{aligned} \tag{13.23}$$

The state vector is

$$\underline{x} = \begin{bmatrix} N_1 \\ M_S \\ N_2 \end{bmatrix} \tag{13.24}$$

where N_1 is the speed of the motor, N_2 the speed of the load and M_S the torque, transmitted by the elastic spring. The input u is the active motor torque (torque at the first mass). This two–mass system has the following data:

$$\begin{aligned} \text{Momentum of inertia, motor:}\quad J_1 &= 0.166\ [kg \cdot m^2] \\ \text{Momentum of inertia, load:}\quad J_2 &= 0.330\ [kg \cdot m^2] \\ \text{Spring stiffness:}\quad c &= 400\ [Nm/rad] \\ \text{Damping constant:}\quad d &= 0.0106\ [Nm \cdot s/rad] \end{aligned}$$

The isolated nonlinearity is a model of a nonlinear friction characteristic at the load–mass, which we want to compensate with the available input u. The control design procedure consists of three main steps:

- Design of the compensation filter $K(s)$
- Design of a linear Luenberger observer
- Computation of the linear state feedback law

Compensation filter $K(s)$
To design the compensation filter $K(s)$, we first calculate the two transfer functions $G_1(s)$ and $G_2(s)$.

$$\begin{aligned} G_1(s) &= \underline{c}^T (sI - A)^{-1} \underline{b} = \frac{s \cdot d + c}{s\,(s^2 \cdot J_1 J_2 + s \cdot d\,(J_1 + J2) + c\,(J_1 + J_2))} \\ G_2(s) &= \underline{c}^T (sI - A)^{-1} \underline{e}_{NL} = \frac{s^2 \cdot J_1 + s \cdot d + c}{s\,(s^2 \cdot J_1 J_2 + s \cdot d\,(J_1 + J2) + c\,(J_1 + J_2))} \end{aligned} \tag{13.25}$$

The compensation filter $K(s)$ can be calculated according to section 13.2.2.2. For $K(s)$ we obtain

$$K(s) = \frac{G_2(s)}{G_1(s)} = \frac{s^2\, J_1 + s\, d + c}{s\, d + c} \tag{13.26}$$

This transfer function shows globally derivative behaviour and has to be divided into a realizable part $K'(s)$ and a derivative part.

$$K(s) = \underbrace{\frac{s\,(d - \frac{J_1 c}{d}) + c}{s\, d + c}}_{K'(s)} + s \cdot \underbrace{\frac{J_1}{d}}_{k_d} \tag{13.27}$$

In this example, the whole state vector was assumed to be measurable. Thus, we can transform the derivative contribution of the compensation algorithm, that is $k_d \cdot \frac{d}{dt}\,(\mathcal{NL}(x_3))$, into

$$\begin{aligned} k_d \cdot \frac{d}{dt}(\mathcal{NL}(x_3)) &= k_d \cdot \frac{\partial}{\partial x_3}(\mathcal{NL}(x_3)) \cdot \dot{x}_3 = \\ &= \frac{J_1}{d} \cdot \frac{100}{1+10^4 x_3^2} \cdot \left[\frac{d}{J_2} \cdot x_1 + \frac{c}{J_2} \cdot x_2 - \frac{d}{J_2} \cdot x_3 + \frac{1}{J_2} \cdot 3 \arctan(10\, x_3)\right] \end{aligned} \tag{13.28}$$

The compensation filter can be calculated without explicit differentiation of $\mathcal{NL}(x_3)$ with respect to time. The complete compensation algorithm, which eliminates the isolated nonlinearity, has the form

$$u = v - K(s) \cdot \mathcal{NL}(x_3) \tag{13.29}$$

with v as new input of the system. The compensated system now behaves like the linear reference model as described in section 13.2.2.2. In a next step, we have to design a linear Luenberger observer to obtain the equivalent linear states for the compensated system.

Linear observer

The input–output behaviour of the compensated system is equivalent to the reference system, but the states are not, because the nonlinearity's influence was only compensated with respect to the system's output y.

To obtain the states of the linear reference model, we implement an observer with input v and error comparison y. Since this is well–known in linear control theory, we will skip the exact design procedure. Damping optimum was used for the error differential equation in this example.

State–space control design

Last, we have to control a system with input v and output y that has linear input–output behaviour. The states $\underline{\hat{x}}_R$ of the equivalent linear system (reference system) are provided by the linear Luenberger observer.

The controller design procedure is equivalent to that of a linear system. We use pole placement and damping optimum with a time constant $T_{ers} = 0.1\,s$ ([11]). The control law has the following form:

$$v = w - \underline{r}^T \cdot \underline{\hat{x}}_R \tag{13.30}$$

w is the reference input of the system (figure 13.5). The results of the simulation are depicted in figures 13.7 and 13.8. Figure 13.7 shows the step response of the reference system, that is our control aim. The same control parameters were applied to the compensated nonlinear system and the reference system, to show the performance of the compensation strategy. Figure 13.8 shows the step response for the compensated system with isolated nonlinearity. The controller and compensation filter are adjusted well, and there is no significant difference between the two step responses. The necessary motor torque is depicted in figure 13.9. No non–realizable high torque peaks are required to perform the compensation strategy. To show the benefit of this control design procedure, we depicted

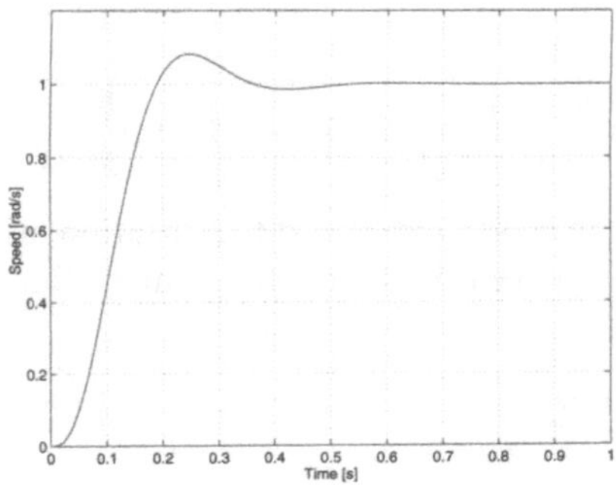

Figure 13.7: Step response $y_R(t)$ of the refer ene system; the desired final value is 1 rad/s

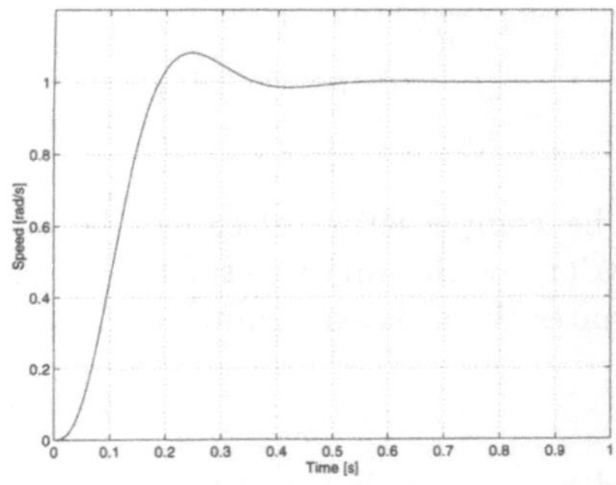

Figure 13.8: Step response $y(t)$ of the compensated nonlinear system

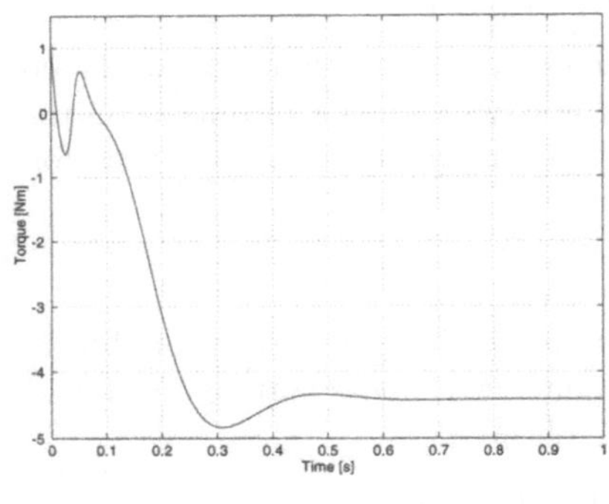

Figure 13.9: Necessary motor torque for the compensation algorithm

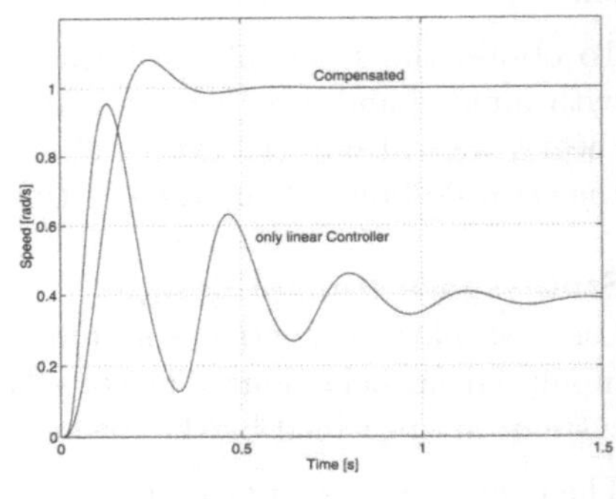

Figure 13.10: Comparision: step response of the nonlinear system, controlled by linear state feedback and the compensated system

the step response of the nonlinear system, controlled by the linear state feedback only. The nonlinearity was not taken into account, and therefore, both dynamics and final value have deteriorated (figure 13.10). The compensation filter takes advantage of the knowledge of the nonlinearity and guarantees an exact final value without increasing the degree of the whole system (e.g. by integrators). The additional states of the compensation filter itself remain unobservable.

13.2.4 Conclusions

In this section we presented a unified approach for controller design for plants containing isolated nonlinearities. The main restriction is the smoothness of $\mathcal{NL}(\underline{x}_M)$. Furthermore, the transfer function $G_1(s)$ has to be *minimum phase.* A major problem in real applications is the differentiation of measured signals. This can be avoided by state measurement or, which is even more realistic, by using the estimated states from the nonlinear observer in chapter 5. The design procedure consists of three main steps

- Design of the compensation filter $K(s) = \dfrac{G_2(s)}{G_1(s)}$
- Design of a linear Luenberger observer for the system with input v and output y. This generates estimates of the states of the equivalent linear system $\hat{\underline{x}}_R$.
- Design of a linear state feedback law, utilizing the estimates $\hat{\underline{x}}_R$ of the equivalent linear system.

The main benefit of this design procedure is the unified approach. Most steps require linear control theory. Nevertheless, the resulting control–algorithm is a nonlinear state–feedback law. It is applicable to a wide class of nonlinear systems, for example in the field of electromechanical drive systems and motion control. A similar procedure for nonlinear state–space controller design is presented in the following section.

Finally, we want to point out the differences and improvements compared to the proposed control strategies in chapters 7 and 11. In chapter 7 the isolated nonlinearity is compensated with an approximation of the exact filter $K(s)$. A PDT_1–transfer function, whose parameters are tuned by trial–and–error, is applied to the plant. Here, we derived a compensation filter such, that the effects of the nonlinearity at the output of the system disappear completely. Additionally, we stated a design procedure not only for the compensation filter $K(s)$, but also for the controller. Compared to the adaptive control concept in chapter 11, compensation filters are only applicable to systems with isolated nonlinearity, not to systems of arbitrary structure. In opposite to *adaptive feedback linearization*, full state measurement is not required for the design of the compensation filter, since the *nonlinear neural observer* can provide estimates of the states.

13.2.5 Alternate Compensation and Control Design

In the last sections, we derived design rules for state–space control of a system with an isolated nonlinearity based on two main steps

1. Design of a linearizing compensation filter
2. Design of a linear state feedback law for the linearized system

In this section, we will swap the two steps: first we will design a linear state feedback. The resulting system is still nonlinear then. In a second step, we will design the linearizing compensation filter, which has different coefficients as in the last sections. This order has the advantage, that we do not need the linear state observer, as we will see later.

The system under consideration is of the following form again:

$$\begin{aligned} \dot{\underline{x}} &= A \cdot \underline{x} + \underline{b} \cdot u + \underline{e}_{NL} \cdot \mathcal{NL}(\underline{x}_M) \\ y &= \underline{c}^T \cdot \underline{x} \end{aligned} \tag{13.31}$$

In the first step, a linear state feedback law is applied to the system with isolated nonlinearity. But how should it be designed? The only nonlinear effect in equation (13.31) is $\mathcal{NL}(\underline{x}_M)$. For the linear controller design, we set $\mathcal{NL}$ to zero and design a linear controller with the desired behaviour for the system in equation (13.31) with $\mathcal{NL} = 0$.

$$\mathcal{NL} = 0 \qquad \text{for controller design} \tag{13.32}$$

Now, we can use any design rule for the remaining linear system, e.g. damping–optimum, pole placement, etc. Here we will use state feedback and damping–optimum as a background theory. The feedback coefficients $\underline{r}$ can be calculated with standard tools using *Matlab*. The feedback law is

$$u = v - \underline{r}^T \cdot \underline{x} \tag{13.33}$$

v is the new input to the controlled system. The state vector $\underline{x}$ is either measured directly or is provided by the *nonlinear observer* as presented in chapter 5. If an observer is necessary, $\hat{\underline{x}}$ instead of $\underline{x}$ is used in the feedback law (13.33). This is possible, since the observer acts independently from the control law. The overall structure of the alternate compensation and control strategy is depicted in figure 13.11.

For controller design, we neglected the isolated nonlinearity and thus a major property of the whole system in equation (13.31). We cannot expect the controlled system to show the desired behaviour. Nevertheless, we can describe its dynamics in state–space notation

$$\begin{aligned} \dot{\underline{x}} &= \underbrace{\left(A - \underline{b} \cdot \underline{r}^T\right)}_{\tilde{A}} \cdot \underline{x} + \underline{b} \cdot v + \underline{e}_{NL} \cdot \mathcal{NL}(\underline{x}_M) \\ y &= \underline{c}^T \cdot \underline{x} \end{aligned} \tag{13.34}$$

The matrix $\tilde{A}$ already describes the desired dynamic behaviour, but still the nonlinearity is present.

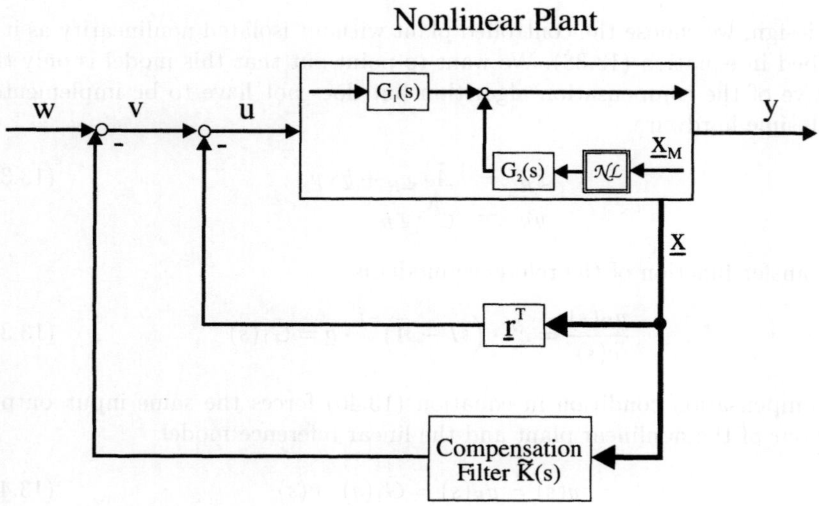

Figure 13.11: Overall structure of the alternate compensation and state–space control strategy; a linear observer is not necessary

In the next step, we will design a nonlinear feedback to compensate the nonlinear influence of $\mathcal{NL}(\underline{x}_M)$. The input $\underline{x}_M$ of the nonlinearity is assumed to be measurable, and the nonlinearity itself is known in advance or is provided by a *nonlinear neural observer*. To calculate the compensation filter, the system of equation (13.34) is first transformed in transfer function description. Similar to section 13.2.2.1, this results in the following equations:

$$\begin{aligned} y(s) &= \underbrace{\left[\underline{c}^T \cdot \left(sI - \tilde{A}\right)^{-1} \cdot \underline{b}\right]}_{\tilde{G}_1(s)} \cdot v(s) + \\ &\quad + \underbrace{\left[\underline{c}^T \cdot \left(sI - \tilde{A}\right)^{-1} \cdot \underline{e}_{NL}\right]}_{\tilde{G}_2(s)} \cdot \mathcal{NL}(\underline{x}_M)(s) \\ &= \tilde{G}_1(s) \cdot v(s) + \tilde{G}_2(s) \cdot \mathcal{NL}(\underline{x}_M)(s) \end{aligned} \tag{13.35}$$

$\tilde{G}_1(s)$ and $\tilde{G}_2(s)$ are defined as

$$\tilde{G}_1(s) = \underline{c}^T \cdot \left(sI - \tilde{A}\right)^{-1} \cdot \underline{b} = \frac{\tilde{\alpha}_{n-1}s^{n-1} + \tilde{\alpha}_{n-2}s^{n-2} + \cdots + \tilde{\alpha}_0}{s^n + \tilde{\gamma}_{n-1}s^{n-1} + \cdots + \tilde{\gamma}_0} \tag{13.36}$$

$$\tilde{G}_2(s) = \underline{c}^T \cdot \left(sI - \tilde{A}\right)^{-1} \cdot \underline{e}_{NL} = \frac{\tilde{\beta}_{n-1}s^{n-1} + \tilde{\beta}_{n-2}s^{n-2} + \cdots + \tilde{\beta}_0}{s^n + \tilde{\gamma}_{n-1}s^{n-1} + \cdots + \tilde{\gamma}_0} \tag{13.37}$$

The compensation goal is to make the nonlinear plant behave like the linear reference model. As a model, that is our goal for the controller and compensation filter design, we choose the controlled plant without isolated nonlinearity as it is described in equation (13.38). We want to point out that this model is only the objective of the compensation algorithm but does not have to be implemented in real–time hardware.

$$\begin{aligned} \dot{\underline{x}}_R &= \tilde{A} \cdot \underline{x}_R + \underline{b} \cdot v \\ y_R &= \underline{c}^T \cdot \underline{x}_R \end{aligned} \tag{13.38}$$

The transfer function of the reference model is

$$\frac{y_R(s)}{v(s)} = \underline{c}^T \cdot \left(sI - \tilde{A}\right)^{-1} \cdot \underline{b} = \tilde{G}_1(s) \tag{13.39}$$

The compensation condition in equation (13.40) forces the same input–output behaviour of the nonlinear plant and the linear reference model.

$$y(s) \stackrel{!}{=} y_R(s) = \tilde{G}_1(s) \cdot v(s) \tag{13.40}$$

The compensation filter $\tilde{K}(s)$ can be derived from equation (13.40). The new input of both, the compensated nonlinear system and the reference model, is w. We will calculate a feedback v such that the nonlinear system's output is equal to the reference model's output.

$$\begin{aligned} y_R(s) = \tilde{G}_1(s) \cdot w(s) &\stackrel{!}{=} y(s) = \tilde{G}_1(s) \cdot v(s) + \tilde{G}_2(s) \cdot \mathcal{NL}(\underline{x}_M)(s) \\ \Longrightarrow v(s) &= w(s) - \frac{\tilde{G}_2(s)}{\tilde{G}_1(s)} \cdot \mathcal{NL}(\underline{x}_M)(s) \end{aligned} \tag{13.41}$$

By introducing the compensation filter $\tilde{K}(s)$, the linearizing feedback law has the following form:

$$\begin{aligned} \tilde{K}(s) &= \frac{\tilde{G}_2(s)}{\tilde{G}_1(s)} = \frac{\tilde{\beta}_{n-1}s^{n-1} + \tilde{\beta}_{n-2}s^{n-2} + \cdots + \tilde{\beta}_0}{\tilde{\alpha}_{n-1}s^{n-1} + \tilde{\alpha}_{n-2}s^{n-2} + \cdots + \tilde{\alpha}_0} \\ \Longrightarrow v(s) &= w(s) - \tilde{K}(s) \cdot \mathcal{NL}(\underline{x}_M)(s) \end{aligned} \tag{13.42}$$

The whole system with input w and output y now shows the desired behaviour of the linear reference model in equation (13.38). The dynamics contributed by the linear state feedback of equation (13.33) is already included in the reference model and therefore in the transfer function $\tilde{G}_1(s)$.

The equivalence of the linear reference model and the controlled and compensated nonlinear system is valid only for the input–output relation, not for internal states, since the influence of the nonlinearity is only compensated with respect to the system's output y.

If $\tilde{K}(s)$ shows globally derivative behaviour, the derivative part of $\tilde{K}(s)$ can be composed of all states of the nonlinear system, as explained in section 13.2.2.3. If the whole state is not accessible for measurement, the *nonlinear neural observer* of chapter 5 will provide the missing states. The general structure, including the *neural observer*, is depicted in figure 13.6.

Note that the controller's coefficients $\underline{r}$ influence the compensation filter $\tilde{K}(s)$ (see equations (13.34) – (13.37)). This can be taken into account when specifying the reference model and thus the controller's coefficients. It may be possible to obtain a more simple compensation filter $\tilde{K}(s)$; pole–zero compensation may be obtained in some applications. To avoid a globally derivative behaviour of $\tilde{K}(s)$, the controller's coefficients may be chosen adequately. Of course, all these possibilities of influencing the compensation filter will cause restrictions in the linear reference model and thus in the dynamics of the controlled nonlinear system. A compromise between realization of $\tilde{K}(s)$ and the desired dynamics of the controlled system has to be found.

A realization of the compensation filter $\tilde{K}(s)$ can be found in the same way as described in section 13.2.2.3. Especially, the derivative of $\mathcal{NL}(\underline{x}_M)$ with respect to time does not have to be calculated explicitly. It can be composed of all states and the nonlinearity of the system. Exact compensation of the isolated nonlinearity is not possible for all cases. All restrictions for the compensation filter and the nonlinearity already mentioned in section 13.2.2.3 also apply to $\tilde{K}(s)$. We just want to point out the most important requirements

- $\mathcal{NL}(\underline{x}_M)$ has to be known (e.g. via nonlinear observer)
- $\tilde{G}_1(s)$ has to be *minimum phase*; all zeros have to be in the negative complex half plane
- If $\tilde{K}(s)$ shows globally derivative behaviour, $\mathcal{NL}(\underline{x}_M)$ has to be smooth

The principle of this control strategy is the same as in the last sections, just the order of necessary steps has changed. The main advantage of this strategy is the unified approach. Most steps only require linear control theory, and no difficult and time consuming calculation of the controller is necessary. The main steps are

- Design of a linear state feedback law, as if the isolated nonlinearity was not present ($\mathcal{NL} = 0$ for controller design)
- Design of the compensation filter $\tilde{K}(s) = \frac{\tilde{G}_2(s)}{\tilde{G}_1(s)}$

A linear observer for the design of the state feedback law is no longer necessary. This reduces the intensity of online computation. If the state of the nonlinear system is not measurable, the *nonlinear neural observer* of chapter 6 will provide information about state and isolated nonlinearity.

This alternate compensation and control design procedure is not as straightforward as the first approach in the last sections. First, we design a linear feedback law for a nonlinear system. This only makes sense in combination with the compensation filter applied afterwards. Nevertheless, the result is the same with less computational intensity, since a linear observer for the state feedback law is no longer necessary.

The last two control approaches were able to linearize a nonlinear plant with an isolated nonlinearity. The compensation filter may cause problems if its poles are very fast. In discrete–time realizations the sample time has to be decreased enormously. To avoid this problem, we want to present a method that generalizes the *linear controllable canonical form* to nonlinear systems and allows the calculation of a linearizing nonlinear state feedback.

13.3 Nonlinear Control with a Controllable Canonical Form

The controllable canonical form for a linear system of order n is a special state–space description of the following form

$$\dot{\underline{x}} = \begin{bmatrix} 0 & 1 & 0 & 0 & \cdots & & 0 \\ 0 & 0 & 1 & 0 & \cdots & & 0 \\ \vdots & & & \ddots & & & \vdots \\ 0 & \cdots & & 0 & 1 & & 0 \\ 0 & \cdots & & & 0 & & 1 \\ -\alpha_0 & -\alpha_1 & \cdots & & & & -\alpha_{n-1} \end{bmatrix} \cdot \underline{x} + \begin{bmatrix} 0 \\ 0 \\ \vdots \\ 0 \\ 0 \\ 1 \end{bmatrix} \cdot u \qquad (13.43)$$

This representation is suitable for designing state–space controllers since the coefficients α_i are also the coefficients of the characteristic polynomial of the system.

According to equation (13.43) the state–space equations are

$$\begin{aligned} \dot{\underline{x}}_1 &= \underline{x}_2 \\ \dot{\underline{x}}_2 &= \underline{x}_3 \\ &\vdots \\ \dot{\underline{x}}_{n-1} &= \underline{x}_n \\ \dot{\underline{x}}_n &= -\alpha_0 x_1 - \alpha_1 x_2 - \cdots - \alpha_{n-1} x_n + u \end{aligned} \qquad (13.44)$$

The objective of the nonlinear controllable canonical form is to find a similar representation for nonlinear systems such that the calculation of the (nonlinear) state feedback law is as simple as in the linear case.

The first $n-1$ differential equations of equation (13.44) remain unchanged; the last equation contains nonlinear functions for the system's dynamics and the coupling of the input u ([13]). The nonlinear controllable canonical form for systems with one input is

$$\underline{\dot{x}} = \underline{f}(\underline{x}) + \underline{g}(\underline{x}) \cdot u \tag{13.45}$$

with

$$\underline{f}(\underline{x}) = \begin{bmatrix} x_2 \\ x_3 \\ \vdots \\ x_n \\ \mathcal{F}(\underline{x}) \end{bmatrix} \qquad \underline{g}(\underline{x}) = \begin{bmatrix} 0 \\ 0 \\ \vdots \\ 0 \\ \mathcal{G}(\underline{x}) \end{bmatrix} \tag{13.46}$$

If the system under consideration is in the controllable canonical form, the control law

$$u = -\frac{1}{\mathcal{G}(\underline{x})} \left(\mathcal{F}(\underline{x}) + \alpha_0 x_1 + \alpha_1 x_2 + \cdots + \alpha_{n-1} x_n - w \right) \tag{13.47}$$

produces the closed loop system

$$\underline{\dot{x}} = A \cdot \underline{x} + \underline{b} \cdot w \tag{13.48}$$

where A and $\underline{b}$ are the matrices from equation (13.43). The resulting linear dynamics can directly be influenced by the choice of the coefficients α_i, which in turn are equal to the coefficients of the characteristic polynonial of the closed loop system. The term $\mathcal{F}(\underline{x})$ in equation (13.47) compensates the nonlinear effect of the original system. An important condition for the applicability of this method is

$$\mathcal{G}(\underline{x}) \neq 0 \tag{13.49}$$

in the whole operation range. Unfortunately, most systems do not have the structure of equation (13.45) from the beginning, but have to be transformed into it. This transformation involves *integrability conditions* and is discussed in detail in [13].

Now, what can neural networks contribute to this control method? First we want to examine the control law in equation (13.47). It depends on all states and therefore requires full state measurement or the ability of state observation. Systems with an isolated nonlinearity are included in equation (13.45), and in this case the control law typically contains the nonlinearity and one or more of its derivatives. With a modification of the *systematic intelligent observer design* of chapter 5 it is possible to learn all existent derivatives of the nonlinearity.

The neural observer provides information about the nonlinearity and its derivatives and about all internal states. Thus, neither full state measurement nor knowledge about the nonlinearity is required. Since the observer adapts to the real nonlinearity, we obtain an adaptive control concept consisting of three steps

1. Design of the neural observer for the system with isolated nonlinearity. It has to be extended for the identification of the nonlinearity's derivatives.

2. Transformation of the original system into the *controllable canonical form* and calculation of the control law.

3. Combination of the previous two steps: the observed states and identified nonlinearity is utilized in the control law. This results in an adaptive nonlinear state–space controller.

This control concept has the advantage that it can adapt to variations of the nonlinearity and does not need extremely low sample times in discrete–time implementations. Practical test runs on simple electrical drive systems have shown robust behaviour even in case of disturbances. This method can also be extended to controllers with integration contribution to guarantee stationary accuracy by a suitable choice of the characteristic polynomial (α_i).

13.4 Nonlinear Control Design with Neural Networks and Numerical Optimization

In the last sections we presented a control design procedure for nonlinear systems enclosing an isolated nonlinearity. Information about the nonlinearity was provided by the *nonlinear neural observer* as described in chapter 5. The resulting controller could be calculated with linear control theory and was non–adaptive in nature. In contrast to this approach we presented in chapter 11 an adaptive controller, performing input–output–linearization. The structure of the system was arbitrary, it only had to be representable by the following equations:

$$\begin{aligned} \dot{\underline{x}} &= \underline{f}(\underline{x}) + \underline{g}(\underline{x}) \cdot u \\ y &= h(\underline{x}) \end{aligned} \tag{13.50}$$

The controller is based on neural networks and special, mathematically proven stable learning laws. The main restriction of the *Stable Model Reference Neurocontrol* is the necessity of full state measurement. The principle of this method is depicted in figure 13.12. A combination of the advantages of the *nonlinear neural observer* and the *stable model reference neuro control* leads to an adaptive control law for systems with isolated nonlinearities. Full state measurement is no longer necessary, but arbitrary plant structure is also no longer possible.

An observation and control concept for plants of arbitrary structure without full state measurement is described in detail in chapter 12. It is based on a nonlinear state transformation, which has to fulfill a diffeomorphism in the operation range. The two main control methods are

1. *Reference control*
   The plant follows a reference model. The controller is based on a nonlinear state transformation and the transformed states.

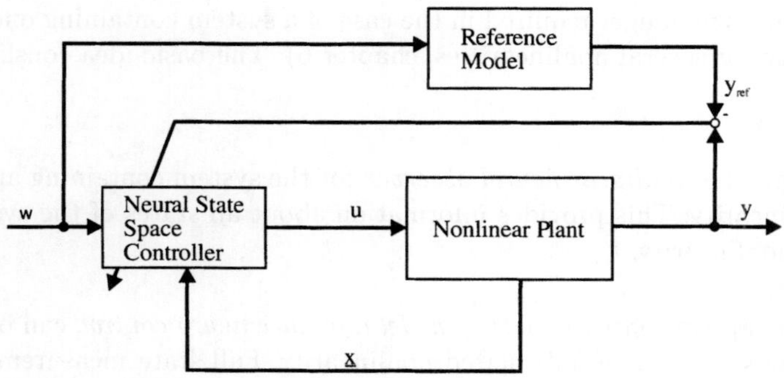

Figure 13.12: Principle of stable model reference neuro control for nonlinear systems; all states are measured

2. *Inverse control*
   The basis is again the nonlinear state transformation, but all nonlinear effects are compensated by the inverse function of the transformed nonlinearity.

In both cases the controller is a neural network (GRNN) and thus is adaptive in nature, which results in very general control concepts. They are applicable to unknown systems of arbitrary structure with varying nonlinearities and parameters.

13.4.1 Model Reference Neuro Control for Systems with Isolated Nonlinearity

Now we want to concentrate on systems with an isolated nonlinearity again. This restriction, compared to chapter 11, will extend the concept of *stable model reference neuro control* to a wider class of real systems. Controllers for plants, where full state measurement is impossible, can now be designed with the following method. The system under consideration has the form

$$\begin{aligned} \dot{\underline{x}} &= A \cdot \underline{x} + \underline{b} \cdot u + \underline{e}_{NL} \cdot \mathcal{NL}(\underline{x}_M) \\ y &= \underline{c}^T \cdot \underline{x} \end{aligned} \qquad (13.51)$$

$\underline{x}_M$ is the measurable input of the nonlinearity. As described in chapter 11, the method of *model reference neuro control* can be applied, if the necessary Lie–derivatives can be learned by a *General Regression Neural Network* or other nonlinear function approximators. In general, this method can also be utilized for systems with an isolated nonlinearity. All important conditions can be found in chapter 11. The most important one is the measurement of all states of the system. This is no longer required in the case of a system containing one isolated nonlinearity (or several nonlinearities, chapter 6). The basic idea consists of two steps

- Design of a *nonlinear neural observer* for the system containing an isolated nonlinearity. This provides information about all states of the system and the nonlinearity.
- The adaptive control concept, *model reference neuro control*, can be applied to the system with an isolated nonlinearity. Full state measurement is no longer required, since the observed states $\underline{\hat{x}}$ of the first step can be used.

The observability of the states of the nonlinear system results in the applicability of *model reference neuro control* to many systems in the field of motion control or electrical drives.

The neural network in the controller approximates the nonlinear state feedback law, which makes the controlled plant follow the linear reference model. Note that the linear reference model has to be implemented in real–time hardware to produce the reference output signal. The main benefit of the proposed method is the applicability to a wide class of nonlinear systems by observing all states with a nonlinear neural observer. Additionally, costly sensors are not required.

The overall structure of *model reference neuro control* for systems with isolated nonlinearities is depicted in figure 13.13. The observed states, instead of the measured ones, are used as an input of the *neural state–space controller.*

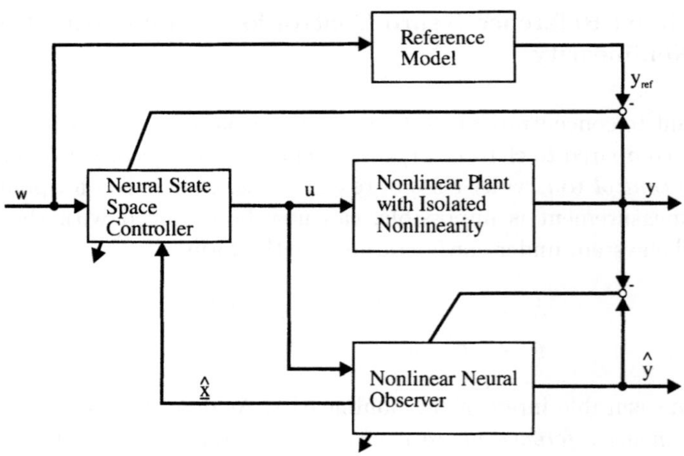

Figure 13.13: Model reference neuro control for systems with isolated nonlinearity; all states are observed

Conclusions
The proposed method improves the *model reference neuro control* concept and makes it applicable to real world plants. The plant's structure is not arbitrary, but has to be representable as a system with an isolated nonlinearity. The observation of the states of the nonlinear system is now mathematically proven stable, and additional expensive sensors are not necessary. This extends the *model reference neuro control* concept to systems with non–measurable states.

13.4.2 Considerations on Numerical Optimization in Nonlinear Control

The aim of *model reference neuro control* is to force the plant to follow a desired behaviour, the reference model. The neural network is an adaptive controller, which can adapt to variations in the nonlinearities or the plant's parameters. But if these variations are very slow (e.g. months or even years), one could identify the nonlinearity first and calculate the control law based on this identification result. The resulting control law is non–adaptive. The following two steps can be repeated from time to time, e.g. when the nonlinearity varies slowly.

- Identification of the isolated nonlinearity.
- Computation of a nonlinear state feedback law based on the identified nonlinearity and the observed states.

The computation of the control law is the main difficulty, since a nonlinear system has to be controlled. In this section, we want to design a controller such that the closed–loop system fulfills any kind of optimality condition, e.g. time–optimal, energy–optimal or exactly like a reference model. Mathematically, these conditions can be described as a *cost function* $J(\underline{x}, u, t)$, which has to be minimized. The result is a control variable $u(t)$ that fulfills the necessary conditions.

We will now shortly discuss how to use the results of optimal control theory in nonlinear control. The optimal control problem can be formulated, according to [8], as follows:

For a given system

$$\dot{\underline{x}} = \underline{f}(\underline{x}, u, t) \tag{13.52}$$

the control variable $u(t)$ and the final time t_f are desired, which minimize the cost function

$$J(u(t), t_f) = \phi(\underline{x}(t_f), t_f) \tag{13.53}$$

Initial conditions

$$\underline{x}(t_0) = \underline{x}_0 \tag{13.54}$$

and final conditions

$$\underline{x}(t_f) = \underline{x}_f \tag{13.55}$$

have to be satisfied. Inequality constraints

$$\underline{g}(\underline{x}(t), u(t), t) \geq 0 \tag{13.56}$$

as well as equality contraints

$$\underline{h}(\underline{x}(t), u(t), t) = 0 \tag{13.57}$$

have to be considered also. Equation (13.56) can be utilized to specify limits of state variables or of the input $u(t)$. Thus, saturation of the control variable u and maximal ranges of the allowed operating condition can be taken into account already in the controller design procedure. If we want to apply optimal control theory to reference model control, we can extend equation (13.52) by the linear state–space equations of the reference model. The cost function is then the difference between the plant's and the reference model's states.

The derivation of the optimal input $u(t)$ is not simple. This dynamic optimization problem can usually only be solved with numerical methods, such as the *direct collocation method* from [15]. An application of this method can be found in [18]. Optimal control theory can also be applied to a system with isolated nonlinearity. Equation (13.52) is then not arbitrary, but the state–space equation of the system under consideration.

The result of the optimization algorithm is the optimal control variable $u(t)$, which minimizes the cost function J. Of course, $u(t)$ depends on the initial and final conditions

$$u(t) = u(t, \underline{x}_0, \underline{x}_f) \tag{13.58}$$

We will not deliver a ready–to–use method for nonlinear controller design in this section, since this approach is still under development. Nevertheless, the idea is as follows: Calculate an optimal control variable $u(t)$, such that the cost function J is minimal. Model reference control, time–optimal and energy–optimal control can be covered with this method, and saturation effects can be taken into consideration.

Usually, controller design involves any kind of optimality condition. In the next section, we will show an example of time–optimal control without explicitly solving an optimization algorithm. The basis of the controller will be a neural network.

13.5 Time–Optimal Tension Control of Production Plants with Continuous Moving Webs

13.5.1 Introduction

In industrial continuous processing plants, the tension of the web is usually controlled in a closed loop cascade control with PI– or PID–controllers; only in some special applications state space control is used. Since these control methods use a linear control algorithm, there is no chance to optimally adapt them to the plant's nonlinear behaviour, irrespective of how the controller parameters are tuned. Therefore, increased demands on the tension control cannot be met by these usual control methods.

Therefore, new control strategies are required that take into account nonlinearities as well as possible limitations of the control variable. Due to their features, neural networks offer a promising chance to realize such control concepts for continuous processing plants.

In this section, we will present a control concept which combines the method of time–optimal control with the advantages of neural networks. At the beginning we will briefly discuss the basic idea of this concept; then we will present the controller design, and in the last section the results of the experimental validation will be shown.

The structure and several problems of continuous processing plants have been discussed already in chapter 2. The system was simplified there by a linearization, which is the usual procedure for the design of a conventional linear controller.

When we consider a system only in its linearized form, though, the problems that arise by its real nonlinear behaviour may not become obvious so easily. Therefore, we shall at this point discuss the nonlinear signal flow graph to get a clearer understanding of the problems a nonlinear control has to cope with.

The possibility that the modular structure of a continuous processing plant can be used for decentralized control design has already been emphasized in chapter 2. The method of time–optimal control proposed in this section also makes use of this advantage.

Since the general structure of a continuous processing plant becomes apparent from figure 2.20, the nonlinear signal flow graph shown in figure 13.14 was reduced to two sucessive subsystems (the term *subsystem*, used in this context describes the behaviour of one nip section and the preceding section of the web). In this figure, for the states and parameters the following notation was used (all quantities are unnormalized):

Nomenclature for figure 13.14 ($x = i, j$)

$F_{(x-1)x}$	:	web–force between rollers $x-1$ and x
M_x	:	motor torque of subsystem x
V_x	:	velocity of the web at roller x
i_{gear}	:	transmission of the gearbox
$J_{x,res}$	:	resultant moment of inertia of subsystem x
$L_{(x-1)x}$	:	length of the web between rollers $x-1$ and x
R_x	:	radius of roller x
$\varepsilon_{(x-1)x}$	:	strain of the web between rollers $x-1$ and x

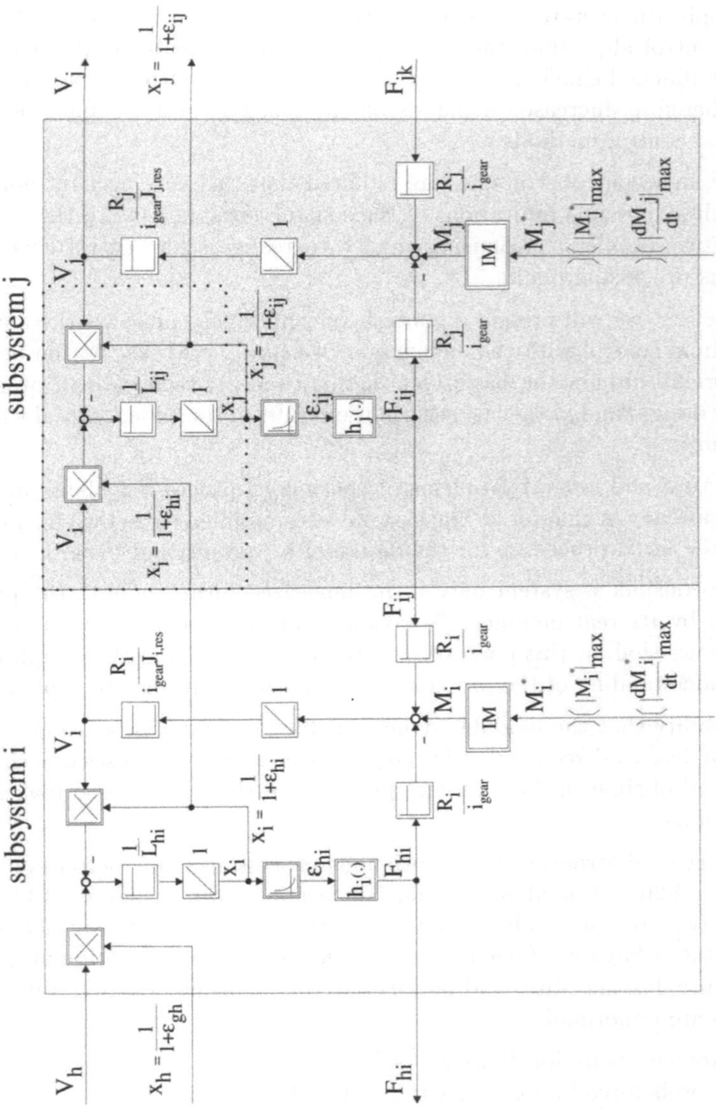

Figure 13.14: Signal flow graph of a continuous processing plant (two sucessive subsystems)

Unlike a linearized one, this signal flow graph represents the system's dynamic behaviour when the variation of the operating point is large, e.g. when the web–

forces are changed rapidly due to technological demands or when the machine is starting up.

Let us now discuss the signal flow in a continuous processing plant: In subsystem j, the electric drive (e.g. IM–drive) receives the set–point value of the motor torque (control variable) M_j^* from the converter–control–system and drives the roller in the nip section. When the forces in the web acting on both sides of the nip section (F_{ij}, F_{jk}) are non–zero, the roller is accelerated by the sum of the motor torque and the load torque caused by the web–forces. In our case, a stiff mechanical coupling of the drive and the roller was assumed. Due to this assumption we can add the moment of inertia of the drive and the roller to a resultant moment of inertia $J_{j,res}$. For a more detailed and realistic representation of the system, of course, the elastic behaviour of the coupling should be taken into account by modelling the drive and the roller as a two– or multi–mass system. There are two reasons why, in our case, the system was simplified in this regard: First of all, as mentioned in chapter 2, in real applications the damping factor d of the shaft is difficult to determine and thus usually not known exactly. Second, in industrial applications, the dimensions of the coupling shaft are usually chosen such that the mechanical–resonance–frequency is very high and the system can be considered as stiff.

Assuming that no slip occurs between the roller and the web, we then obtain the velocity V_j of the web at the nip section by integration and weighting of the accelerating torque. The subsequent section of the signal flow graph, framed by the dotted line, represents the law of conservation of mass, which states that the temporal change of mass in the control volume is the difference of the input and output of the mass of this volume. From the equation describing this law, we obtain ε_{ij}, the strain of the web, and also the web–force F_{ij} in subsystem j.

Due to the material transported in the plant, all subsystems are coupled against and in the direction of transportation (i.e. in upstream and downstream direction).

In *upstream* direction the web–force of subsystem j acts without any time–delay as a load torque on the roller of subsystem i. In *downstream* direction the velocity V_i and $x_i = f(\varepsilon_{hi})$ of subsystem i influence the strain and thus the web–force of subsystem j.

In contrast to the linearized representation shown in figure 2.20, the nonlinear elements in figure 13.14 (marked with double–framed boxes) can no longer be linearized when the large–signal behaviour of the system is examined. Limitating the absolute value of the maximum and the time derivative of the control variable M_x^* ($x = i, j$) adds another nonlinearity to the signal flow graph (not marked with a framebox in figure 13.14). In real applications, the parameters of the controller usually are assigned such that the mentioned limits are never reached by the control variable; thus a limitation is not necessary. This does not hold in cases when the dynamics of the control have to be optimal, e.g. due to higher demands on the product quality. The control method proposed in this section combines the demand for limitation of the control variable with the idea of time–optimality.

Following the signal flow in subsystem j once more, we find further nonlinear connections in the part representing the law of conservation of mass. The strain ε_{ij} can be determined by the difference of the products of the velocities of the actual and the preceding subsystem and the strains in both subsystems (functions of the strains, respectively, with $x_x = \dfrac{1}{1+\varepsilon_{(x-1)x}}$ $(x = i, j)$).

The relation between the stress and the strain is usually modelled by Hooke's law. This law is valid only, though, under the restriction that the web is elastic and thin. As long as this assumption cannot be made, the stress–strain relation has to be described by a nonlinear function, e.g. $h_x(.)$, $(x = i, j)$.

Not shown in figure 13.14 is the friction in the plant which acts as an additional load torque on the roller of each nip section. Friction, in continuous processing plants, is a nonlinear function e.g. of the web–force, the velocity or the temperature in the considered subsystem.

In the field of control design for continuous processing plants, the coupling and the nonlinear dynamic behaviour are the major challenges to be met in the future. Up to now the design of a control is done assuming a linearized system, and it is optimized without regarding the influence of the coupling (see [17]). Under special conditions, such a control can lead to acceptable control results. However, when the demand on a tension control rises, e.g. due to higher production speed in the plant, measures for nonlinear control design and decoupling must be considered. A promising approach is *decentralized decoupling* ([16]), a method which is based on the decentralized state space control of continuous processing plants. The goal of the method is the minimization of the influence of the remaining system on the considered subsystem. The first advantage of the method is that no measurement of the coupling quantities is required; the second, that it is robust against variations of the parameters in a wide area. Since this control concept is implemented at the experimental plant of the institute, which was also used for the validation of the time–optimal control, a comparison of the two methods' dynamic behaviour was possible.

A control like decentralized decoupling cannot be optimal, though, as the system is nonlinear. Therefore, a decentralized control by fuzzy logic was designed and was tested also at the aforementioned experimental plant. Here, promising results regarding the decoupling behaviour and the dynamic response were obtained ([9]).

The mentioned approaches attack some of the problems we have discussed up to now, and we learned that in some cases this can lead to acceptable control results. Results optimal in *every* respect, though, could be achieved only with a control taking into account, theoretically, all the problematic features of a processing plant with a continuous moving web.

In this section we will not present an idea that fulfills this demand. The proposed control concept is only another step in this direction. Like the approaches mentioned above, the concept will focus on one of the problems and will, for the time being, disregard the rest. The problem we will concentrate on is the limitation of the absolute value of the maximum and the time derivative of the control variable in order to avoid an overload of the motors on the one hand but, exploit the system's dynamics as best as possible on the other. In a first step, we will therefore do the controller design on the basis of the linearized signal flow graph of an isolated subsystem. Afterwards, the controller can be extended by degrees in the attack of th remaining problems.

13.5.2 Controller Design

It was mentioned in the last section that the fundamental idea of the proposed control concept is the principle of time–optimal control; it can be expressed in a short and compact form

$$\underline{x}(0) = \underline{x}_0 \stackrel{t_{min}}{\longrightarrow} \underline{x}_{end} \tag{13.59}$$

where it is assumed that the control variable of the system, u(t), is constrained in magnitude by the relation

$$u(t) \leq |M| \qquad \text{for all } t \tag{13.60}$$

and can be switched between the boundary values $\{+M, \ -M\}$ in zero time. The magnitude constraint is the result of physical limitations on the amount of, in our case, torque, which may be realized in practice by equipment ([1]). Figure 13.15 shows one possible curve shape of the control when the system is to be brought from an initial state with $u_0 = +M$ to a final state with $u_{end} = 0$ in minimum time.

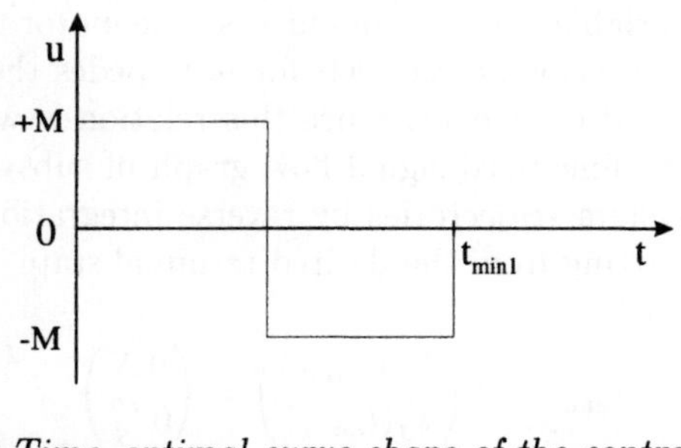

Figure 13.15: Time–optimal curve shape of the control variable u(t)

If this control sequence now was applied to a real electromechanical system, the result would be, as may easily be understood, vibrations and, consequently, noise and increased wear of the mechanical components. Moreover, in the context of time–optimal control, two problematic features are known that have similar effects: The chattering and the limit cycle, both characterized by a rapid switching of the control between $\pm M$.

Most of these problems can be avoided if the vertical parts of the curve shape are replaced by a finite gradient that is tolerable for the mechanical parts and does not impair the system's dynamics. The solid line in figure 13.16 is a possible curve shape for the control u(t) that fulfills this restriction. It must be noted that in this case the control of the system cannot be completely time–optimal anymore, since $t_{min2} > t_{min1}$. In view of the mentioned advantages, we will accept this. When the controller in the sequel is referred to as time–optimal, the reader may bear in mind that this means only quasi–time–optimal.

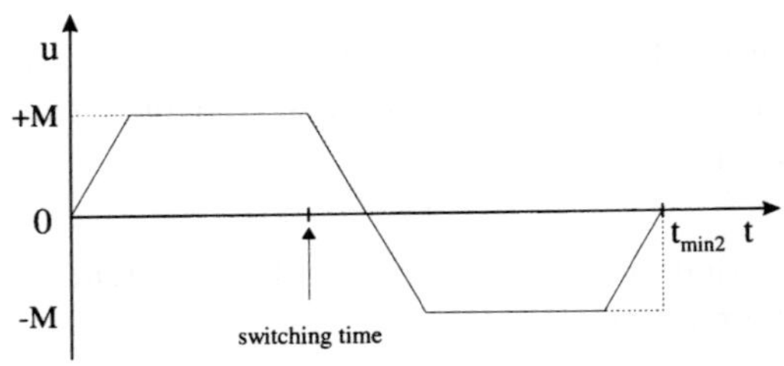

Figure 13.16: Curve of the control variable u(t) modified for practical application

Two problems arise when the optimal–control law for one subsystem of a continuous processing plant shall be determined. First of all, a curve shape like the solid line shown in figure 13.16 cannot be expressed anymore in the form required to compute the system trajectories analytically. And second, the relation between the control variable, which is in our case the motor torque, and the two states in the subsystem, velocity and web–force, impedes the calculation of the trajectories by mathematical means. Since this relation is well known, though, (figure 13.17 shows the linearized signal flow graph of subsystem j) we can use it to determine the system trajectories by reverse integration (reverse meaning backwards in time) starting from the desired terminal state

$$\underline{x}_{end} = \begin{pmatrix} F_{ij}(t_{min2}) \\ V_j(t_{min2}) \end{pmatrix} = \begin{pmatrix} 0\,N \\ 0\,\frac{m}{s} \end{pmatrix} \tag{13.61}$$

$$u_{end} = M_j(t_{min2}) = 0\,Nm \tag{13.62}$$

This terminal state is chosen to avoid that for every possible operating point the system trajectories have to be calculated anew; the steady state value of the motor torque is superimposed by addition afterwards. Hence, the controller's input values are $\Delta F_{ij} = F_{ij} - F_{ij}^*$ and $\Delta V_j = V_j - V_j^*$; the output is $\Delta M_j = M_j^* - M_j$.[1]

Note: Since we use the linearized signal flow graph of one subsystem for the controller design, the values F, V and M are small deviations from the steady state. This is why the desired terminal state zero is chosen in equations (13.61) and (13.62). In practical applications, though, the control is fed with the actual values of the web–force and the velocity from the measuring instruments with the goal to achieve $F_{ij} = F_{ij}^*$ and thus $\Delta F_{ij} = 0$. For the simulation of the system, the quantities in the linearized signal flow graph will be treated as in practical application.

By a variation of both the switching time and t_{min2}, a large number of curves then is generated such that the main part of the plant's operating range is covered. To avoid any ambiguous coherence between the control and the states, curve shapes like the dashed line in figure 13.16 are generated, neglecting both the slope at the beginning and the end. Figure 13.20 shows how the absolute value of the time derivative of the controller's output can be limited by an appropriate device for one subsystem of the plant. The slope at the end, which determines the behaviour of the control surface around the origin of the state space, is additionally smoothed afterwards by the neural net.

Two restrictions have to be made for the use of the method of reverse integration:

- **Neglect of the coupling**
  First of all, the system trajectories cannot be determined for the *total* system as in this case we would have to deal with coherences too complex. As mentioned already in section 13.5.1, the controller design is therefore done for the *isolated* subsystem j as shown in figure 13.17; the coupling will be taken into account in a next step.

- **Simplification of the IM-drive**
  The second restriction is that the simple time delaying behaviour of the induction motor, that we will assume as the electrical drive of subsystem j, will be reduced to a proportional element with unity gain to avoid another state in the signal flow graph.

Figure 13.18 shows several trajectories of the considered subsystem j that were generated with curve shapes of the control $u(t) = M_j^* = M_j$ like the dashed line in figure 13.16. Since the trajectories for negative curve shapes (not shown in figure 13.18) are centrosymmetric to the shown trajectories, they can easily be determined.

[1] Though the equipment with load cells for the measurement of the web–force is not state of the art in most industrial continuous processing plants, we will assume in this section that the current values of the web–force, the velocity and the motor torque are available for control.

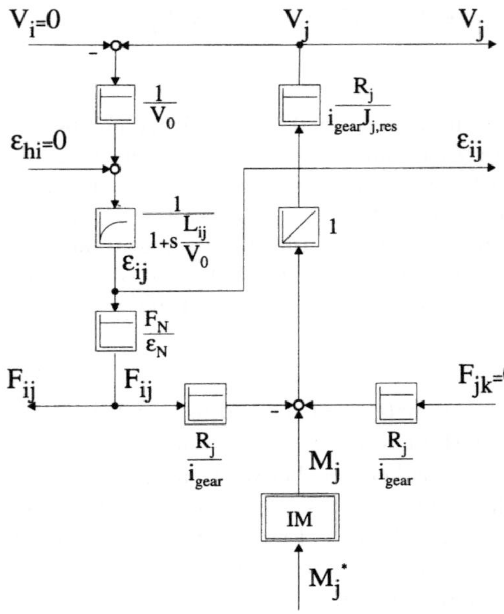

Figure 13.17: Linearized signal flow graph of the isolated subsystem j

Let us assume an initial state with e.g. $\Delta F_{ij} > 0$, $\Delta V_j < 0$: From this point in the state plane, the controller will drive the system along the corresponding trajectory to the desired terminal state described by equations (13.61) and (13.62), first of all accelerating the system with $\Delta M_j = +M$ (in figure 13.18 $+M = 15\ Nm$).

The control surface made up of the system trajectories could be used in principle for the control of one subsystem of a continuous processing plant. However, such a control surface has two severe drawbacks:

- First of all, it is impossible to completely cover the operating range of the plant by trajectories, since gaps between the individual curves cannot be avoided.
- And secondly, even a moderate number of trajectories results in a huge data volume, which is not sensible to use in real–time processing.

These problems can be solved when the control surface is approximated by a neural network which uses the whole data volume as sample points. Thus, we are able to determine the value of ΔM_j at arbitrary positions in the state space. In this case, equally spaced values are chosen, which enables the application of a special algorithm for numerical efficiency. By variation of the space between

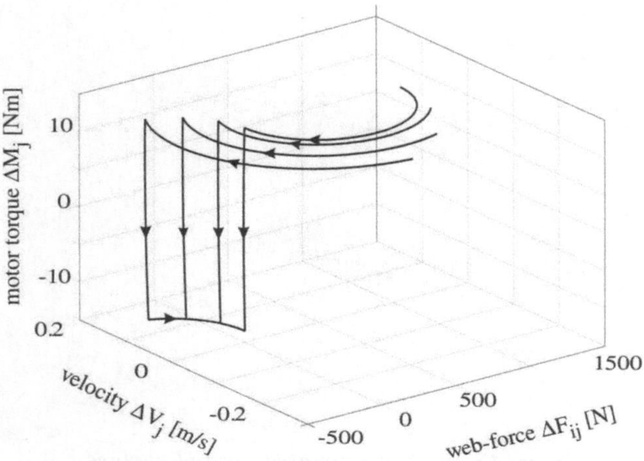

Figure 13.18: Three–dimensional time–optimal trajectories generated by reverse integration; the arrows indicate the motion of the states for positive time

the values, we are able to adjust the density of meshes such that we get the best possible balance between the number of sample points and the created, though reduced, data volume. Together with the corresponding coordinates ΔF_{ij} and ΔV_j, the values of ΔM_j are then stored e.g. as a matrix and themselves used afterwards as sample points for the neural net implemented in the time–optimal controller. Due to its advantages described in chapter 4, a *General Regression Neural Network* (GRNN) was chosen both for the approximation of the control surface and for the neural net in the controller. In the preceding chapters neural networks, and especially the GRNN, were used for identification, since they are able to reproduce (learn) unknown functions or parameters by weight adjustment. For the controller design in this section, though, the GRNN is used only as a function approximator. The ability of the GRNN to learn certain functions will be of interest again when, as mentioned in section 13.5.1, the remaining problems like the nonlinear dynamic behaviour or time–variant parameters of a subsystem shall be tackled.

Figure 13.19 shows the approximated control surface. The gridlike structure caused by the equally spaced sample points can easily be seen. Additionally shown are the trajectories from figure 13.18, which helps to see the difference between the original control surface and the approximated one. By the smoothing factor σ in the equation of the GRNN, the slope around the origin of the state space can be tuned to a value that is acceptable for the mechanical components. The smoothness of the transition from the slope to the parts where the controller output is $+M$ or $-M$ can also be varied by σ.

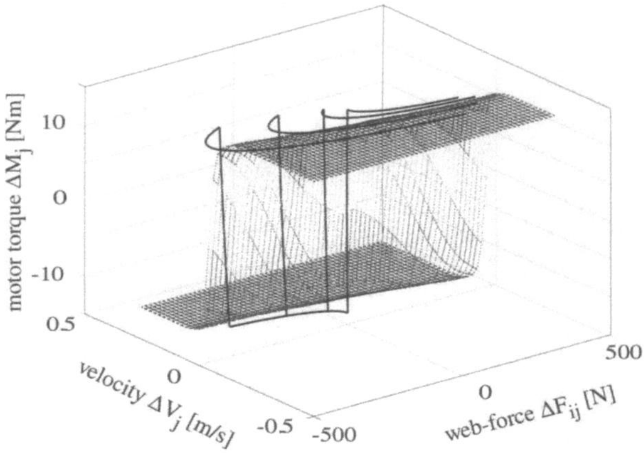

Figure 13.19: Approximated control surface

As mentioned above, the approximated control surface is used as a set of sample points for the neural net implemented in the controller. Figure 13.20 shows the controller structure applied to subsystem j. The controller's input variables, ΔF_{ij} and ΔV_j are calculated during operation of the plant; depending upon their values, the controller determines the actual value ΔM_{jGRNN} of ΔM_j using the algorithm for numerical efficiency (described e.g. in [2, 4]). The setpoint value F_{ij}^* depends on the demands from the process, but has to lie within the limits $[0; F_N]$, whereas V_j^* is a dependent quantity that can be determined by the equation

$$V_j^* \quad = \quad V_i^* + V_0 \frac{\varepsilon_N}{F_N} F_{ij}^* \tag{13.63}$$

with

F_N		nominal web–force
V_i^*	$=$	0 (no deviation from the reference value in the steady state)[2]
V_0		average velocity of the web in the actual operating point
ε_N		nominal strain

The value of the control in the stationary state, M_{jstat}, is a function of the setpoint value F_{ij}^*. The additional torques M_{jcd} and M_{jcu} take into account the coupling in downstream and upstream direction. M_{jcu} compensates the influence of the subsequent subsystem k by a simple addition of the corresponding web–force F_{jk}. A variation of the web–force of a subsystem, though, not only influences its direct neighbours but is passed on downstream. Therefore, a simple compensation can be done by the addition (M_{jcu}) of the weighted web–force

[2]This, of course, only holds for the linearized signal flow graph which is used in the simulation of the system. When the controller is later on implemented at a real plant for the experimental validation, V_i^* is set to V_0.

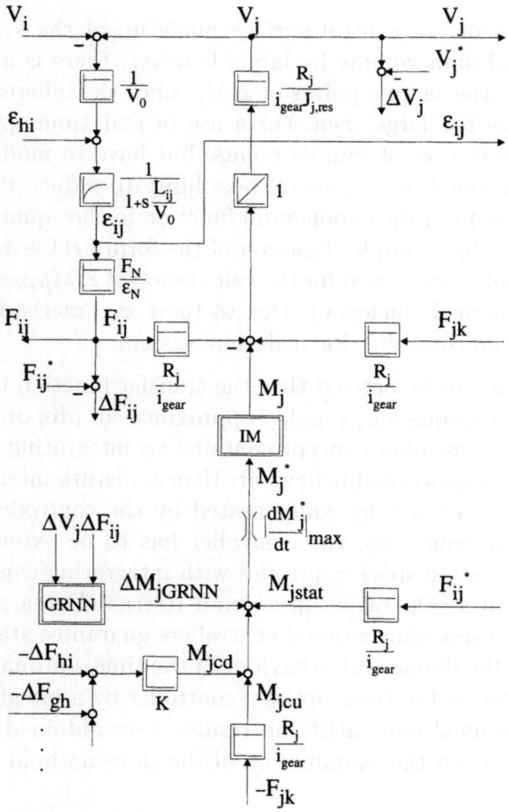

Figure 13.20: Subsystem j with time–optimal controller

differences of all preceding subsystems i, g, h etc. to the reference value of the torque of subsystem j. This method is one possibility to *centrally* decouple the subsystems; it was proposed first in [2]. The drawback of this simple and most obvious method, though, is that quantities from other subsystems (web–forces), provided by measuring instruments or observers, are required; this is a contrast to the idea of decentralized control. One approach for a decentralized decoupling controller based on the idea of time–optimal control was proposed in [5]. Here, the obtained results were not satisfying. However, in [5] there also was a decentralized observer for the nonlinear stress–strain relation designed (for the general procedure of intelligent observer design see chapter 5) by which a very exact reproduction of the web-force is possible. Intelligent observers can also be used for the identification of other parameters in a subsystem of a continuous processing plant.

The approximation of the control surface made up of the system trajectories reduces the original data volume by large; however, there is an upper limit for the spaces between the sample points of ΔM_j, since the inherent approximation error would become too large then. For a use in real–time applications, we do not have to reduce the set of sample points, but have to modify the algorithm ΔM_{jGRNN} is determined by. A general possibility to reduce the computational time is the replacement of the exponential function in the equation of the GRNN (e.g. equation (4.7)) by a simpler function of the form $1/(1+ax^2)$. Additionally, the number of sample points used for the calculation of ΔM_{jGRNN} can be reduced (algorithm for numerical efficiency). Due to these two methods, we are able to use the time–optimal controller for real–time systems.

From figure 13.20 it can be derived that the transfer function between the web–force and the motor torque has, roughly approximated, proportional time delaying behaviour. The controller concept contains no integrating contribution and is therefore also of proportional behaviour. Hence, disturbances occurring in the system (e.g. friction) cannot be compensated by the controller and may cause a control offset. To avoid this, the controller has to be extended for example by a parallel PI– or state space controller with integrating contribution or even by a simple integrator. Operating in a small restricted area around the origin of the state space, these conventional controllers guarantee stationary accuracy without impairing the dynamical behaviour of the time–optimal controller. In [2] and [4] the extension of the time–optimal controller by a integrating state space controller was examined and satisfying results were obtained. In this chapter, though, we will focus on the examination of the time–optimal controller alone.

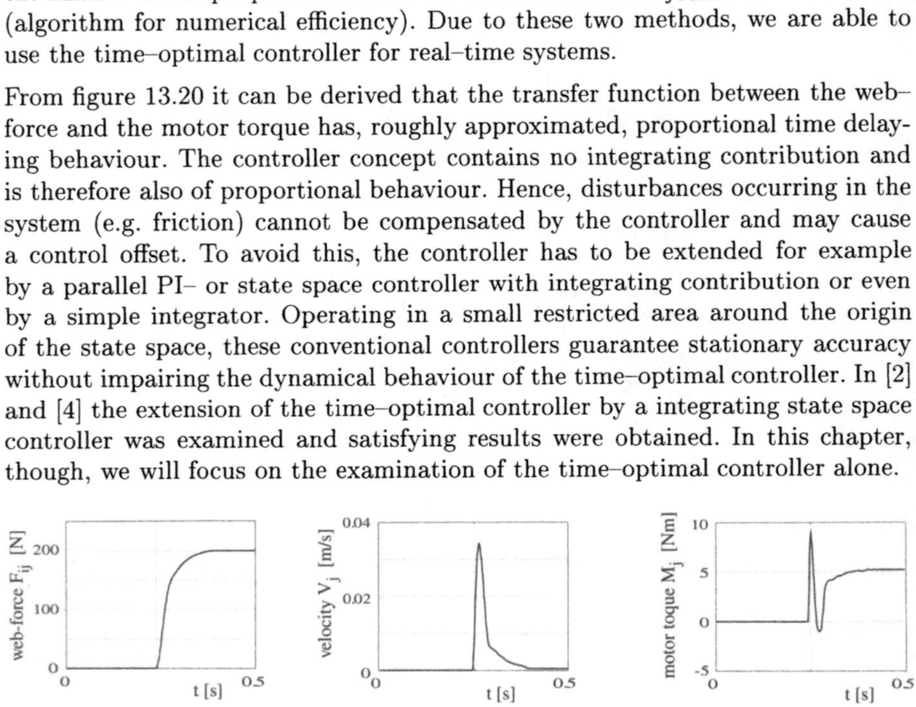

Figure 13.21: Simulation of the transient response to a step change of the web–force of a time–optimal controlled isolated subsystem

Figure 13.21 shows the simulated transients F_{ij}, V_j and M_j of an isolated subsystem for a step change of the reference web–force F_{ij}^* by 200 N. Due to the sudden change of ΔF_{ij}, the controller's input variables leave the origin of the control surface in negative directions, and the controller's output changes from zero to a positive torque. From this point in the control surface, ΔF_{ij} and ΔV_j are driven back to the origin along one time–optimal trajectory within the control surface. As can be derived from the peak in the transient of the velocity, the time–optimal control is a very dynamic control concept by which the steady state is reached only after 125 ms. In the simulation the time–optimal controller alone is able to reach a steady state without any offset. This, of course, is possible because friction or any other disturbances were not implemented in the simulation model.

13.5.3 Experimental Validation

For validation, the time–optimal controller was implemented at subsystem 3 of the experimental plant whose structure is shown in figure 13.22.

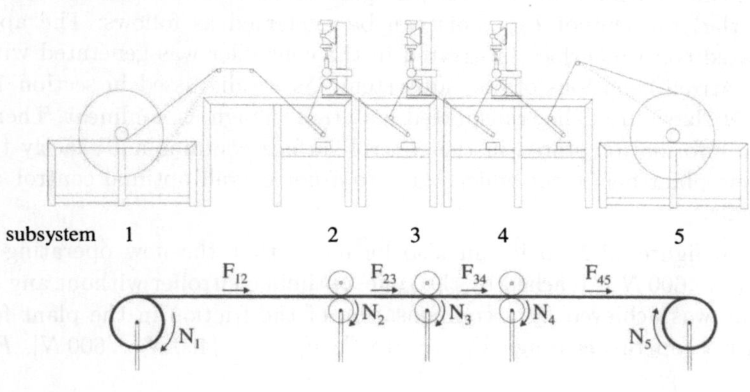

Figure 13.22: Experimental plant

The algorithm of the time–optimal controller was realized in the programming language C and implemented on a PC (for features see table 8.1) equipped with an AD/DA-board to read the actual and reference values of the web–force F_{23} and the velocity V_3 from the plant's control system. During the operation of the plant, the controller accesses the data of the approximated control surface stored in the RAM of the PC. The actual value of ΔM_{3GRNN} is then determined from the control surface and transferred back to the control system. The centralized decoupling as shown in figure 13.20 is done in the plant's control system, since the required signals are more easily available there.

Since the time–optimal controller was implemented only at subsystem 3, a different control concept had to be chosen for the 4th one (Subsystem 2 is used to impress the actual value of the average velocity V_0 of the web upon the total system and is therefore only controlled in speed). For conventional control, different strategies (e.g. cascaded control, state space control, decentralized decoupling) are available in the plant's control system. In [4], all possible combinations of the time–optimal controller (subsystem 3) with the conventional concepts (subsystem 4) were tested. In this section we want to restrict ourselves to the arrangement

subsystem 3: time–optimal neural controller
subsystem 4: decentralized decoupling

Dynamic response to a change of the reference value

Figure 13.23 shows the response of subsystem 3 to a step change of F_{23}^* by 200 N for the above mentioned controller arrangement (13.23 b) and for the control of both subsystems by decentralized decoupling (13.23 a). As can be noted, a shorter rise time of the web–force is achievable with the time–optimal control. Yet, in figure 13.23 b there is a slight overshoot in the transient of F_{23}. This can be explained as follows: The approximated control surface integrated in the controller was generated with the linearized equations of one subsystem. As we discussed in section 13.5.1, the behaviour of the transported material, though, is nonlinear. Therefore, the information stored in the control surface contains not exactly the inverse plant behaviour which leads to a not overall optimal control–action result.

From figure 13.23 b it can also be noted that the new operating point $F_{23} = 600\,N$ is reached by the time–optimal controller without any offset. This was achieved by a compensation of the friction in the plant for the narrow operating range $V_0 = 0.8\,\frac{m}{s}$, $F_{23} = [400\,N \ldots 600\,N]$, $F_{34} = 400\,N$.

The difference between the two control strategies best becomes clear when we look at the transients of the motor torque M_3 of figures 13.23 a and 13.23 b. When the step change of the reference value F_{23}^* is applied, the control by decentralized decoupling raises the torque to the future steady state value, thus achieving a steady change of F_{23} without any overshoot. The time–optimal controller achieves the change of the web–force by a very dynamic rise of the torque with a peak value of 10 Nm.

Decoupling behaviour

For the investigation of the decoupling behaviour of the time–optimal controller, the same step change of 200 N was impressed upon subsystem 4. The results are shown in figure 13.24 b for the above mentioned controller arrangement and in figure 13.24 a for the control by decentralized decoupling. In the transient of F_{23} (13.24 b) no dynamic deviation due to the rise of F_{34} can be noted; however, the change of F_{34} causes a steady state offset. This can be explained by the changed friction torque acting upon the subsystem 3. Since the friction compensation was done only for a small operating range, the offset cannot be suppressed any longer.

As mentioned above, subsystem 2 of the experimental plant has a special function and is only controlled in speed. Thus, a change of F_{12} is not achieved as usual by a temporary change of V_2 (N_2, respectively) but of the winder speed N_1. The decoupling behaviour regarding disturbances caused by the preceding subsystem 2 could therefore not be examined.

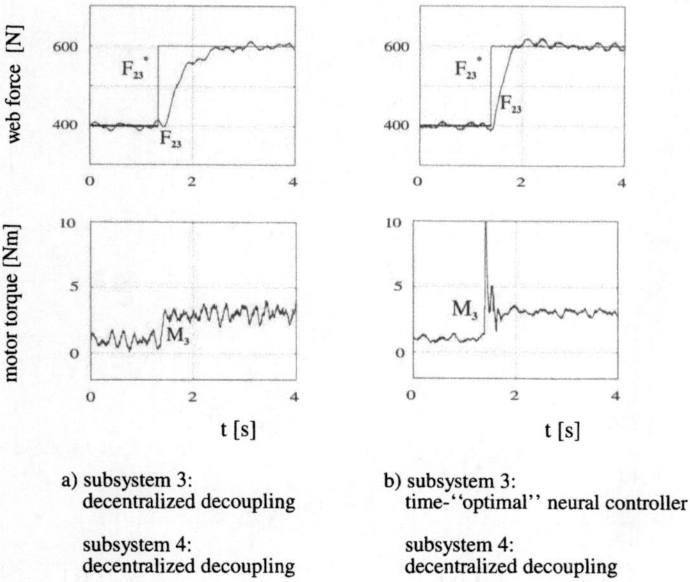

Figure 13.23: Web–forces and motor torque at subsystem 3 for a step change of F_{23}^* *by 200 N* ($V_0 = 0.8\frac{m}{s}$)

13.5.4 Conclusion

Generally it can be stated that the results achieved with the time–optimal control are very promising. Concerning the response to a setpoint change, the neural controller is dynamically superior to the control by decentralized decoupling, whereas the decoupling results are equal. However, a final estimation of the control quality cannot be done before the controller has been implemented and tested at two successive subsystems. In such a case, we will also be able to investigate the downstream decoupling behaviour.

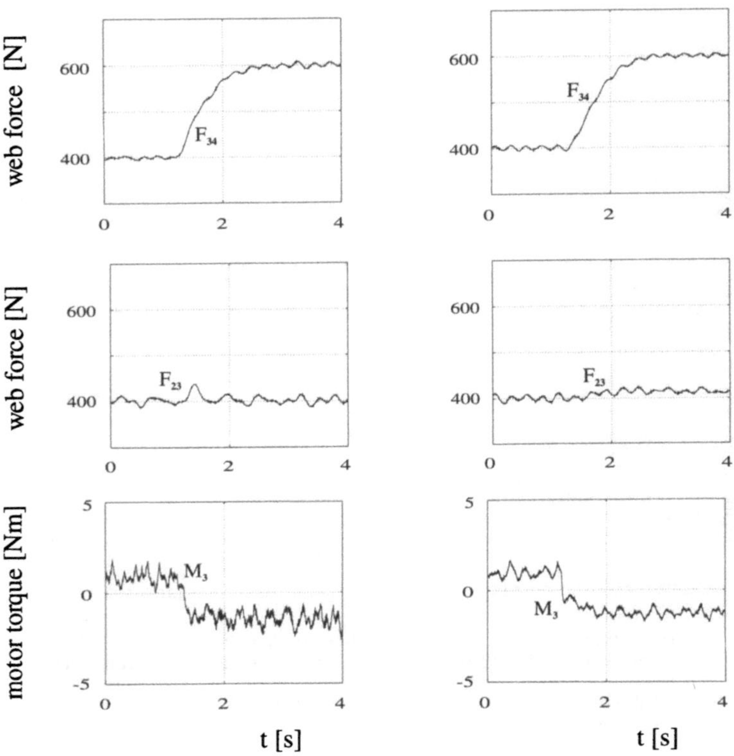

a) subsystem 3:
decentralized decoupling

subsystem 4:
decentralized decoupling

b) subsystem 3:
time-optimal neural controller

subsystem 4:
decentralized decoupling

Figure 13.24: Web–forces and motor torque at subsystem 3 and 4 for a step change of F_{34}^* *by 200 N* ($V_0 = 0.8\,\frac{m}{s}$)

Yet, the time–optimal controller does not take into account the nonlinear dynamic behaviour of the subsystems. A next step to the goal of control optimal in every respect (as defined in section 13.5.1) will be the determination of the control surface using the *nonlinear* signal flow graph of a subsystem where also the elastic coupling of the electrical drive and the roller is modelled. By doing so, an improvement of the control–action behaviour can be expected.

Further problems are the friction and time–varying parameters (the average velocity V_0 of the plant or the strain ε). The friction in a continuous processing plant depends on several quantities and also varies when these quantities change. In conventional controls, friction in the system is of no importance because of the usual integrating part in the controller. It was mentioned above that in our case the time–optimal controller has proportional behaviour, and therefore, an extension was discussed in section 13.5.2. Nonetheless, it seems sensible to compensate the influence of the friction since it is a disturbance of the system.

When parameters vary, the behaviour of the plant also changes, which will lead to a non–optimal control–action result as long as the data in the control surface remains constant. If there is the possibility to gather information about these parameters the control surface can be adapted online to the changed situation. The calculation of the trajectories and the approximation may be done as a background process which does not interfere with the control algorithm. At regular intervals, the actual control surface can then be updated, starting from the mainly used area around the origin.

In most systems, usually only few quantities are available for measurement. With the systematic intelligent observer design, however, we have a powerful tool for the identification of nonlinearities and unknown parameters. Implemented at a continuous processing plant, such observers can provide the information required for optimal adaptation of the control surface.

Since a control concept like this is able to adjust itself to changing plant behaviour or nonlinearities online, the user will be allowed to put the system into operation more or less automatically.

13.6 References

[1] Athans, M., Falb, P.
Optimal Control. McGraw–Hill, New York, 1966.

[2] Frenz, T.
Entwurf und Simulation von nichtlinearen Regelungen mit neuronalen Netzen für kontinuierliche Fertigungsanlagen. Diplomarbeit, TU München, 1993.

[3] Frenz, T.
Stabile neuronale online Identifikation und Kompensation statischer Nichtlinearitäten am Beispiel von Werkzeugmaschinenvorschubantrieben. Dissertation, TU München, 1998.

[4] Klor, A.
Entwurf und Implementierung einer neuronalen Regelung für kontinuierliche Fertigungsanlagen. Diplomarbeit, TU München, 1997.

[5] Lenz, U.
Entwurf von lernfähigen Beobachtern und nichtlinearen Regelungsverfahren für kontinuierliche Fertigungsanlagen. Diplomarbeit, TU München, 1993.

[6] Lenz, U.
Lernfähige neuronale Beobachter für eine Klasse nichtlinearer dynamischer Systeme und ihre Anwendung zur intelligenten Regelung von Verbrennungsmotoren. Dissertation, TU München, 1998.

[7] Omidvar, O., Elliott, D. L.: *Neural Systems For Control.* Academic Press, 1997.

[8] Papageorgiou, M.: *Optimierung.* R. Oldenbourg Verlag, München, 1991.

[9] Pecher, H.
Entwurf dezentraler Regelungen und Beobachter mit Zustands- und Fuzzyreglern bei kontinuierlichen Fertigungsanlagen mit hoher Teilsystemzahl. Diplomarbeit, TU München, 1994.

[10] Schäffner, C.
Analyse und Synthese neuronaler Regelungsverfahren. Dissertation, TU München, 1996.

[11] Schröder, D.
Elektrische Antriebe 2. Springer–Verlag, Berlin, Heidelberg, 1995.

[12] Slotine, J., Weiping, L.
Applied Nonlinear Control. Prentice–Hall, 1991.

[13] Sommer, R.
Synthese nichtlinearer Systeme mit Hilfe einer kanonischen Form. VDI–Verlag, 1981.

[14] Specht, D.
A General Regression Neural Network. IEEE Transactions on Neural Networks, Vol. 2, 6. November 1991, pp. 568–576.

[15] von Stryk, O.
User's Guide for DIRCOL, A Direct Collocation Method for the Solution of Optimal Control Problems. Lehrstuhl für Höhere Mathematik und Numerische Mathematik, TU München, 1997.

[16] Wolfermann, W., Schröder, D.
New Decentralized Control in Processing Machines with Continuous Moving Webs Proc. of the Second International Conference on Web Handling, Oklahoma, 1993.

[17] Wolfermann, W.
Tension Control of Webs - A Review of the Problems and Solutions in the Present and Future. Proc. of the Third International Conference on Web Handling, Oklahoma, 1995.

[18] Zoelch, U., Schröder, D.
Dynamic Optimization Method for Design and Rating of the Components of a Hybrid Vehicle. International Journal of Vehicle Design, Vol. 19, No. 1, 1998, pp. 1–13.

List of Figures

1.1 Fundamental structures of control methods: linear plants 2

1.2 Fundamental structures of control methods: nonlinear plants; aspects of this book (shaded boxes) 3

1.3 Self–Tuning Controller (STC, indirect control method) 6

1.4 Model Reference Adaptive Control (MRAC, direct control method) 6

2.1 Signal flow graph of a two–mass system with friction and backlash effects 21

2.2 Bode plot of the open $\dot{\varphi}_1$ control loop (solid line: one–mass system, dashed line: two–mass system) and corresponding signal flow graph 23

2.3 Bode plot of the open $\dot{\varphi}_2$ control loop (solid line: one–mass system, dashed line: two–mass system) and corresponding signal flow graph 24

2.4 Step response with $\Omega_{0N} = 0,1\, \omega_D$ (— $\dot{\varphi}_1$, - - - $\dot{\varphi}_2$) 25

2.5 Step response with $\Omega_{0N} = \omega_D$ (— $\dot{\varphi}_1$, - - - $\dot{\varphi}_2$) 26

2.6 Step response with $\Omega_{0N} = 10\, \omega_D$ (— $\dot{\varphi}_1$, - - - $\dot{\varphi}_2$) 26

2.7 State–space control of $\dot{\varphi}_2$ without integrating contribution ($T_N = 1 sec$) 28

2.8 State–space control without integrating contribution with $\Omega_{0N} = 628.32\, s^{-1}$ 31

2.9 State–space control without integrating contribution with $\Omega_{0N} = 62.832\, s^{-1}$ 32

2.10 State–space control without integrating contribution with $\Omega_{0N} = 6.2832\, s^{-1}$ 33

2.11 Signal flow graph of a two–mass system with state–space control with integrating contribution 34

2.12 State–space control with integrating contribution with $\Omega_{0N} = 628.32\, s^{-1}$ 36

2.13 State–space control with integrating contribution with $\Omega_{0N} = 62.832\,s^{-1}$ 37
2.14 State–space control with integrating contribution with $\Omega_{0N} = 6.2832\,s^{-1}$ 38
2.15 Position state–space controlled two mass system with friction . . 40
2.16 Three–mass system 41
2.17 Bode plot of a two–mass of inertia system 43
2.18 Bode plot of a three–mass system with $(\Theta_1+\Theta_2) = \Theta_L$ and $\Theta_1 < \Theta_2$ 44
2.19 Bode plot of a three–mass system with $(\Theta_1+\Theta_2) = \Theta_L$ and $\Theta_1 > \Theta_2$ 45
2.20 Multi variable process system with continuous processing of material 47
2.21 Mechanical spring and mass system representing the system of figure 2.20 48
2.22 Step response of f_{23} and the coupling to f_{34} and vice versa . . . 48
2.23 Reduced decoupling observer 49
2.24 Step response of n_3 due to n_3^* (a) and response of n_3 due to a f_{34}–variation (b and c) 50
2.25 Schematic feedforward control of axial winders 51
2.26 Step responses due to reference input 52
2.27 Step responses due to disturbance 53
2.28 Web force and winder speed for a noncircular winder 53
2.29 Structure of the control with optimal output feedback 55
2.30 Step responses of the speed with optimal output feedback; above: step response to reference value, below: step response to disturbance 56
2.31 Step responses with decentralized control 58
2.32 Comparison of fuzzy and decentralized decoupling control 62

3.1 The brain consists of connected neurons. The associations reflected in the connections combine input scheme and action scheme. . . 68
3.2 Simple structure of open–loop control 69
3.3 Simple control structure. The controller diminishes the deviation of the systems output from the intended behaviour. 70
3.4 Conditional feedback structure: the responses to variations of the reference signal and to disturbances are separately optimizable. 71
3.5 Model reference adaptive control (MRAC), SISO–system 73
3.6 Signal flow chart representation of a rigid drive system with speed dependent friction 75

3.7 Direct and indirect approach for adaptive control of a rigid drive system with speed dependent friction 76
3.8 Stability of an equilibrium point 77
3.9 General classification of technical systems 78

4.1 Approximation and interpolation using given sample points . . 84
4.2 Examples of localized basis functions 86
4.3 Topology of a radial basis function network 89
4.4 Topology of a general regression neural network 90
4.5 Improved interpolation and extrapolation of the GRNN compared to the RBF network 90
4.6 Extrapolation of the GRNN using Gaussian activation compared to polynomial activation of the DANN for two different smoothing factors σ 92
4.7 Neural network representation of the singleton defuzzification method 95
4.8 Function approximation by a fuzzy controller 97
4.9 Example function $y = \arctan(x)$ 99
4.10 Approximation results of the RBF 100
4.11 Approximation results of the GRNN 100
4.12 Approximation results of the DANN 100
4.13 Approximation result of the fuzzy controller 101

5.1 Schematic depiction of a nonlinear SISO–plant "electrically driven throttle" (single input: torque, single output: position) 105
5.2 Friction depending on speed as an example for an isolated nonlinearity in a feedback path 106
5.3 Possible structural locations of a nonlinearity within linear parts of the plant 107
5.4 Signal flow chart of the state space representation of a SISO–system enclosing an isolated nonlinearity $\underline{\mathcal{NL}}(\underline{x}, u)$ 109
5.5 Isolated nonlinearity $\underline{\mathcal{NL}}(\underline{x})$ in state space signal flow graph representation 109
5.6 State observer for a linear dynamic system, applicable e.g. when noise occurs 112
5.7 Intelligent observer for SISO–systems with isolated nonlinearity; applicable for identification and state observation (e.g. noise) . . 114

5.8 The observer structure leads to the observer error defined as the difference over $H(s)$ between the nonlinearity regarded as input of a linear plant (depending on $\underline{x}$and u) and the neural net $\hat{\mathcal{NL}}$ 115

5.9 Bode plots for a SPR–transfer function (left side) and a not-SPR–transfer function (right side) 117

5.10 Signal flow graph of the error model with delayed activation; in the lower part, the extended error is depicted 119

5.11 Signal flow graph of the transformed error model 121

5.12 State observer for a linear SISO–system according to Luenberger, state reconstruction based on the system's output 125

5.13 Intelligent observer for a dynamic SISO–system containing an isolated nonlinearity in the signal flow graph state space representation; application for state observation and identification of the nonlinearity based on the measured system output 127

6.1 Plant with separable nonlinearities 136

6.2 The nonlinear observer with accessible states 138

6.3 The time–variant dynamic $H'(s)$ 146

7.1 Experimental machine 149

7.2 Two–mass model 150

7.3 Simulated and measured data — linear model 151

7.4 Identification of friction characteristic with different learning rates 153

7.5 Simulated and measured data – nonlinear model 154

7.6 Simulated compensation of friction influence 155

7.7 Hardware concept 156

7.8 Evaluated slide speed 158

7.9 Evaluated and observed slide speed 158

7.10 Error between evaluated and observed slide speed 159

7.11 Learning process 159

7.12 Learned friction characteristic 160

7.13 Signal flow chart of the feedforward compensation controller . . 161

7.14 Evaluated slide speed without friction compensation 162

7.15 Evaluated slide speed with friction compensation 162

7.16 Signal for compensation and slide speed 163

8.1 Backlash model 167
8.2 Elastic two–mass system with backlash and friction 168
8.3 Normalized signal flow graph of the elastic two–mass system with backlash and friction 169
8.4 Transformed signal flow graph of the elastic two–mass system without differentiation 170
8.5 Backlash model in elastic two–mass systems 173
8.6 Load–side subsystem 174
8.7 Complete signal flow graph of backlash identification with load–side observer 177
8.8 Motor–side subsystem 178
8.9 Complete signal flow graph of backlash identification with motor–side observer 181
8.10 Learning result of MBO in a stiff system 182
8.11 Learning result of LBO in a flexible plant 182
8.12 Experimental set–up for backlash identification 183
8.13 Photograph of the coupling of the two motors 184
8.14 Learning result at the real system 186
8.15 Learning procedure at the real system (current output of the neural network) 186
8.16 Learning procedure at the real system (evaluation of the neural network over the whole input space) 187

9.1 Simplified structure of a one stand cold rolling mill 190
9.2 General control concept (indirect control method) 190
9.3 Possible form of eccentricity 191
9.4 Observer concept for the identification of the eccentricity and signal flow chart of tension dynamics 193
9.5 Bode-plot of the error transfer function $H(s)$: SPR–condition not fulfilled 194
9.6 Observer structure with delayed activation and additional error (augmented error) 195
9.7 Simulation results of eccentricity identification and compensation 197
9.8 The aims of roll bite identification 199
9.9 Signal flow chart of the roll bite 200
9.10 Observer structure of roll bite identification 201

9.11 Possible concept of neural control of roll bite 202
9.12 Simulation results of roll bite identification 205
9.13 Experimental plant 206
9.14 Results of friction identification 208
9.15 Three–dimensional representation of friction with respect to tension force and speed 209
9.16 Results of eccentricity identification 210
9.17 Control concept 211
9.18 Step responses of a conventional and a state space controller . . 212
9.19 Learning results and compensation effect 213
9.20 Simultaneous learning and compensation: comparison of simulation and experimental results 214

10.1 Two–mass system with nonlinear friction characteristic 220
10.2 Transformed two–mass system with nonlinear friction characteristic 221
10.3 Two–mass system with nonlinear spring characteristic 223
10.4 Control of a second order plant by input–output linearization . 225
10.5 Control of two–mass system with nonlinear friction characteristic by input–output–linearization ($y = x_1$); top: reference signal y_m (- - -), load speed x_1 (· · · ·), motor speed x_3 (——); bottom: torque command u 226
10.6 Control of two–mass system with nonlinear friction characteristic by input–output–linearization ($y = x_3$); top: reference signal y_m (- - -), load speed x_1 (· · · ·), motor speed x_3 (——); bottom: torque command u 229
10.7 Control of a second order plant by input–output linearization (simplified control law) 231

11.1 Stable Model Reference Neurocontrol: basic system structure . 236
11.2 Stable Model Reference Neurocontrol for example plant 241
11.3 Simulation results: plant output y (left, continuous), reference signal y_m (left, dashed) and plant input u (right) 245
11.4 Simulation results: Development of the controller parameters u^i 245
11.5 Simulation results: ideal control law $f_c^*(w = 0, x_1, x_2)$ (left) and learned control law $f_c(w = 0, x_1, x_2)$ after 10^5 s (right) 246
11.6 Experimental setup 247

11.7 Experimental results: plant output y (left, continuous), modified reference signal y_{mm} according to section 11.4.1 (left, dotted) and plant input u (right) 248
11.8 Experimental results: ideal control law $f_c^*(w = 0, x_1, x_2)$ (left) and learned control law $f_c(w = 0, x_1, x_2)$ after 750 s (right) 248

12.1 Indirect and direct control methods 257
12.2 Possible net structures 258
12.3 Discrete and continuous dynamic neural networks with external coupling 259
12.4 Observer and identificator 261
12.5 Observer structures 262
12.6 Feedforward and recurrent observer structures 264
12.7 Differences between observer and identificators 266
12.8 Whole identification structure 272
12.9 Possible control structures based on dynamic identification . . . 273
12.10 Simulation results of the first example 276
12.11 Two–mass system and dynamic neural net structure 277
12.12 Results of the identification of a unknown two–mass system . . 278
12.13 Simulation results of the identification and control of the system described by equation (12.57) 280

13.1 Signal flow chart of a system enclosing an isolated nonlinearity $\mathcal{NL}$ in state–space description 286
13.2 Signal flow chart of a system enclosing an isolated nonlinearity $\mathcal{NL}$ in transfer function description 288
13.3 Signal flow chart of the compensation algorithm for a system with isolated nonlinearity $\mathcal{NL}$ 289
13.4 Signal flow chart of the separation of the compensation filter $K(s)$ 290
13.5 State–space control structure for the compensated system with an isolated nonlinearity; explicit differentiation is not necessary . . 292
13.6 State–space control structure for the compensated system with isolated nonlinearity including a nonlinear observer; explicit differentiation is not necessary 293
13.7 Step response $y_R(t)$ of the reference system; the desired final value is $1\,rad/s$ 296
13.8 Step response $y(t)$ of the compensated nonlinear system 296

13.9 Necessary motor torque for the compensation algorithm 296
13.10Comparision: step response of the nonlinear system, controlled by linear state feedback and the compensated system 296
13.11Overall structure of the alternate compensation and state–space control strategy; a linear observer is not necessary 299
13.12Principle of stable model reference neuro control for nonlinear systems; all states are measured 305
13.13Model reference neuro control for systems with isolated nonlinearity; all states are observed 306
13.14Signal flow graph of a continuous processing plant (two sucessive subsystems) 310
13.15Time–optimal curve shape of the control variable u(t) 313
13.16Curve of the control variable u(t) modified for practical application 314
13.17Linearized signal flow graph of the isolated subsystem j 316
13.18Three–dimensional time–optimal trajectories generated by reverse integration; the arrows indicate the motion of the states for positive time 317
13.19Approximated control surface 318
13.20Subsystem j with time–optimal controller 319
13.21Simulation of the transient response to a step change of the web–force of a time–optimal controlled isolated subsystem 320
13.22Experimental plant 321
13.23Web–forces and motor torque at subsystem 3 for a step change of F^*_{23} by 200 N ($V_0 = 0.8 \frac{m}{s}$) 323
13.24Web–forces and motor torque at subsystem 3 and 4 for a step change of F^*_{34} by 200 N ($V_0 = 0.8 \frac{m}{s}$) 324

Index

a–priori knowledge, 144
– additive, 145
– multiplicative , 146
activation, 86, 92
adaptation law, 115, 127
adaptive fuzzy control, 96
approximation, 84
approximation error, 86
Artificial Intelligence, 67
augmented error, 196, 208, 237, 271

backlash, 20, 167
– description for neural identification, 172
– experimental validation, 183
– identification, 167, 180, 183
– model, 167
backpropagation, 93
basis function, 85, 96

cascaded control, 20
Cerebellar Model Articulation Controller, 93
closed–loop control, 70
CMAC, 93
compensation, 152, 189, 261
– algorithm, 287
– condition, 286, 287
– filter, 288
– of eccentricities, 209
– of isolated nonlinearities, 285
– of non–circularity, 198
conditional feedback, 71
continuous processing of material, 46
control
– direct, 273
– indirect, 190, 209, 256
– inverse, 273
– reference, 273
– time–optimal, 313
control law, 96
control surface, 316
– approximation, 316
controllable canonical form, 302
convergence, 86, 212
cost function, 307
coulomb friction, 98

damping optimization criterion, 29, 35
DANN, 92, 99
deadzone, 169
decentralized decoupling, 57, 312, 322
decoupling filter, 142
defuzzification, 94
delayed activation, 195
diffeomorphism, 269, 304
direct collocation method, 308
Distance Activation Neural Network, 92, 99
dynamic approximator, 270
dynamic identificator, 255, 265
dynamic optimization, 308

eccentricity, 191, 209
error matrix, 141
error model, 117
error transfer function, 114, 126, 196, 270
estimated nonlinearity, 87
estimation error, 87
external feedback, 258
extrapolation, 85

factorization, 91
feed drive, 149

feedback linearization, 267
feedforward control, 50, 150
feedforward structure, 152, 257
finite element analysis, 40
friction, 20, 98, 149, 207
function approximation, 83, 94
fuzzy basis function, 96
fuzzy control, 62, 312
fuzzy controller, 94, 96, 99
fuzzy neural network, 94
fuzzy set, 94, 99

Gaussian radial basis function, 85
gear play, 167
General Regression Neural Network, 89, 99, 317
GRNN, 89, 99, 317

ideal control law, 224
identification
– of eccentricities, 207
– of friction, 207
– with accessible states, 136
– with unknown states, 139
inherent approximation error, 86, 192, 270
input–output behaviour, 261
input–output–linearization, 217, 304
– examples, 224
– mathematical requirements, 226
– without differentiation of the plant output, 228
internal dynamics, 227
internal states, 258
interpolation, 84, 87

Kohonen Neural Network, 93

Lagrangian–approach, 41
LBO, 172, 174
learning in control engineering, 73
learning rate, 152
Lie derivatives, 218
– examples, 219
load–side observer, 174
localized basis function, 85
Luenberger observer, 124

mass flow condition, 201
mass flow continuum, 192
MBO, 172, 178
McCulloch–Pitts–neuron, 95
membership function, 95
model
– linear, 150
– nonlinear, 152
motion control, 59
motor–side observer, 178
multidimensional input–space, 91
Multilayer Perceptron, 93

natural frequency, 150
natural mode, 150
neural control, 255
neural networks
– dynamic, 255, 257
– feedforward, 258
– recurrent, 258
– static, 257
neural observer, 190, 207
neuro–fuzzy system, 94
non–circularity, 191, 194
nonlinear function approximation, 83
nonlinear observer, 265
nonlinear state feedback, 290, 297
nonlinearity, 83
– isolated, 107, 191, 262, 285
– separable, 110, 135
plants with, 135
– static, 110
– structural, 110
– visibility of, 111
numerical accuracy, 92

observer approach, 113
open–loop control, 69
optimal constant output feedback, 54

parameter convergence, 87, 118
parameter error, 87
perceptron, 93
persistent excitation, 87, 117, 122
PI–controller, 209
polynomial activation, 92
processing plants with continuous moving webs
– coupling of subsystems, 311

– nonlinearsignal flow graph, 310
– time–optimal control, 308

Radial Basis Function Network, 99
Radial Basis Function network, 88
RBF, 88, 99
recurrent structure, 260, 271
relative degree, 219
– examples, 219
resonant mechanical system, 21
restricted update area, 92
roll bite, 189, 198
– control of, 198, 204
roll force, 198, 199

sigmoidal activation function, 93
singleton defuzzification, 94
sliding zone, 201
smoothing parameter, 88
SMRNC, 235
SPR–condition, 152
SPR–transfer function, 116
Stable Model Reference Neurocontrol, 235, 304
– examples, 240
– experimental results, 244
– mathematical requirements, 238
– plants with control saturation, 249
– reduction of learning times, 250
– simplified control law, 249
– with differentiation of the plant output, 250
state observation, 261
state–space control, 20, 27, 291
– integrating contribution, 34
state–space controller, 209
state–space observer, 39
stick–slip, 39
structural knowledge, 262
subsystem observer, 172
systematic observer design, 111, 126

three–mass system, 41
time–optimal control, 313
tool machine, 149
transfer function description, 286
transformed plant, 269
two–mass system, 20, 150, 168, 191, 207
– signal flow graph, 169
– state–space equations, 171
– transformed representation, 170

universal approximator, 260
universal function approximator, 86

zone of advance, 202

MIX
Papier aus verantwortungsvollen Quellen
Paper from responsible sources
FSC® C105338

If you have any concerns about our products,
you can contact us on
ProductSafety@springernature.com

In case Publisher is established outside the EU,
the EU authorized representative is:
Springer Nature Customer Service Center GmbH
Europaplatz 3, 69115 Heidelberg, Germany

Printed by Libri Plureos GmbH
in Hamburg, Germany